# Hardware/Software Co-Design: Principles and Practice

# Hardware/Software Co-Design: Principles and Practice

edited by

**Jørgen Staunstrup**
Technical University,
Lyngby, Denmark

and

**Wayne Wolf**
Princeton University,
Princeton, NJ, U.S.A.

KLUWER ACADEMIC PUBLISHERS
BOSTON / DORDRECHT / LONDON

A C.I.P. Catalogue record for this book is available from the Library of Congress.

ISBN 978-1-4419-5018-5

Published by Kluwer Academic Publishers,
P.O. Box 17, 3300 AA Dordrecht, The Netherlands.

Sold and distributed in the U.S.A. and Canada
by Kluwer Academic Publishers,
101 Philip Drive, Norwell, MA 02061, U.S.A.

In all other countries, sold and distributed
by Kluwer Academic Publishers Group,
P.O. Box 322, 3300 AH Dordrecht, The Netherlands.

*Printed on acid-free paper*

Contents

## Contributing Authors

*Thomas Benner*
Institut für Datenverarbeitungsanlagen
Hans-Sommer-Straße 66
D-38106 Braunschweig
Germany

*A. Changuel*
System-Level Synthesis Group
TIMA/INPG
46, ave Felix Viallet
F-38031 Grenoble cedex
France

*J. M. Daveau*
System-Level Synthesis Group
TIMA/INPG
46, ave Felix Viallet
F-38031 Grenoble cedex
France

*Rainer Dömer*
Department of Information and Computer Science
University of California at Irvine
Irvine CA 92691-3425
doemer@ics.uci.edu

*Rolf Ernst*

Institut für Datenverarbeitungsanlagen
Hans-Sommer-Straße 66
D-38106 Braunschweig
Germany
ernst@ida.ing.tu-bs.de

*Dirk Hermann*
Institut für Datenverarbeitungsanlagen
Hans-Sommer-Straße 66
D-38106 Braunschweig
Germany

*F. Hessel*
System-Level Synthesis Group
TIMA/INPG
46, ave Felix Viallet
F-38031 Grenoble cedex
France

*Daniel D. Gajski*
Department of Information and Computer Science
University of California at Irvine
Irvine CA 92691-3425
gajski@ics.uci.edu

*Jesper Grode*
Department of Information Technology
Technical University of Denmark
DK-2800, Lyngby
Denmark

*Peter V. Knudsen*
Department of Information Technology
Technical University of Denmark
DK-2800, Lyngby
Denmark

*Ph. Le Marrec*
System-Level Synthesis Group
TIMA/INPG
46, ave Felix Viallet

F-38031 Grenoble cedex
France

*Ahmed Amine Jerraya*
System-Level Synthesis Group
TIMA/INPG
46, ave Felix Viallet
F-38031 Grenoble cedex
France
jerraya@tima.inpg.fr

*Clifford Liem*
Laboratorie TIMA
Institut National Polytechnique de Grenoble
46, ave Felix Viallet
F-38031 Grenoble cedex
France
liem@tima.inpg.fr

*Jan Madsen*
Department of Information Technology
Technical University of Denmark
DK-2800, Lyngby
Denmark
jan@it.dtu.dk

*G. F. Marchioro*
System-Level Synthesis Group
TIMA/INPG
46, ave Felix Viallet
F-38031 Grenoble cedex
France

*Achim Österling*
Institut für Datenverarbeitungsanlagen
Hans-Sommer-Straße 66
D-38106 Braunschweig
Germany

*Pierre Paulin*
SGS-Thomson Microelectronics

850, rue Jan Monnet
BP 16, 38921 Crolles cedex
France
pierre.paulin@st.com

*M. Romdhani*
System-Level Synthesis Group
TIMA/INPG
46, ave Felix Viallet
F-38031 Grenoble cedex
France

*Wolfgang Rosenstiel*
Universität Tübingen
Technische Informatik
Sand 13
72076 Tübingen
Germany
rosenstiel@informatik.uni-tuebingen.de

*Thomas Scholz*
Institut für Datenverarbeitungsanlagen
Hans-Sommer-Straße 66
D-38106 Braunschweig
Germany

*Jørgen Staunstrup*
Department of Information Technology
Technical University of Denmark
DK-2800, Lyngby
Denmark
jst@it.dtu.edu

*C. A. Valderrama*
System-Level Synthesis Group
TIMA/INPG
46, ave Felix Viallet
F-38031 Grenoble cedex
France

*Wayne Wolf*

Department of Electrical Engineering
Princeton University
Princeton NJ 08544 USA
wolf@princeton.edu

*Wei Ye*
Institut für Datenverarbeitungsanlagen
Hans-Sommer-Straße 66
D-38106 Braunschweig
Germany

*Jianwen Zhu*
Department of Information and Computer Science
University of California at Irvine
Irvine CA 92691-3425
jzhu@ics.uci.edu

# Preface

This book presents a number of issues of fundamental importance for the design of integrated hardware software products such as embedded, communication, and multimedia systems. Co-design is still a new field but one which has substantially matured over the past few years. The book is inteded to provide:

- material for an advanced course,
- an overview of fundamental concepts in hardware/software co-design, and
- the necessary background for practioners who wants to get, an overview of the area.

As an interdisciplinary field, co-design touches on a number of established disciplines. We assume at various points in the book that the reader has some familiarity with a variety of topics ranging from concurrent programming languages through field-programmable gate arrays. Hopefully, our demands are not too strenuous and the bibliography will supply readers with necessary background.

This book is the result of a series of Ph.D. courses on hardware/software co-design held over the past two-and-a-half years. The first of these courses were held at the Technical University of Denmark in August, 1995 and was sponsored in part by the Danish Technical Research Council and the EUROPRACTICE program. The second was taught at the Institut National Polytechnique de Grenoble in October of 1996 and was sponsored in part by the European Commission under project EC-US-045. The third course will be held in Tokyo just after the publication of this book, in December, 1997 and is sponsored by the IEICE and Institute of System and Information Technologies, Kyushu. We would like to thank the sponsors of these courses for their generous support. We would also like to thank the students for their patience and insight.

The Authors and Editors

# 1 ESSENTIAL ISSUES IN CODESIGN

Daniel D. Gajski,
Jianwen Zhu,
and Rainer Dömer

Department of Information and Computer Science
University of California, Irvine
Irvine, California, USA

## 1.1 MODELS

In the last ten years, VLSI design technology, and the CAD industry in particular, have been very successful, enjoying an exceptional growth that has been paralleled only by the advances in IC fabrication. Since the design problems at the lower levels of abstraction became humanly intractable earlier than those at higher abstraction levels, researchers and the industry alike were forced to devote their attention first to lower-level problems such as physical and logic design. As these problems became more manageable, CAD tools for logic simulation and synthesis were developed successfully and introduced into the design process. As design complexities have grown and time-to-market requirements



*J. Staunstrup and W. Wolf (eds.), Hardware/Software Co-Design: Principles and Practice*, 1-45.

have shrunk drastically, both industry and academia have begun to focus on system levels of design since they reduce the number of objects that a designer needs to consider by an order of magnitude and thus allow the design and manufacturing of complex application specific integrated circuits (ASIC) quickly.

This chapter introduces key concepts in codesign: models of computations, hardware architectures, required language features for describing embedded systems and a generic methodology for codesign. The other chapters in this book give a more detailed survey and explore models (Chapter 6), languages (Chapter 7), architectures (Chapter 4), synthesis algorithms (Chapter 2), compilers (Chapter 5), and prototyping and emulation of embedded systems (Chapter 3). The last three chapters provide details of three research systems that demonstrate several aspects of codesign methodology.

### *1.1.1 Models and Architectures*

System design is the process of implementing a desired functionality using a set of physical components. Clearly, the whole process of system design must begin with specifying the desired functionality. This is not, however, an easy task.

The most common way to achieve the level of precision we need in specification is to think of the system as a collection of simpler subsystems, or pieces, and the method or the rules for composing these pieces to create system functionality. We call such a method a **model**. To be useful, a model should possess certain qualities. First, it should be formal so that it contains no ambiguity. It should also be complete, so that it can describe the entire system. In addition, it should be comprehensible to the designers who need to use it, as well as being easy to modify, since it is inevitable that, at some point, they will wish to change the system's functionality. Finally, a model should be natural enough to aid, rather than impede, the designer's understanding of the system.

It is important to note that a model is a formal system consisting of objects and composition rules, and is used for describing a system's characteristics. Typically, we would use a particular model to decompose a system into pieces, and then generate a specification by describing these pieces in a particular language. A language can capture many different models, and a model can be captured in many different languages (see Chapter 7).

Designers choose different models in different phases of the design process, in order to emphasize those aspects of the system that are of interest to them at that particular time. For example, in the specification phase, the designer knows nothing beyond the functionality of the system, so he will tend to use a model that does not reflect any implementation information. In the implementation phase, however, when information about the system's components is available, the designer will switch to a model that can capture the system's structure.

Once the designer has found an appropriate model to specify the functionality of a system, he can describe in detail exactly how that system will work. At that point, however, the design process is not complete, since such a model has still not described exactly how that system is to be manufactured. The next step, then, is to transform the system functionality into an **architecture**, which defines the system implementation by specifying the number and types of components as well as the connections between them.

In summary, models describe how a system works, while architectures describe how it will be manufactured. The **design process** or **methodology** is the set of design tasks that transform a model into an architecture. In the rest of this section we will describe several basic computational models whose variants and extensions are frequently used in system design.

### *1.1.2 Finite-State Machines*

A **finite-state machine** (FSM) [1, 2] is the most popular model for describing control systems, since the temporal behavior of such systems is most naturally represented in the form of states and transitions between states.

Basically, the FSM model consists of a set of **states**, a set of **transitions** between states, and a set of **actions** associated with these states or transitions.

The finite state machine can be defined abstractly as the quintuple

$$\langle S, I, O, f, h \rangle$$

where $S, I$, and $O$ represent a set of states, set of inputs and a set of outputs, respectively, and $f$ and $h$ represent the next-state and the output functions. The next state function $f$ is defined abstractly as a mapping $S \times I \rightarrow S$. In other words, $f$ assigns to every pair of state and input symbols another state symbol. The FSM model assumes that transitions from one state to another occur only when input symbols change. Therefore, the next-state function $f$ defines what the state of the FSM will be after the input symbols change.

The output function $h$ determines the output values in the present state. There are two different types of finite state machine which correspond to two different definitions of the output function $h$. One type is a **state-based** or **Moore-type**, for which $h$ is defined as a mapping $S \rightarrow O$. In other words, an output symbol is assigned to each state of the FSM and outputed during the time the FSM is in that particular state. The other type is an **input-based** or **Mealy-type** FSM, for which $h$ is defined as the mapping $S \times I \rightarrow O$. In this case, an output symbol in each state is defined by a pair of state and input symbols and it is outputed while the state and the corresponding input symbols persist.

According to our definition, each set $S, I$, and $O$ may have any number of symbols. However, in reality we deal only with binary variables, operators and memory elements. Therefore, $S, I$, and $O$ must be implemented as a cross-product of binary signals or memory elements, whereas functions $f$ and $h$ are defined by Boolean expressions that will be implemented with logic gates.

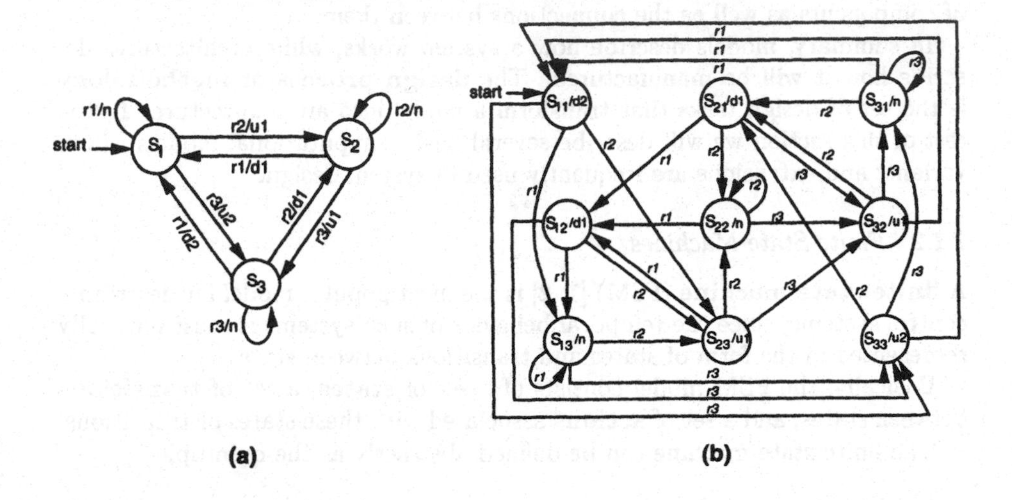


**Figure 1.1** FSM model for the elevator controller: (a) input-based, (b) state-based. (†)

In Figure 1.1(a), we see an input-based FSM that models an elevator controller in a building with three floors. In this model, the set of inputs $I = \{r1, r2, r3\}$ represents the floor requested. For example, $r2$ means that floor 2 is requested. The set of outputs $O = \{d2, d1, n, u1, u2\}$ represents the direction and number of floors the elevator should go. For example, $d2$ means that the elevator should go down 2 floors, $u2$ means that the elevator should go up 2 floors, and $n$ means that the elevator should stay idle. The set of states represents the floors. In Figure 1.1(a), we can see that if the current floor is 2 (i.e., the current state is $S_2$), and floor 1 is requested, then the output will be $d1$.

In Figure 1.1(b) we see the state-based model for the same elevator controller, in which the value of the output is indicated in each state. Each state has been split into three states representing each of the output signals that the state machine in Figure 1.1(a) will output when entering that particular state.

In practical terms, the primary difference between these two models is that the state-based FSM may require quite a few more states than the input-based model. This is because in a input-based model, there may be multiple arcs pointing to a single state, each arc having a different output value; in the state-based model, however, each different output value would require its own state, as is the case in Figure 1.1(b).

### *1.1.3 Dataflow Graph*

A **dataflow graph** (DFG) is the most popular model for describing computational intensive systems, since mathematical equations can be naturally represented by a directed graph in which the nodes represent operations or functions and the arcs represent the order in which the nodes are executed.

The dataflow model of computation is based on two principles: asynchrony and functionality. The asynchrony principle states that all operations are executed when and only when the required operands are available. The functionality principle states that all operations behave as functions which do not have any side effects. This implies that any two enabled operations can be executed in either order or concurrently.

Formally a DFG can be described as a quintuple

$$\langle N, A, V, v^0, f \rangle$$

where

$N = \{n_1, n_2, \ldots, n_M\}$ is the set of nodes;

$A = \{a_1, a_2, \ldots, a_L\} \subseteq N \times N$ is the set of arcs between the nodes;

$V = \{\langle v_1, v_2, \ldots v_L \rangle\} \subseteq V_1 \times V_2 \times \ldots V_L$ represents the set of values associated with the arcs, where $v_i \in V_i$ is the value at arc $A_i$;

$v^0 \in V$ represents the initial values at the arcs;

$f = \{f_{n_i} : \prod_{a_j \in I(n_i)} V_j \mapsto \prod_{a_k \in O(n_i)} V_k \mid n_i \in N\}$ defines the function performed by each node $n_i \in N$, where $I(n_i)$ and $O(n_i)$ are the set of incoming and outgoing arcs for each node $n_i \in N$.

At any point in time, a node with data values on all its inputs can be executed and will produce a value at its output after some finite amount of time. In Figure 1.2 we can see that the computation of $\sqrt{a^2 + b^2}$ can be represented by a DFG with four nodes performing squaring, addition and square root computation. Initially, at time $t_1$, only the values 3 and 4 are available at the input of the squaring nodes ($v^{t_1} = \langle 3, 4, \perp, \perp, \perp, \perp \rangle$). The squaring nodes compute

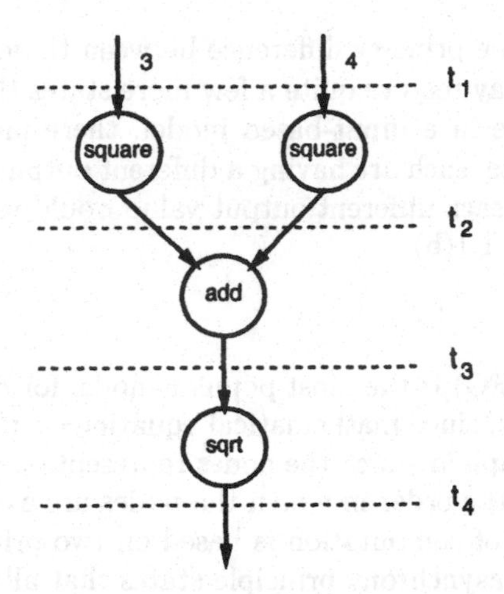


**Figure 1.2** Example of a dataflow graph.

the values 9 and 16 at time $t_2$ ($v^{t_2} = \langle 3, 4, 9, 16, \perp, \perp \rangle$), after which the addition is executed ($v^{t_3} = \langle 3, 4, 9, 16, 25, \perp \rangle$). Finally the square root is computed ($v^{t_4} = \langle 3, 4, 9, 16, 25, 5 \rangle$) at time $t_4$.

Dataflow graphs are excellent for representing computations described by complex functions, but not suitable for representing a control part which is found in most programming languages. For this reason, they are very popular in describing DSP components and systems. Many variations of dataflow models are used in DSP research such as the synchronous dataflow graph (SDFG) [3], where the set $V$ models FIFOs of tokens. Several extensions of DFGs such as task graphs and control/dataflow graphs (CDFGs) can be found in Chapter 2 and Chapter 6.

### *1.1.4 Finite-State Machine with Datapath*

Most of the real systems combine the features of control and computation. Thus, we must combine the features of the FSM and the DFG models. One solution is to divide time into equal time intervals, called states, and allocate one or more states for each node in the DFG. Since DFG computations are executed in a datapath we call this model a finite-state machine with datapath (FSMD).

In order to formally define a FSMD[2, 4], we must extend the definition of a FSM introduced in Section 1.1.2. A FSM is a quintuple

$$\langle S, I, O, f, h \rangle$$

In order to include a datapath, we must extend this definition by adding the set of datapath variables, inputs and outputs. More formally, we define a variables set $V$ which defines the state of the datapath by defining the values of all variables in each state.

In the same fashion, we can separate the set of FSMD inputs into a set of FSM inputs $I_C$ and a set of datapath inputs $I_D$. Thus, $I = I_C \times I_D$. Similarly, the output set consists of FSM outputs $O_C$ and datapath outputs $O_D$. In other words, $O = O_C \times O_D$. Except for very trivial cases, the size of the datapath variables, and ports makes specification of functions $f$ and $h$ in a graphical or tabular form very difficult. In order to be able to specify variable values in an efficient and understandable way in the definition of an FSMD, we will specify variable values with arithmetic expressions.

We define the set of all possible expressions, $Expr(V)$, over the set of variables $V$, to be the set of all constants $K$ of the same type as variables in $V$, the set of variables $V$ itself and all the expressions obtained by combining two expressions with arithmetic, logic, or rearrangement operators.

Using $Expr(V)$ we can define the values of the status signals as well as transformations in the datapath. Let $STAT = \{stat_k = e_i \Delta e_j \mid e_i, e_j, \in Expr(V), \Delta \in \{\leq, <, =, \neq, >, \geq\}\}$ be the set of all status signals which are described as relations between variables or expressions of variables. Examples of status signals are $Data \neq 0$, $(a-b) > (x+y)$ and $(counter = 0) AND (x > 10)$. The relations defining status signals are either true, in which case the status signal has value 1 or false in which case it has value 0.

With formal definition of expressions and relations over a set of variables we can simplify function $f : (S \times V) \times I \rightarrow S \times V$ by separating it into two parts: $f_C$ and $f_D$.. The function $f_C$ defines the next state of the control unit

$$f_C : S \times I_C \times STAT \rightarrow S$$

while the function $f_D$ defines the values of datapath variables in the next state

$$f_D : S \times V \times I_D \rightarrow V$$

In other words, for each state $s_i \in S$ we compute a new value for each variable $V_j \in V$ in the datapath by evaluating an expression $e_j \in Expr(V)$.

Similarly, we can decompose the output function $h : S \times V \times I \rightarrow O$ into two different functions, $h_C$ and $h_D$ where $h_C$ defines the external control outputs $O_C$ as in the definition of an FSM and $h_D$ defines external datapath outputs.

Therefore,

$$h_C : S \times I_C \times STAT \rightarrow O_C$$

and

$$h_D : S \times V \times I_D \rightarrow O_D$$

Note, again that variables in $O_C$ are Boolean variables and that variables in $O_D$ are Boolean vectors.

Using this kind of FSMD, we could model the elevator controller example in Figure 1.1 with only one state, as shown in Figure 1.3. This reduction in the number of states is possible because we have designated a variable *cfloor* to store the state value of the FSM in Figure 1.1(a) and *rfloor* to store the values of $r1$, $r2$ and $r3$ (A further useful extension to the FSMD model can be found in Chapter 7).

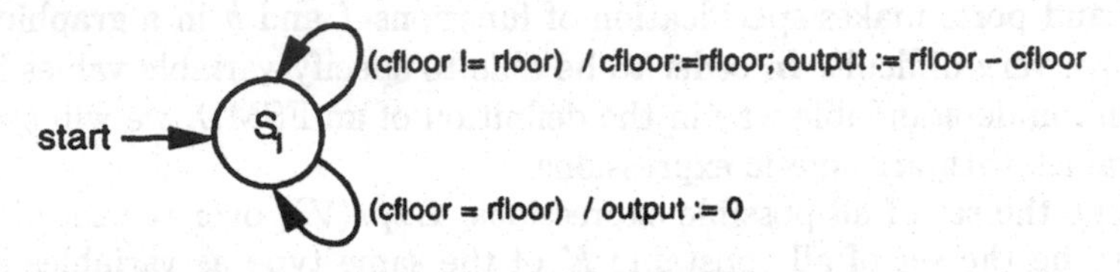


**Figure 1.3** FSMD model for the elevator controller. (†)

In general, the FSM is suitable for modeling control-dominated systems, and the DFG for computation-dominated systems, while the FSMD can be suitable for both control- and computation-dominated systems. However, it should be pointed out that neither the FSM nor the FSMD model is suitable for complex systems, since neither one explicitly supports concurrency and hierarchy. Without explicit support for concurrency, a complex system will precipitate an explosion in the number of states. Consider, for example, a system consisting of two concurrent subsystems, each with 100 possible states. If we try to represent this system as a single FSM or FSMD, we must represent all possible states of the system, of which there are $100 \times 100 = 10,000$. At the same time, the lack of hierarchy would cause an increase in the number of arcs. For example, if there are 100 states, each requiring its own arc to transition to a specific state for a particular input value, we would need 100 arcs, as opposed to the single arc required by a model that can hierarchically group those 100 states into one state. The problem with such models, of course, is that once they reach several hundred states or arcs, they become incomprehensible to humans.

### *1.1.5 Hierarchical Concurrent Finite-State Machines*

The **hierarchical concurrent finite-state machine** (HCFSM) [5] is essentially an extension of the FSM model, which adds support for **hierarchy** and **concurrency**, thus eliminating the potential for state and arc explosion that occurred when describing hierarchical and concurrent systems with FSM models.

Like the FSM, the HCFSM model consists of a set of **states** and a set of **transitions**. Unlike the FSM, however, in the HCFSM each state can be further decomposed into a set of **substates**, thus modeling hierarchy. Furthermore, each state can also be decomposed into **concurrent substates**, which execute in parallel and communicate through global variables. The transitions in this model can be either structured or unstructured, with structured transitions allowed only between two states on the same level of hierarchy, while unstructured transitions may occur between any two states regardless of their hierarchical relationship.

One language that is particularly well-adapted to the HCFSM model is Statecharts [6], since it can easily support the notions of hierarchy, concurrency and communication between concurrent states. Statecharts uses unstructured transitions and a broadcast communication mechanism, in which events emitted by any given state can be detected by all other states.

The Statecharts language is a graphic language. Specifically, we use rounded rectangles to denote states at any level, and encapsulation to express a hierarchical relation between these states. Dashed lines between states represent concurrency, and arrows denote the transitions between states, each arrow being labeled with an event and, optionally, with a parenthesized condition and/or action.

Figure 1.4 shows an example of a system represented by means of Statecharts. In this figure, we can see that state $Y$ is decomposed into two concurrent states, $A$ and $D$; the former consisting of two further substates, $B$ and $C$, while the latter comprises substates $E$, $F$, and $G$. The bold dots in the figure indicate the starting points of states. According to the Statecharts language, when event $b$ occurs while in state $C$, $A$ will transfer to state $B$. If, on the other hand, event $a$ occurs while in state $B$, $A$ will transfer to state $C$, but only if condition $P$ holds at the instant of occurrence. During the transfer from $B$ to $C$, the action $c$ associated with the transition will be performed.

Because of its hierarchy and concurrency constructs, the HCFSM model is well-suited to represent complex control systems. Even when the FSM model is replaced with the FSMD model it can only associate very simple actions, such as assignments, with its transitions or states. As a result, the HCFSMD is not suitable for modeling certain characteristics of complex systems, which may require complex data structures or may perform in each state an arbitrarily complex activity. For such systems, this model alone would probably not suffice.

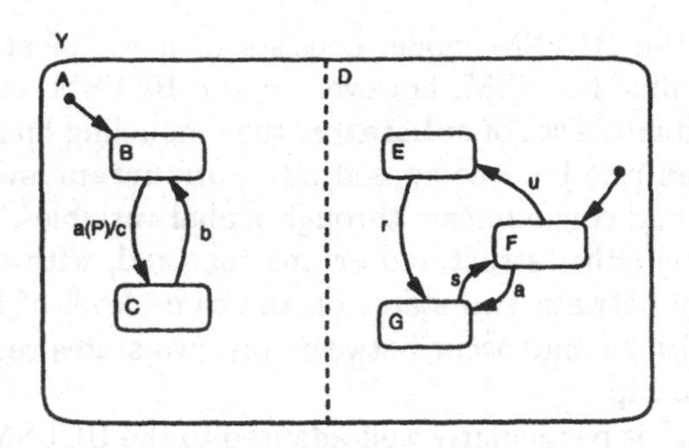


**Figure 1.4** Statecharts: hierarchical concurrent states. (†)

### *1.1.6 Programming Languages*

**Programming languages** provide a heterogeneous model that can support data, activity and control modeling. Unlike the structure chart, programming languages are presented in a textual, rather than a graphic, form.

There are two major types of programming languages: imperative and declarative. The **imperative** class includes languages like C and Pascal, which use a control-driven model of execution, in which statements are executed in the order written in the program. LISP and PROLOG, by contrast, are examples of **declarative** languages, since they model execution through demand-driven or pattern-driven computation. The key difference here is that declarative languages specify no explicit order of execution, focusing instead on defining the target of the computation through a set of functions or logic rules.

In the aspect of data modeling, imperative programming languages provide a variety of data structures. These data structures include, for example, **basic** data types, such as integers and reals, as well as **composite** types, like arrays and records. A programming language would model small activities by means of **statements**, and large activities by means of **functions** or **procedures**, which can also serve as a mechanism for supporting hierarchy within the system. These programming languages can also model control flow, by using control constructs that specify the order in which activities are to be performed. These control constructs can include **sequential** composition (often denoted by a semicolon), **branching** (*if* and *case* statements), **looping** (*while*, *for*, and *repeat*), as well as **subroutine calls**.

The advantage to using an imperative programming language is that this paradigm is well-suited to modeling computation-dominated behavior, in which some problem is solved by means of an algorithm, as, for example, in a case when we need to sort a set of numbers stored in an array.

The main problem with programming languages is that, although they are well-suited for modeling the data, activity, and control mechanism of a system, they do not explicitly model the system's states, which is a disadvantage in modeling embedded systems.

### *1.1.7 Program-State Machines*

A **program-state machine** (PSM) is an instance of a heterogeneous model that integrates an HCFSM with a programming language paradigm. This model basically consists of a hierarchy of **program-states**, in which each program-state represents a distinct mode of computation. At any given time, only a subset of program-states will be active, i.e., actively carrying out their computations.

Within its hierarchy, the model would consist of both composite and leaf program-states. A **composite** program-state is one that can be further decomposed into either **concurrent** or **sequential** program-substates. If they are concurrent, all the program-substates will be active whenever the program-state is active, whereas if they are sequential, the program-substates are only active one at a time when the program-state is active. A sequentially decomposed program-state will contain a set of transition arcs, which represent the sequencing between the program-substates. There are two types of transition arcs. The first, a **transition-on-completion arc (TOC)**, will be traversed only when the source program-substate has completed its computation and the associated arc condition evaluates to true. The second, a **transition-immediately arc (TI)**, will be traversed immediately whenever the arc condition becomes true, regardless of whether the source program-substate has completed its computation. Finally, at the bottom of the hierarchy, we have the **leaf** program-states whose computations are described through programming language statements.

When we are using the program-state machine as our model, the system as an entity can be graphically represented by a rectangular box, while the program-states within the entity will be represented by boxes with curved corners. A concurrent relation between program-substates is denoted by the dotted line between them. Transitions are represented with directed arrows. The starting state is indicated by a triangle, and the completion of individual program-states is indicated by a transition arc that points to the *completion point*, represented as a small square within the state. TOC arcs are those that originate from a square inside the source substate, while TI arcs originate from the perimeter of the source substate.

Figure 1.5 shows an example of a program-state machine, consisting of a root state $Y$, which itself comprises two concurrent substates, $A$ and $D$. State

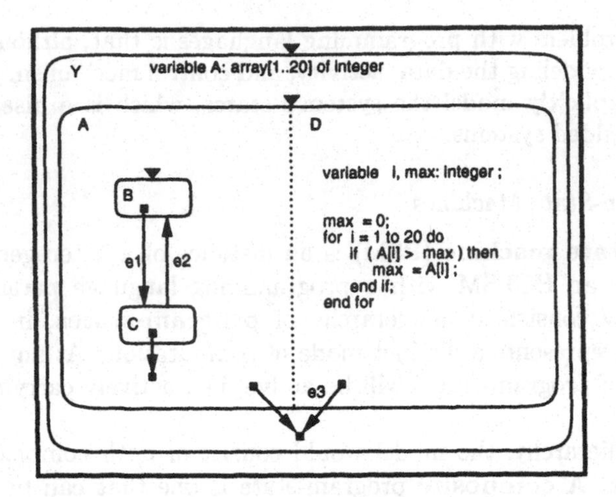


**Figure 1.5** An example of program-state machine. (†)

*A*, in turn, contains two sequential substates, *B* and *C*. Note that states *B*, *C*, and *D* are leaf states, though the figure shows the program only for state *D*. According to the graphic symbols given above, we can see that the arcs labeled *e*1 and *e*3 are TOC arcs, while the arc labeled *e*2 is a TI arc. The configuration of arcs would mean that when state *B* finishes and condition *e*1 is true, control will transfer to state *C*. If, however, condition *e*2 is true while in state *C*, control will transfer to state *B* regardless of whether *C* finishes or not.

Since PSMs can represent a system's states, data, and activities in a single model, they are more suitable than HCFSMDs for modeling systems which have complex data and activities associated with each state. A PSM can also overcome the primary limitation of programming languages, since it can model states explicitly. It allows a modeler to specify a system using hierarchical state-decomposition until he/she feels comfortable using program constructs. The programming language model and HCFSMD model are just two extremes of the PSM model. A program can be viewed as a PSM with only one leaf state containing language constructs. A HCFSMD can be viewed as a PSM with all its leaf states containing no language constructs.

In this section we presented the main models to capture systems. Obviously, there are more models used in codesign, mostly targeted at specific applications. For example, the codesign finite state machine (CFSM) model [7], which is based on communicating FSMs using event broadcasting, is targeted at reactive real-time systems.

## 1.2 ARCHITECTURES

To this point, we have demonstrated how a model is used to describe a system's functionality, data, control and structure. An architecture is intended to supplement these descriptive models by specifying how the system will actually be implemented. The goal of an architecture design is to describe the number of components, the type of each component, and the type of each connection among these various components in a system.

Architectures can range from simple controllers to parallel heterogeneous processors, but most architectures nonetheless fall into a few distinct classes: **application-specific architectures**, such as DSP systems; **general-purpose processors**, such as RISCs; and **parallel processors**, such as VLIW, SIMD and MIMD machines (see also Chapter 4).

### *1.2.1 Controller Architecture*

The simplest of the application-specific architectures is the **controller** variety, which is a straight-forward implementation of the FSM model presented in Section 1.1.2 and defined by the quintuple $< S, I, O, f, h >$. A controller consists of a register and two combinational blocks, as shown in Figure 1.6. The register, usually called the *State register*, is designed to store the states in $S$, while the two combinational blocks, referred to as the *Next-state logic* and the *Output logic*, implement functions $f$ and $h$. *Inputs* and *Outputs* are representations of Boolean signals that are defined by sets $I$ and $O$.

As mentioned in Section 1.1.2, there are two distinct types of controllers, those that are input-based and those that are state-based. These types of controllers differ in how they define the output function, $h$. For input-based controllers, $h$ is defined as a mapping $S \times I \rightarrow O$, which means that the *Output logic* is dependent on two parameters, namely, *State register* and *Inputs*. For state-based controllers, on the other hand, $h$ is defined as the mapping $S \rightarrow O$, which means the *Output logic* depends on only one parameter, the *State register*. Since the inputs and outputs are Boolean signals, in either case, this architecture is well-suited to implementing controllers that do not require data manipulation.

### *1.2.2 Datapath Architecture*

Datapaths can be used for implementation of DFGs in many applications where a fixed computation must be performed repeatedly on different sets of data, as is the case in the digital signal processing (DSP) systems used for digital filtering, image processing, and multimedia. A datapath architecture often consists of

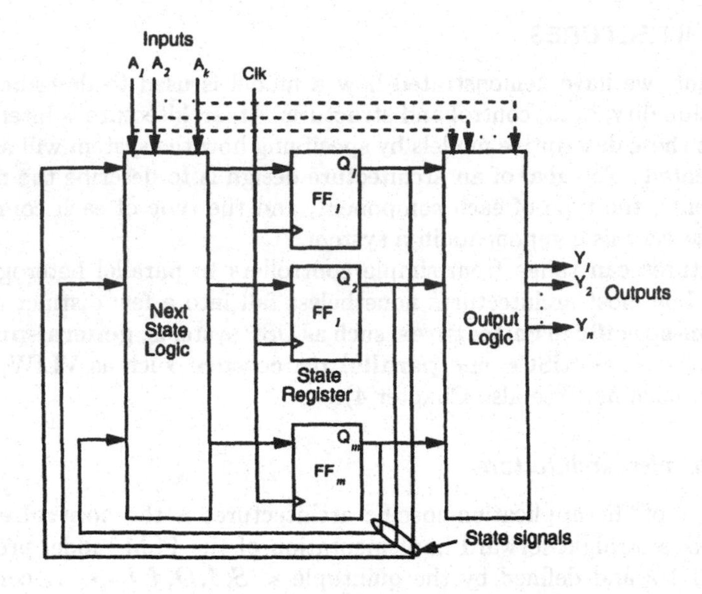


**Figure 1.6** A generic controller design. (‡)

high-speed arithmetic units, connected in parallel and heavily pipelined in order to achieve a high throughput.

Such a datapath is one implementation of the DFG model in which a computation is devided into pipeline stages of equal length and the values between the stages are stored in registers controlled by a common clock.

In Figure 1.7, we can see two different datapaths, both of which are designed to implement a finite-impulse-response (FIR) filter, which is defined by the expression $y(i) = \sum_{k=0}^{N-1} x(i-k)b(k)$ where $N$ is 4. Note that the datapath in Figure 1.7(a) performs all its multiplications concurrently, and adds the products in parallel by means of a summation tree. The datapath in Figure 1.7(b) also performs its multiplications concurrently, but it will then add the products serially. Further, note that the datapath in Figure 1.7(a) has three pipeline stages, each indicated by a dashed line, whereas the datapath in Figure 1.7(b) has four similarly indicated pipeline stages. Although both datapaths use four multipliers and three adders, the datapath in Figure 1.7(b) is regular and easier to implement in ASIC technologies.

A general-purpose datapath can compute DFGs in which each node is scheduled into one or more control steps or clock cycles. A datapath may include counters, registers, register-files and memories with a varied number of ports that are connected with several buses. Note that these same buses can be used

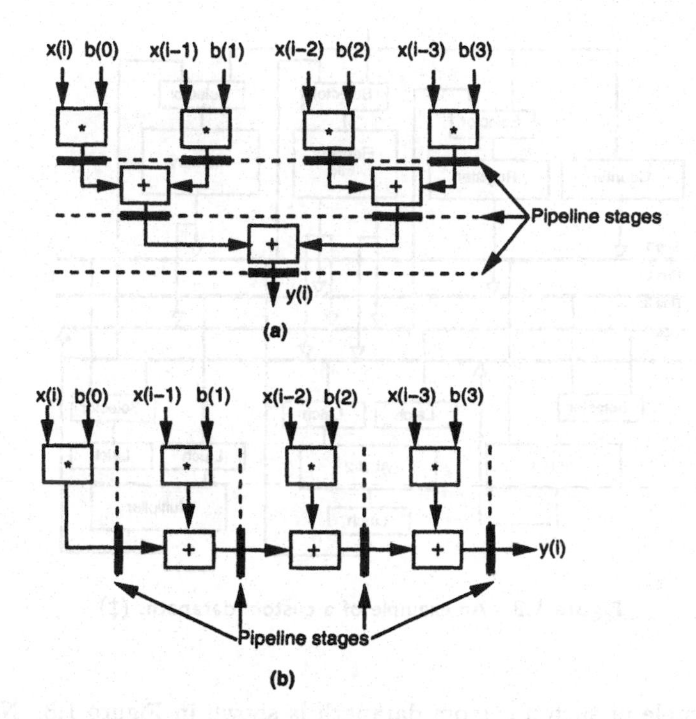


**Figure 1.7** Two different datapaths for FIR filter: (a) with three pipeline stages, (b) with four pipeline stages. (†)

to supply operands to functional units as well as to supply results back to storage units. It is also possible for the functional units to obtain operands from several buses, though this would require the use of a selector in front of each input. It is also possible for each unit to have input and output latches which are used to temporarily store the input operands or results. Such latching can significantly shorten the amount of time that the buses will be used for operand and result transfer, and thus can increase the traffic over these buses.

On the other hand, input and output latching requires a more complicated control unit since each operation requires more than one clock cycle. Namely, at least one clock cycle is required to fetch operands from registers, register files or memories and store them into input latches, at least one clock cycle to perform the operation and store a result into an output latch, and at least one clock cycle to store the result from an output latch back to a register or memory.

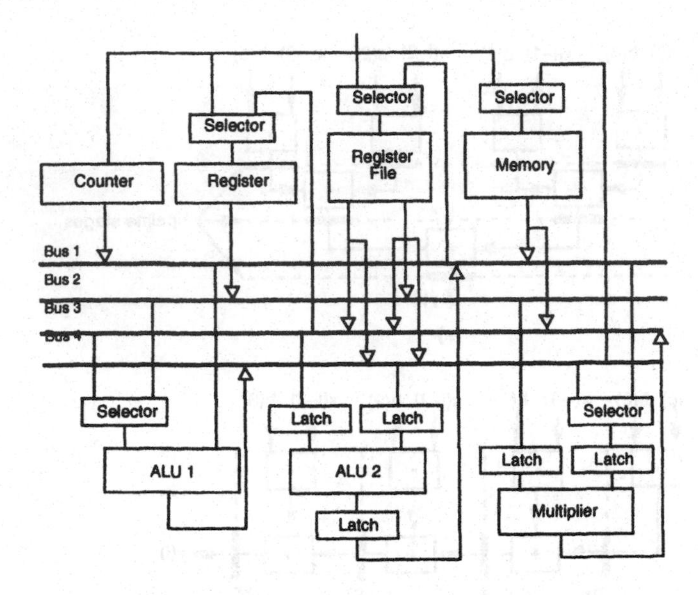


**Figure 1.8** An example of a custom datapath. (‡)

An example of such a custom datapath is shown in Figure 1.8. Note that it has a counter, a register, a 3-port register file and a 2-port memory. It also has four buses and three functional units: two ALUs and a multiplier. As you can see, ALU1 does not have any latches, while ALU2 has latches at both the inputs and the outputs and the single multiplier has only the inputs latched. With this arrangement, ALU1 can receive its left operand from buses 2 and 4, while the multiplier can receive its right operand from buses 1 and 4. Similarly, the storage units can also receive data from several buses. Such custom datapaths are frequently used in application specific design to obtain the best performance-cost ratio.

We have seen in the previous example that as long as each operation in a DFG is implemented by its own unit, as in Figure 1.7, we do not need a control for the system, since data simply flows from one unit to the next, and the clock is used to load pipeline registers. Sometimes, however, it may be necessary to share units to save silicon area, in which case we would need a simple controller to steer the data among the units and registers, and to select the appropriate arithmetic function for those units that can perform different functions at different times. Another situation would be to implement more than one algorithm (DFG) with the same datapath, with each algorithm executing at a different time. In this case, since each algorithm requires a unique flow of data through the datapath, we would need a controller to regulate the flow. Such controllers are usually simple and without conditional branches.

In general any computation can be executed in serial-parallel manner where the parallelism is constrained by resources. In such a case we can use the datapath in Figure 1.8 and add a controller to serialize the computation which leads us to the FSMD architecture.

### *1.2.3 FSMD Architecture*

A FSMD architecture implements the FSMD model by combining a controller with a datapath. As shown in Figure 1.9, the datapath has two types of I/O ports. One type of I/O ports are data ports which are used by the outside environment to send and receive data to and from the ASIC. The data could be of type integer, floating-point, or characters and it is usually packed into one or more words. The data ports are usually 8, 16, 32 or 64 bits wide. The other type of I/O ports are control ports which are used by the control unit to control the operations performed by the datapath and receive information about the status of selected registers in the datapath.

As shown in Figure 1.9, the datapath takes the operands from storage units, performs the computation in the combinatorial units and returns the results to storage units during each state, which is usually equal to one clock cycle.

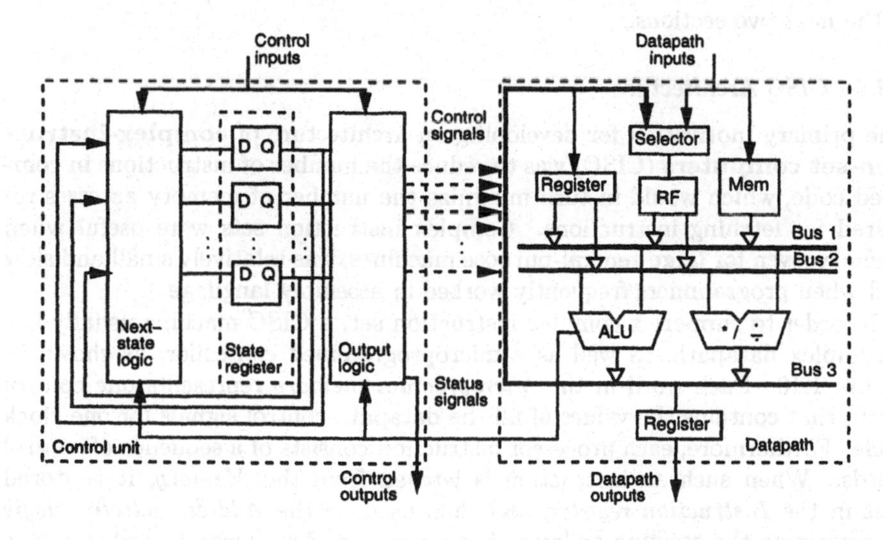


**Figure 1.9** Design model: register-transfer-level block diagram. (‡)

As mentioned in the previous section the selection of operands, operations and the destination for the result is controlled by the control unit by setting proper values of datapath control signals. The datapath also indicates through status signals when a particular value is stored in a particular storage unit or when a particular relation between two data values stored in the datapath is satisfied.

Similar to the datapath, a control unit has a set of input and a set of output signals. There are two types of input signals: external signals and status signals. External signals represent the conditions in the external environment on which the FSMD architecture must respond. On the other hand, the status signals represent the state of the datapath. Their value is obtained by comparing values of selected variables stored in the datapath. There are also two types of output signals: external signals and datapath control signals. External signals identify to the environment that a FSMD architecture has reached certain state or finished a particular computation. The datapath controls, as mentioned before, select the operation for each component in the datapath.

FSMD architectures are used for the design of general-purpose processors as well as ASICs. Each ASIC design consists of one or more FSMD architectures, although two implementations may differ in the number of control units and datapaths, the number of components and connections in the datapath, the number of states in the control unit and the number of I/O ports. The FSM controller and DSP datapath mentioned in the previous sections are two special cases of this kind of architecture. So are CISC and RISC processors as described in the next two sections.

### *1.2.4 CISC Architecture*

The primary motivation for developing an architecture of **complex-instruction-set computers** (CISC) was to reduce the number of instructions in compiled code, which would in turn minimize the number of memory accesses required for fetching instructions. Complex instruction sets were useful when memory, even for large general-purpose machines, was relatively small and slow and when programmers frequently worked in assembly language.

In order to support a complex instruction set, a CISC machine usually has a complex datapath, as well as a microprogrammed controller, as shown in Figure 1.10. Each word in the *Microprogram memory* represents one control word, that contains the values of all the datapath control signals for one clock cycle. Furthermore, each processor instruction consists of a sequence of control words. When such an instruction is fetched from the *Memory*, it is stored first in the *Instruction register*, and then used by the *Address selection logic* to determine the starting address of the corresponding control-word sequence

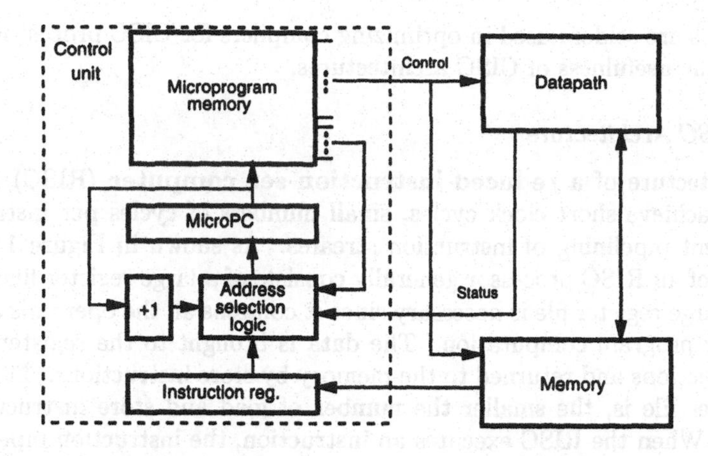


**Figure 1.10** CISC with microprogrammed control. (†)

in the *Microprogram memory.* After this starting address has been loaded into the *MicroPC,* the corresponding control word will be fetched from the *Microprogram memory,* and used to transfer the data in the datapath from one register to another. Since the *MicroPC* is concurrently incremented to point to the next control word, this procedure will be repeated for each control word in the sequence. Finally, when the last control word is being executed, a new instruction will be fetched from the *Memory,* and the entire process will be repeated.

From this description, we can see that the number of control words, and thus the number of clock cycles can vary for each instruction. As a result, instruction pipelining can be difficult to implement in CISCs. In addition, relatively slow microprogram memory requires a clock cycle to be longer than necessary. Since instruction pipelines and short clock cycles are necessary for fast program execution, CISC architectures may not be well-suited for high-performance processors.

Although a variety of complex instructions could be executed by a CISC architectures, program-execution statistics have shown that the instructions used most frequently tend to be simple, with only a few addressing modes and data types. Statistics have also shown that the most complex instructions were seldom or never used. This low usage of complex instructions can be attributed to the slight semantic differences between programming language constructs and available complex instructions, as well as the difficulty in mapping language constructs into such complex instructions. Because of this difficulty, complex instructions are seldom used in optimizing compilers for CISC processors, thus reducing the usefulness of CISC architectures.

### *1.2.5 RISC Architecture*

The architecture of a **reduced-instruction-set computer** (RISC) is optimized to achieve short clock cycles, small numbers of cycles per instruction, and efficient pipelining of instruction streams. As shown in Figure 1.11, the datapath of an RISC processor generally consists of a large register file and an ALU. A large register file is necessary since it contains all the operands and the results for program computation. The data is brought to the register file by load instructions and returned to the memory by store instructions. The larger the register file is, the smaller the number of load and store instructions in the code. When the RISC executes an instruction, the instruction pipe begins by fetching an instruction into the *Instruction register.* In the second pipeline stage the instruction is then decoded and the appropriate operands are fetched from the *Register file.* In the third stage, one of two things occurs: the RISC either executes the required operation in the *ALU*, or, alternatively, computes the address for the *Data cache.* In the fourth stage the data is stored in either the *Data cache* or in the *Register file.* Note that the execution of each instruction takes only four clock cycles, approximately, which means that the instruction pipeline is short and efficient, losing very few cycles in the case of data or branch dependencies.

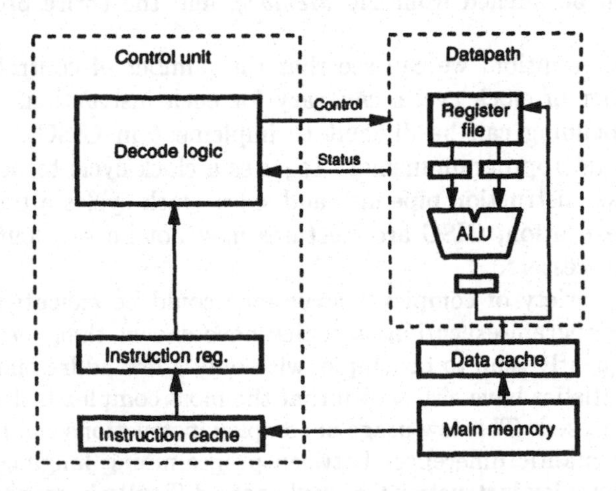


**Figure 1.11** RISC with hardwired control. (†)

We should also note that, since all the operands are contained in the register file, and only simple addressing modes are used, we can simplify the design of the datapath as well. In addition, since each operation can be executed in one clock cycle and each instruction in four, the control unit remains simple and can be implemented with random logic, instead of microprogrammed control. Overall, this simplification of the control and datapath in the RISC results in a short clock cycle, and, ultimately, higher performance.

It should also be pointed out, however, that the greater simplicity of RISC architectures require a more sophisticated compiler. Furthermore, due to the fact that the number of instructions is reduced, the RISC compiler will need to use a sequence of RISC instructions in order to implement complex operations.

### *1.2.6 VLIW Architecture*

A **very-long-instruction-word computer** (VLIW) exploits parallelism by using multiple functional units in its datapath, all of which execute in a lock step manner under one centralized control. A VLIW instruction contains one field for each functional unit; each field of a VLIW instruction specifies the addresses of the source and destination operands, as well as the operation to be performed by the functional unit. As a result, a VLIW instruction is usually very wide, since it must contain approximately one standard instruction for each functional unit.

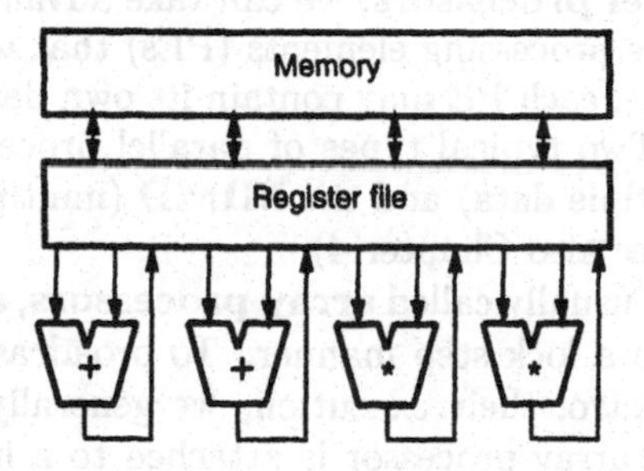


**Figure 1.12** An example of VLIW datapath. (†)

In Figure 1.12, we see an example of a VLIW datapath, consisting of four functional units: two ALUs and two multipliers, a register file and a memory. The register file requires 16 ports to keep all the function units busy: eight output ports which supply operands to the functional units; four input ports which store the results obtained from functional units; and four input/output ports to allow communication with the memory. Ideally, this architecture provides four times the performance we could get from a processor with a single functional unit, under the assumption that the code executing on the VLIW had four-way parallelism, which enables the VLIW to execute four independent instructions in each clock cycle. In reality, however, most code has a large amount of parallelism interleaved with code that is fundamentally serial. As a result, a VLIW with a large number of functional units might not be fully utilized. The ideal conditions would also require us to assume that all the operands were in the register file, with eight operands being fetched and four results stored back on every clock cycle, in addition to four new operands being brought from the memory to be available for use in the next clock cycle. It must be noted, however, that this computation profile is not easy to achieve, since some results must be stored back to memory and some results may not be needed in the next clock cycle. Under these conditions, the efficiency of a VLIW datapath will be less than ideal.

Finally, we should point out that there are two technological limitations that can affect the implementation of a VLIW architecture. First, while register files with 8–16 ports can be built, the efficiency and performance of such register files tend to degrade quickly when we go beyond that number. Second, since VLIW program and data memories require a high communication bandwidth, these systems tend to require expensive high-pin packaging technology as well.

### *1.2.7 Parallel Architecture*

In the design of **parallel processors**, we can take advantage of spatial parallelism by using multiple processing elements (PEs) that work concurrently. In this type of architecture, each PE may contain its own datapath with registers and a local memory. Two typical types of parallel processors are the **SIMD** (single instruction multiple data) and the **MIMD** (multiple instruction multiple data) processors (see also Chapter 4).

In SIMD processors, usually called **array processors**, all of the PEs execute the same instruction in a lock step manner. To broadcast the instructions to all the PEs and to control their execution, we generally use a single global controller. Usually, an array processor is attached to a host processor, which means that it can be thought of as a kind of hardware accelerator for tasks that are computationally intensive. In such cases, the host processor would load the data into each PE, and then collect the results after the computations are finished. When it is necessary, PEs can also communicate directly with their nearest neighbors.

The primary advantage of array processors is that they are very convenient for computations that can be naturally mapped on a rectangular grid, as in the case of image processing, where an image is decomposed into pixels on a rectangular grid, or in the case of weather forecasting, where the surface of the globe is decomposed into $n$-by-$n$-mile squares. Programming one grid point in the rectangular array processor is quite easy, since all the PEs execute the same instruction stream. However, programming any data routing through the array is very difficult, since the programmer would have to be aware of all the positions of each data for every clock cycle. For this reason, problems, like matrix triangulations or inversions, are difficult to program on an array processor.

An MIMD processor, usually called a **multiprocessor system**, differs from an SIMD in that each PE executes its own instruction stream. Each processor can communicate with every other processor within the multiprocessor system, using one of the two communication mechanisms. In a **shared-memory** mechanism, all the processors are connected to a common shared memory through an interconnection network, which means that each processor can access any data in the shared memory. In a **message-passing** mechanism, on the other hand, each processor tends to have a large local memory, and sends data to other processors in the form of messages through an interconnection network.

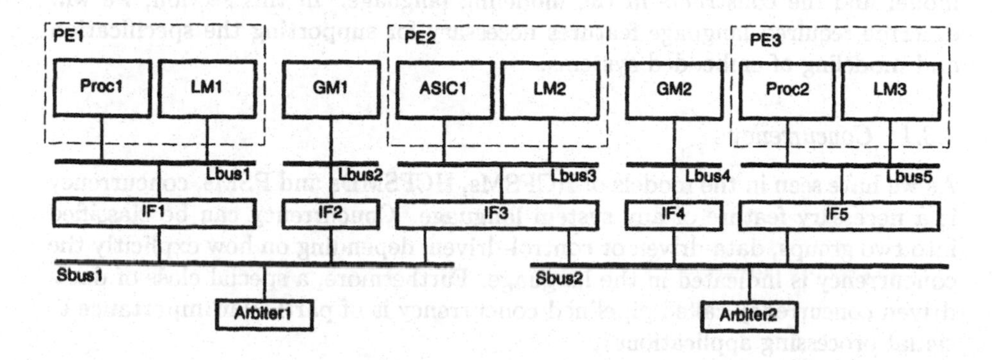


**Figure 1.13** A heterogeneous multiprocessor

In a heterogeneous processor both of the above mechanisms can be combined. Such a heterogeneous multiprocessor, in which the interconnection network consists of several buses, is shown in Figure 1.13. Each **processing element** (PE) consists of a processor or ASIC and a local memory connected by the local bus. The shared or global memory may be either single port, dual port, or special purpose memory such as FIFO. The PEs and global memories are connected by one or more **system buses** via corresponding **interfaces**. The system bus is associated with a well-defined **protocol** to which the components on the bus have to respect. The protocol may be standard, such as VME bus, or custom. An **interface** bridges the gap between a local bus of a PE/memory and system buses.

## 1.3 LANGUAGES

In order to map models of computation into an architecture we need languages to specify the initial system functionality and then to describe the system on different levels of abstraction at each design step.

In order to validate designs through simulation after each design step, we need an **executable modeling** language (see also Chapter 7 and 6). Such a language should also be able to describe design artifacts from previous designs and **intellectual properties** (IPs) provided by other sources.

Since different abstraction levels possess different characteristics, any given modeling language can be well or poorly suited for that particular abstraction level, depending on whether it supports all or just a few of the design characteristics. To find the language that can capture a given model directly, we need to establish a one-to-one correlation between the characteristics of the model and the constructs in the modeling language. In this section, we will describe required language features necessary for supporting the specification and modeling of embedded systems.

### *1.3.1 Concurrency*

As we have seen in the models of HCFSMs, HCFSMDs and PSMs, concurrency is a necessary feature of any system language. Concurrency can be classified into two groups, data-driven or control-driven, depending on how explicitly the concurrency is indicated in the language. Furthermore, a special class of data-driven concurrency called pipelined concurrency is of particular importance to signal processing applications.

**Data-driven concurrency** As we have seen in the DFG model, operation execution depends only upon the availability of data, rather than upon the physical location of the operation or statement in the language. Dataflow representations can be easily described from programming languages using the **single assignment rule**, which means that each variable can appear exactly once on the left hand side of an assignment statement.

**Pipelined concurrency** Since in a dataflow description the execution of each operation is determined by the availability of its input data, the degree of concurrency that can be exploited is limited by data dependencies. However, when the same dataflow operations are applied to a stream of data samples, we can use **pipelined** concurrency to improve the throughput, that is, the rate at which the system is able to process the data stream. Such throughput improvement is achieved by dividing operations into groups, called pipeline **stages**, which operate on different data sets in the stream. By operating on different data sets, pipeline stages can run concurrently. Note that each stage will take the same amount of time, called a **cycle**, to compute its results.

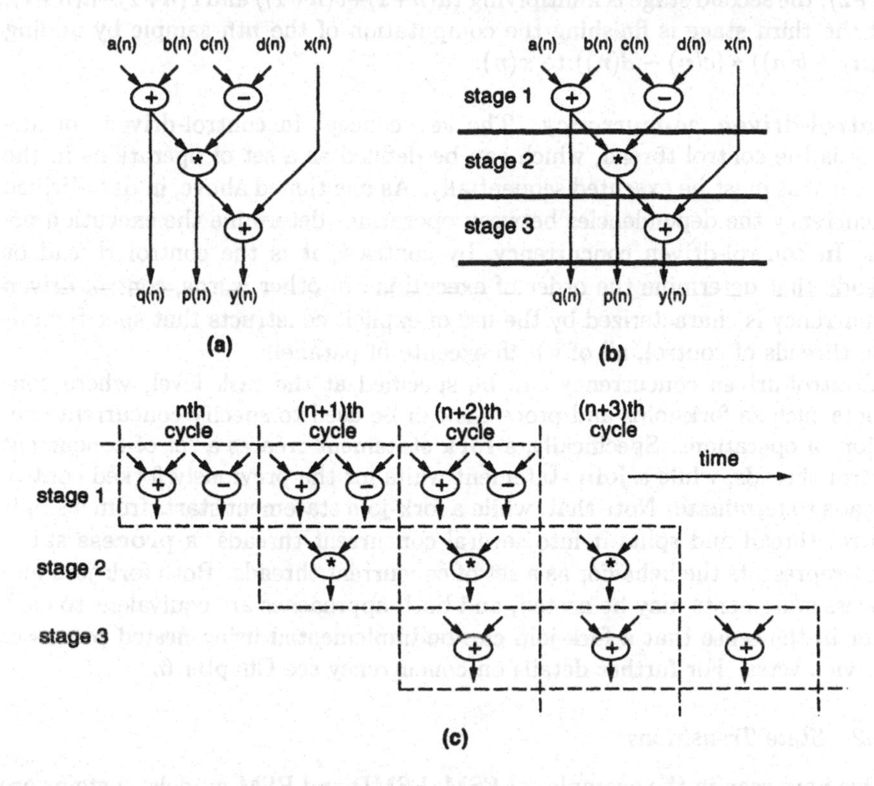


**Figure 1.14** Pipelined concurrency: (a) original dataflow, (b) pipelined dataflow, (c) pipelined execution.

For example, Figure 1.14(a) shows a dataflow graph operating on the data set $a(n), b(n), c(n), d(n)$ and x(n), while producing the data set $q(n), p(n)$ and $y(n)$, where the index $n$ indicates the $n$th data in the stream, called **data sample** $n$. Figure 1.14(a) can be converted into a pipeline by partitioning the graph into three stages, as shown in Figure 1.14(b).

In order for the pipeline stages to execute concurrently, storage elements such as registers or FIFO queues have to be inserted between the stages (indicated by thick lines in Figure 1.14(b)). In this way, while the second stage is processing the results produced by the first stage at the previous cycle, the first stage can simultaneously process the next data sample in the stream. Figure 1.14(c) illustrates the pipelined execution of Figure 1.14(b), where each row represents a stage, each column represents a cycle. In the third column, for example, while the first stage is adding $a(n+2)$ and $b(n+2)$, and subtracting $c(n+2)$ and $d(n+2)$, the second stage is multiplying $(a(n+1)+b(n+1))$ and $c(n+1)-d(n+1)$, and the third stage is finishing the computation of the $n$th sample by adding $((a(n)+b(n)) * (c(n)-d(n))$ to $x(n)$.

**Control-driven concurrency** The key concept in control-driven concurrency is the control thread, which can be defined as a set of operations in the system that must be executed sequentially. As mentioned above, in data-driven concurrency the dependencies between operations determine the execution order. In control-driven concurrency, by contrast, it is the control thread or threads that determine the order of execution. In other words, control-driven concurrency is characterized by the use of explicit constructs that specify multiple threads of control, all of which execute in parallel.

Control-driven concurrency can be specified at the task level, where constructs such as fork-joins and processes can be used to specify concurrent execution of operations. Specifically, a **fork** statement creates a set of concurrent control threads, while a **join** statement waits for the previously forked control threads to terminate. Note that, while a fork-join statement starts from a single control thread and splits it into several concurrent threads, a **process** statement represents the behavior as a set of concurrent threads. Both fork-join and process statements may be nested, and both approaches are equivalent to each other in the sense that a fork-join can be implemented using nested processes and vice versa. For further details on concurrency see Chapter 6.

### *1.3.2 State Transitions*

As we have seen in the examples of FSM, FSMD and PSM models, systems are often best conceptualized as having various **modes**, or states, of behavior. For example, a traffic-light controller [5] might incorporate different modes for day and night operation, for manual and automatic functioning, and for the status of the traffic light itself.

In systems with various modes, the transitions between these modes sometimes occur in an unstructured manner, as opposed to a linear sequencing through the modes. Such arbitrary transitions are akin to the use of *goto* statements in programming languages.

In systems like this, transitions between modes can be triggered by the detection of certain events or certain conditions. Furthermore, actions can be associated with each transition, and a particular mode or state can have an arbitrarily complex behavior or computation associated with it. In the case of the traffic-light controller, for example, in one state it may simply be sequencing between the red, yellow and green lights, while in another state it may be executing an algorithm to determine which lane of traffic has a higher priority based on the time of the day and the traffic density. For example, in the FSMD (Section 1.1.2) and HCFSMD (Section 1.1.5) models, simple assignment statements, such as $x = y+1$, can be associated with a state. In the PSM model (Section 1.1.7), any arbitrary program with iteration and branching constructs can be associated with a state.

### *1.3.3 Hierarchy*

Hierarchy is frequently used in modeling systems as we have seen in Section 1.1.5. First, since hierarchical models allow a system to be conceptualized as a set of smaller subsystems, the system modeler is able to focus on one subsystem at a time. This kind of modular decomposition of the system greatly simplifies the development of a conceptual view of the system. Furthermore, once we arrive at an adequate conceptual view, the hierarchical model greatly facilitates our comprehension of the system's functionality. Finally, a hierarchical model provides a mechanism for scoping objects, such as declaration types, variables and subprogram names. Since a lack of hierarchy would make all such objects global, it would be difficult to relate them to their particular use in the model, and could hinder our efforts to reuse these names in different portions of the same model.

There are two distinct types of hierarchy – structural hierarchy and behavioral hierarchy – where the first is used in architectural descriptions and the second in modeling descriptions.

**Structural hierarchy** A structural hierarchy is one in which a system specification is represented as a set of interconnected components. Each of these components, in turn, can have its own internal structure, which is specified with a set of lower-level interconnected components, and so on. Each instance of an interconnection between components represents a set of communication channels connecting the components.

This kind of structural hierarchy in systems can be specified at several different levels of abstraction. For example, a system can be decomposed into a set of processors and ASICs communicating over buses in a parallel architecture. Each of these chips may consist of several blocks, each representing a FSMD architecture. Finally, each RT component in the FSMD architecture can be further decomposed into a set of gates while each gate can be decomposed

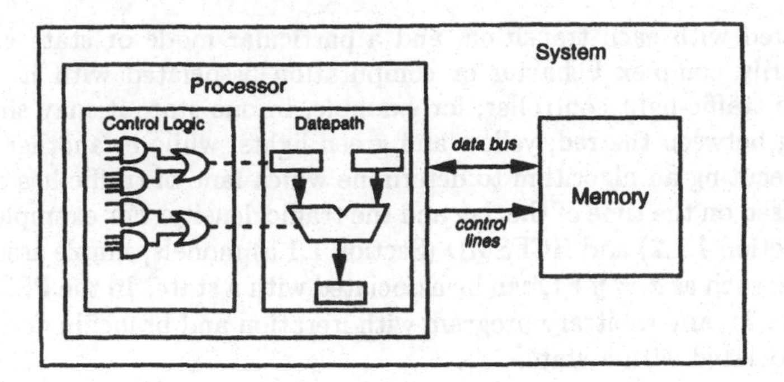


**Figure 1.15** Structural hierarchy. (†)

into a set of transistors. In addition, we should note that different portions of the system can be conceptualized at different levels of abstraction, as in Figure 1.15, where the processor has been structurally decomposed into a datapath represented as a set of RT components, and into its corresponding control logic represented as a set of gates.

**Behavioral hierarchy** The specification of a **behavioral hierarchy** is defined as the process of decomposing a behavior into distinct subbehaviors, which can be either sequential or concurrent.

The **sequential decomposition** of a behavior may be represented as either a set of procedures or a state machine. In the first case, a **procedural sequential decomposition** of a behavior is defined as the process of representing the behavior as a sequence of procedure calls. A procedural sequential decomposition of behavior $P$ is shown in Figure 1.16(a), where behavior $P$ consists of a sequential execution of the subbehaviors represented by procedures $Q$ and $R$. Behavioral hierarchy would be represented here by nested procedure calls.

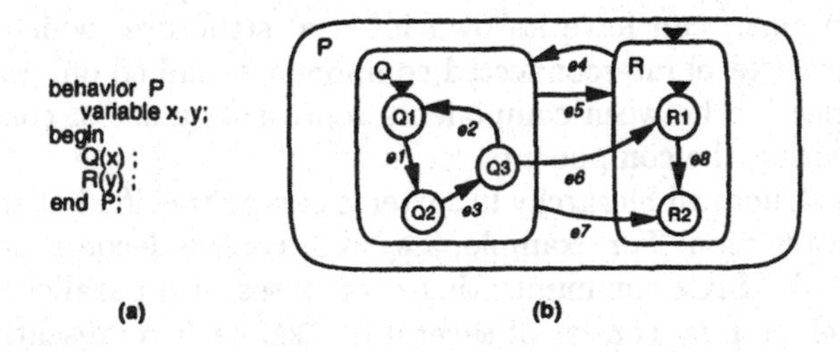


**Figure 1.16** Sequential behavioral decomposition: (a) procedures, (b) state-machines. (†)

Figure 1.16(b) shows a **state-machine sequential decomposition** of behavior $P$. In this diagram, $P$ is decomposed into two sequential subbehaviors $Q$ and $R$, each of which is represented as a state in a state-machine. This state-machine representation conveys hierarchy by allowing a subbehavior to be represented as another state-machine itself. Thus, $Q$ and $R$ are state-machines, so they are decomposed further into sequential subbehaviors. The behaviors at the bottom level of the hierarchy, including *Q1, ... R2*, are called **leaf behaviors**.

The **concurrent decomposition** of behaviors allows subbehaviors to run in parallel or in pipelined fashion.

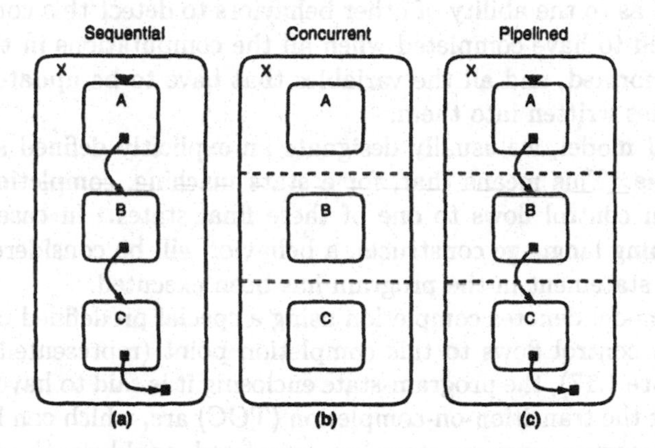


**Figure 1.17** Behavioral decomposition types: (a) sequential, (b) parallel, (c) pipelined.

Figure 1.17 shows a behavior $X$ consisting of three subbehaviors $A$, $B$ and $C$. In Figure 1.17(a) the subbehaviors are running sequentially, one at a time, in the order indicated by the arrows. In Figure 1.17(b), $A, B$ and $C$ run in parallel, which means that they will start when $X$ starts, and when all of them finish, $X$ will finish, just like the fork-join construct discussed in Section 1.3.1. In Figure 1.17(c), $A, B$ and $C$ run in pipelined mode, which means that they represent pipeline stages which run concurrently where $A$ supplies data to $B$ and $B$ to $C$ as discussed in Section 1.3.1.

### *1.3.4 Programming Constructs*

Many behaviors can best be described as sequential algorithms. Consider, for example, the case of a system intended to sort a set of numbers stored in an array, or one designed to generate a set of random numbers. In such cases the functionality can be most directly specified by means of an algorithm.

The advantage of using programming constructs to specify a behavior is that they allow the system modeler to specify an explicit sequencing for the computations in the system. Several notations exist for describing algorithms, but programming language constructs are most commonly used. These constructs include assignment statements, branching statements, iteration statements and procedures. In addition, data types such as records, arrays and linked lists are usually helpful in modeling complex data structures.

### *1.3.5 Behavioral Completion*

Behavioral completion refers to a behavior's ability to indicate that it has completed, as well as to the ability of other behaviors to detect this completion. A behavior is said to have completed when all the computations in the behavior have been performed, and all the variables that have to be updated have had their new values written into them.

In the FSM model, we usually designate an explicitly defined set of states as **final states**. This means that, for a state machine, completion will have occurred when control flows to one of these final states. In cases where we use programming language constructs, a behavior will be considered complete when the last statement in the program has been executed.

The PSM model denotes completion using a special predefined **completion point**. When control flows to this completion point (represented by a black square in Figure 1.17), the program-state enclosing it is said to have completed, at which point the transition-on-completion (TOC) arc, which can be traversed only when the source program-state has completed, could now be traversed.

The specification of behavioral completion has two advantages. First, in hierarchical specifications, completion helps designers to conceptualize each hierarchical level, and to view it as an independent module, free from interference from inter-level transitions. The second advantage of specifying behavioral completion is in the implementation of join constructs where a system has to wait for all concurrent behaviors to complete.

### *1.3.6 Exception Handling*

Often, the occurrence of a certain event can require that a behavior or mode be interrupted immediately, thus prohibiting the behavior from updating values further. Since the computations associated with any behavior can be complex, taking an indefinite amount of time, it is crucial that the occurrence of the event, or **exception**, should terminate the current behavior immediately rather than having to wait for the computation to complete. When such exceptions arise, the next behavior to which control will be transferred is indicated explicitly.

Depending on the direction of transferred control the exceptions can be further divided into two groups: (a) **abortion**, when the behavior is terminated,

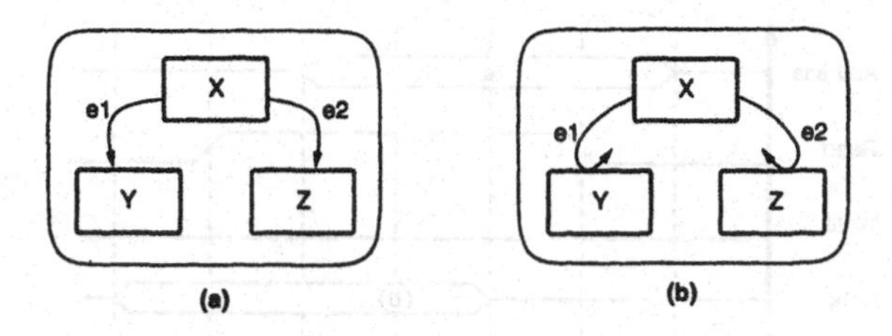


**Figure 1.18** Exception types: (a) abortion, (b) interrupt.

and (b) **interrupt**, where control is temporarily transferred to other behaviors. An example of an abortion is shown in Figure 1.18(a) where behavior *X* is terminated after the occurrence of events *e1* or *e2*. An example of interrupt is shown in Figure 1.18(b) where control from behavior *X* is transferred to *Y* or *Z* after the occurrence of *e1* or *e2* and is returned after their completion.

Examples of such exceptions include resets and interrupts in many computer systems.

### 1.3.7 *Timing*

Although computational models do not explicitly include timing, there may be need to specify detailed timing relations in system specification, when a component receives or generates events in specific time ranges, which are measured in real time units such as nanoseconds.

In general, a timing relation can be described by a 4-tuple $T = (e1, e2, min, max)$, where event e1 preceeds e2 by at least *min* time units and at most *max* time units. When such a timing relation is used with real components it is called **timing delay**, when it is used with component specifications it is called **timing constraint**.

Such timing information is especially important for describing parts of the system which interact extensively with the environment according to a predefined **protocol**. The protocol defines the set of timing relations between signals, which both communicating parties have to respect.

A protocol is usually visualized by a **timing diagram**, such as the one shown in Figure 1.19 for the read cycle of a static RAM. Each row of the timing diagram shows a waveform of a signal, such as *Address, Read, Write* and *Data* in Figure 1.19. Each dashed vertical line designates an occurrence of an event, such as *t1, t2* through *t7*. There may be timing delays or timing constraints associated with pairs of events, indicated by an arrow annotated by $x/y$, where $x$ stands for the *min* time, $y$ stands for the *max* time. For example, the arrow between *t1* and *t3* designates a timing delay, which says that *Data*

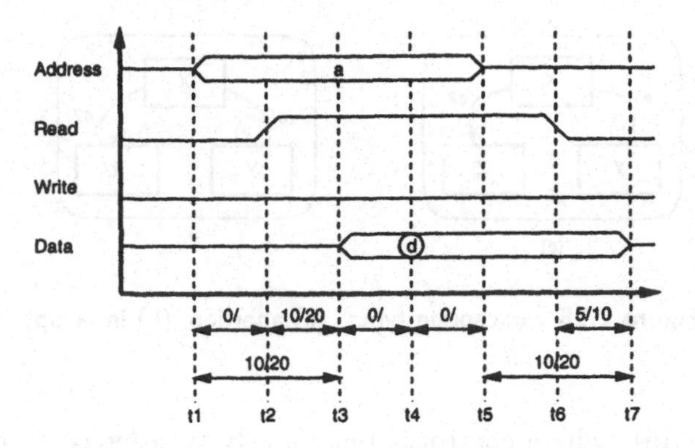


**Figure 1.19** Timing diagram

will be valid at least 10, but no more than 20 nanoseconds after *Address* is valid.

The timing information is very important for the subset of embedded systems known as real time systems, whose performance is measured in terms of how well the implementation respects the timing constraints. A favorite example of such systems would be an aircraft controller, where failure to respond to an abnormal event in a predefined timing limit will lead to disaster.

### *1.3.8 Communication*

In general, systems consist of several interacting behaviors which need to communicate with each other. In traditional programming languages the communication is predefined and hidden from the programmer. For example, functions communicate through global variables, which share a common memory space, or via parameter passing. In case of local procedure calls, parameter passing is implemented by exchanging information on the stack or through processor registers. In the case of remote procedure calls, parameters are passed via the complex protocol of marshaling/unmarshaling and sending/receiving data through a network.

While these mechanisms are sufficient for standard programming languages, they poorly address the needs for embedded systems, where it is necessary to (a) separate the description of computation and communication, (b) declare abstract communication functions, and (c) define a custom communication implementation.

In order to satisfy these requirements the language must include the concept of a **channels**. While the behavior specifies how the computation is performed and when the communication is started, the channels encapsulate the communication implementation.

Each behavior contains a set of **ports** through which the behavior can communicate. Each channel contains a set of communication functions and a set of **interfaces**. An interface declares a subset of the functions of the channel, which can be used by the connected behaviors. So while the declaration of the communication functions is given in the interfaces, the implementation of these functions is specified in the channel.

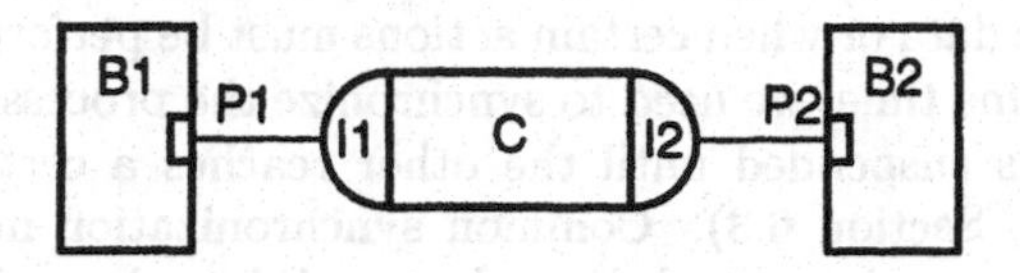


**Figure 1.20** Communication model.

For example, the system shown in Figure 1.20 contains two behaviors $B1$ and $B2$, and a channel $C$. Behavior $B1$ communicates with the left interface $I1$ of channel $C$ via its port $P1$. Similarly behavior $B2$ accesses the right interface $I2$ of channel $C$ through its port $P2$. Note that behaviors $B1$ and $B2$ can be easily replaced by other behaviors as long as the port types stay the same. Similarly channel $C$ can be exchanged with any other channels that provides compatible interfaces.

More specifically, a channel serves as an encapsulator of a set of variables, and a set of functions that operate on these variables. The functions specify how data is transferred over the channel. All accesses to the channel are restricted to these functions.

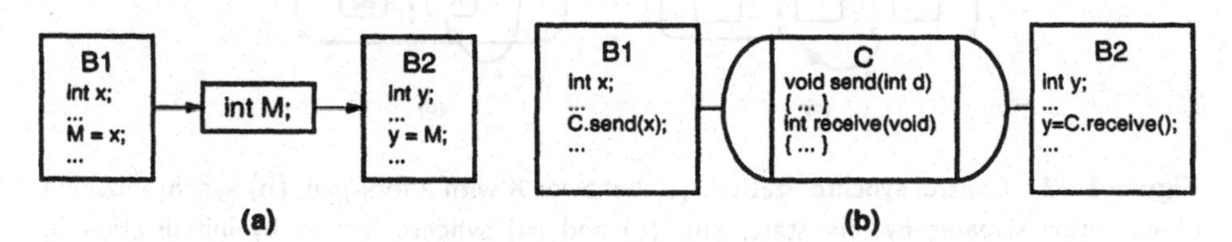


**Figure 1.21** Examples of communication: (a) shared memory, (b) channel.

For example, Figure 1.21 shows two communication examples. Figure 1.21(a) shows two behaviors communicating via a shared variable $M$. Figure 1.21(b) shows a similar situation using the channel model. In fact, communication through shared memory is just a special case of the general channel model.

A channel can also be hierarchical. For example, a channel may implement a high level communication protocol which breaks a stream of data packets into a byte stream, and in turn uses a lower level channel, for example a synchronous bus, which transfers the byte stream one bit at a time.

### *1.3.9 Process Synchronization*

In modeling concurrent processes, each process may generate data and events that need to be recognized by other processes. In cases like these, when the processes exchange data or when certain actions must be performed by different processes at the same time, we need to synchronize the processes in such a way that one process is suspended until the other reaches a certain point in its execution (see also Section 6.3). Common synchronization methods fall into two classifications, namely control-dependent and data-dependent schemes.

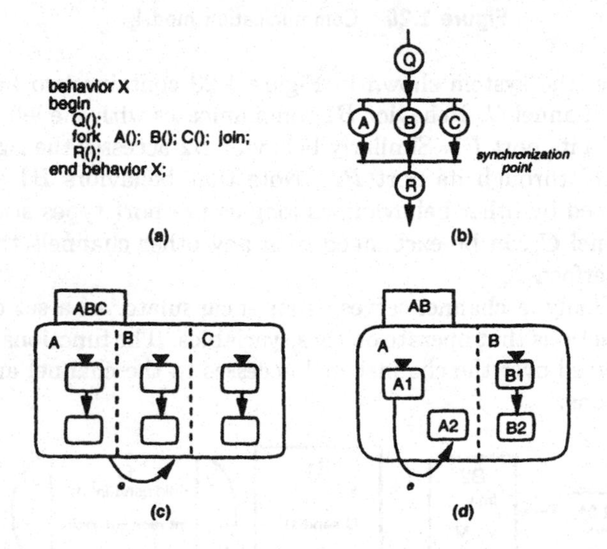


Figure 1.22 Control synchronization: (a) behavior X with a fork-join, (b) synchronization of execution streams by join statement, (c) and (d) synchronization by initialization in Statecharts. (†)

**Control-dependent synchronization** In control-dependent synchronization techniques, it is the control structure of the behavior that is responsible for synchronizing two processes in the system. For example, the ***fork-join*** statement introduced in Section 1.3.1 is an instance of such a control construct. Figure 1.22(a) shows a behavior $X$ which forks into three concurrent subprocesses, *A*, *B* and *C*. In Figure 1.22(b) we see how these distinct execution streams for the behavior *X* are synchronized by a *join* statement, which ensures that the three processes spawned by the fork statement are all complete before *R* is executed. Another example of control-dependent synchronization is the technique of **initialization**, in which processes are synchronized to their initial states either the first time the system is initialized, as is the case with most HDLs, or during the execution of the processes. In the Statecharts [5] of Figure 1.22(c), we can see how the event *e*, associated with a transition arc that reenters the boundary of *ABC*, is designed to synchronize all the orthogonal states *A*, *B* and *C* into their default substates. Similarly, in Figure 1.22(d), event *e* causes *B* to initialize to its default substate *B1* (since *AB* is exited and then reentered), at the same time transitioning *A* from *A1* to *A2*.

**Data-dependent synchronization** In addition to these techniques of control-dependent synchronization, processes may also be synchronized by means of one of the methods for interprocess communication: shared memory or message passing as mentioned in Section 1.3.8.

**Shared-memory based synchronization** works by making one of the processes suspend until the other process has updated the shared memory with an appropriate value. In such cases, the variable in the shared memory might represent an event, a data value or the status of another process in the system, as is illustrated in Figure 1.23 using the Statecharts language.

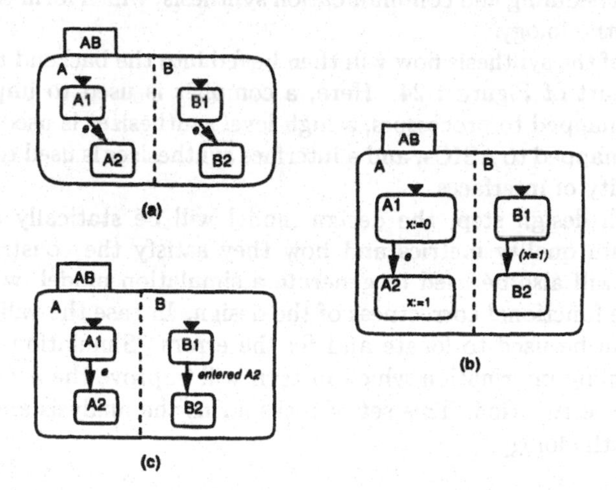


**Figure 1.23** Data-dependent synchronization in Statecharts: (a) synchronization by common event, (b) synchronization by common data, (c) synchronization by status detection. (†)

**Synchronization by common event** requires one process to wait for the occurrence of a specific event, which can be generated externally or by another process. In Figure 1.23(a), we can see how event *e* is used for synchronizing states *A* and *B* into substates *A2* and *B2*, respectively. Another method is that of **synchronization by common variable**, which requires one of the processes to update the variable with a suitable value. In Figure 1.23(b), *B* is synchronized into state *B2* when we assign the value "1" to variable $x$ in state *A2*.

Still another method is **synchronization by status detection**, in which a process checks the status of other processes before resuming execution. In a case like this, the transition from *A1* to *A2* precipitated by event *e*, would cause *B* to transition from *B1* to *B2*, as shown in Figure 1.23(c).

## 1.4 A GENERIC CO-DESIGN METHODOLOGY

In this section we present a methodology that converts an initial specification encapsulating one or more models of computation into an architecture leading to manufacturing by use of standard methods and CAD tools.

As shown in Figure 1.24, co-design may start from a high-level specification which specifies the functionality as well as the performance, power, cost, and other constraints of the intended design. During the co-design process, the designer will go through a series of well-defined design steps that include , partitioning, scheduling and communication synthesis, which form the synthesis flow of the methodology.

The result of the synthesis flow will then be fed into the backend tools, shown in the lower part of Figure 1.24. Here, a compiler is used to implement the functionality mapped to processors, a high-level synthesizer is used to map the functionality mapped to ASICs, and a interface synthesizer is used to implement the functionality of interfaces.

During each design step, the design model will be statically analyzed to estimate certain quality metrics and how they satisfy the constraints. This design model will also be used to generate a simulation model, which is used to validate the functional correctness of the design. In case the validation fails, a debugger can be used to locate and fix the errors. Simulation is also used to collect profiling information which in turn will improve the accuracy of the quality metrics estimation. This set of tasks forms the analysis and validation flow of the methodology.

### *1.4.1 System Specification*

We have described the language features needed for specifying systems in Section 1.3. The system specification should describe the functionality of the

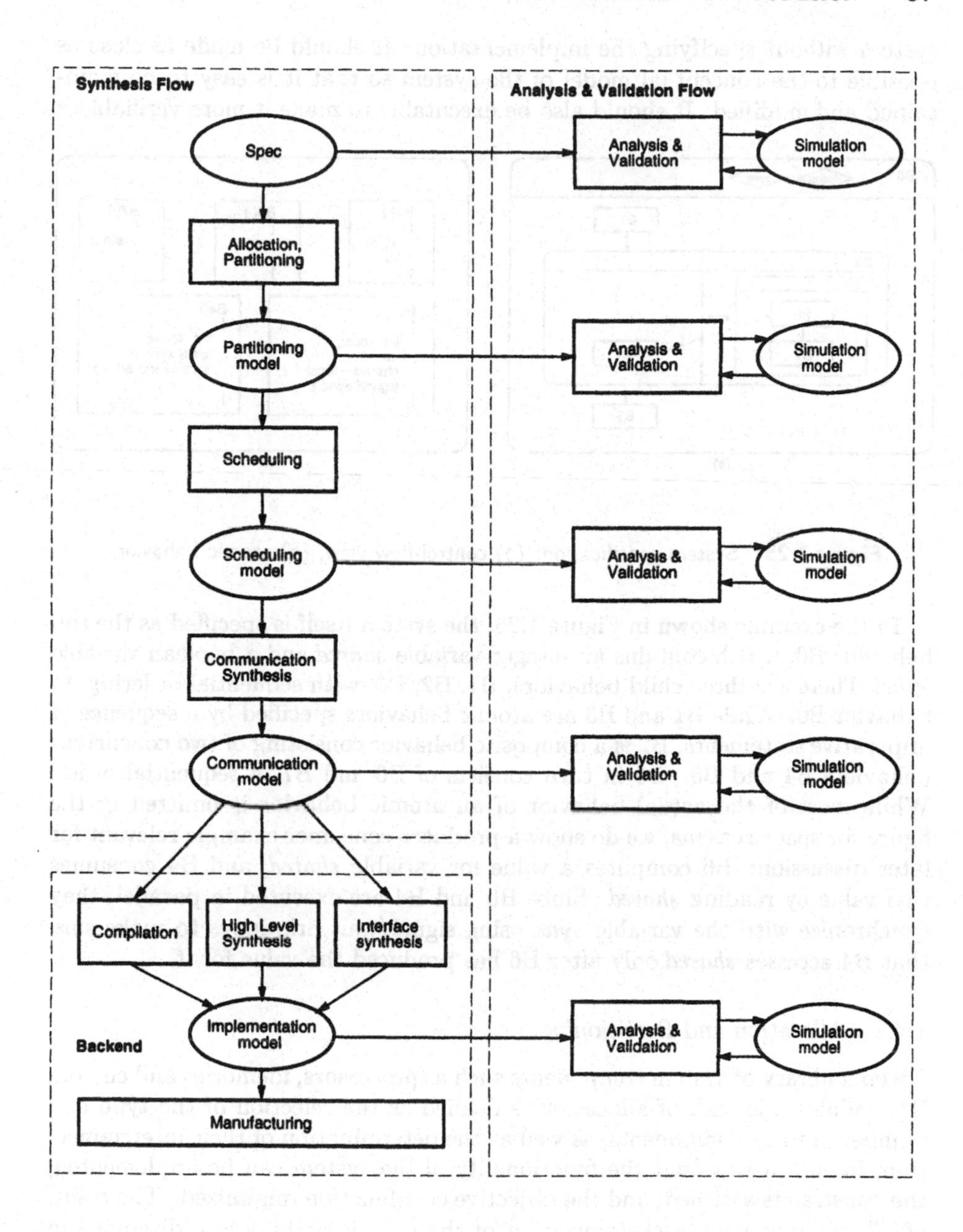


**Figure 1.24** A generic co-design methodology.

system without specifying the implementation. It should be made as close as possible to the conceptual model of the system so that it is easy to be maintained and modified. It should also be executable to make it more verifiable.

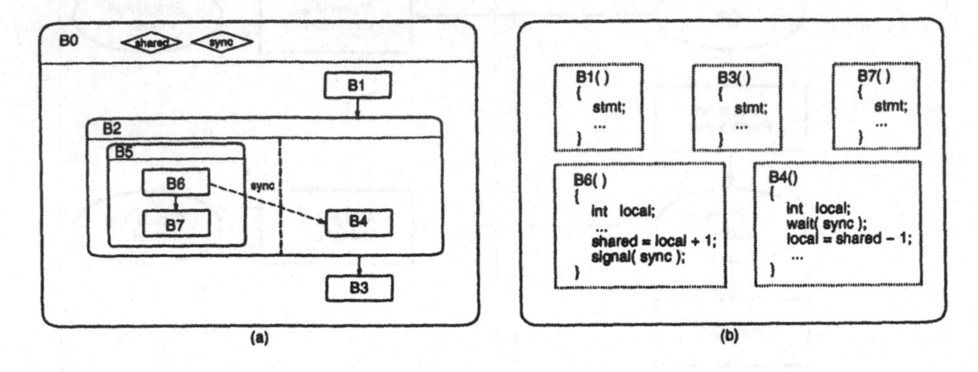


**Figure 1.25** System specification: (a) control-flow view, (b) atomic behaviors.

In the example shown in Figure 1.25, the system itself is specified as the top behavior B0, which contains an integer variable *shared* and a boolean variable *sync*. There are three child behaviors, B1, B2, B3, with sequential ordering, in behavior B0. While B1 and B3 are atomic behaviors specified by a sequence of imperative statements, B2 is a composite behavior consisting of two concurrent behaviors B4 and B5. B5 in turn consists of B6 and B7 in sequential order. While most of the actual behavior of an atomic behavior is omitted in the figure for space reasons, we do show a producer-consumer example relevant for later discussion: B6 computes a value for variable *shared*, and B4 consumes this value by reading *shared*. Since B6 and B4 are executed in parallel, they synchronize with the variable *sync* using signal/wait primitives to make sure that B4 accesses *shared* only after B6 has produced the value for it.

### *1.4.2 Allocation and Partitioning*

Given a library of system components such as processors, memories and custom IP modules, the task of allocation is defined as the selection of the type and number of these components, as well as the determination of their interconnection, in such a way that the functionality of the system can be implemented, the constraints satisfied, and the objective cost function minimized. The result of allocation can be a customization of the generic architecture discussed in Section 1.2.2. Allocation is usually carried out manually by designers and is the starting point of **design exploration**.

Partitioning defines the mapping between the set of behaviors in the specification and the set of allocated components in the selected architecture. The quality of such a mapping is determined by how well the result can meet the design constraints and minimize the design cost (see Chapter 2 for several partioning algorithms).

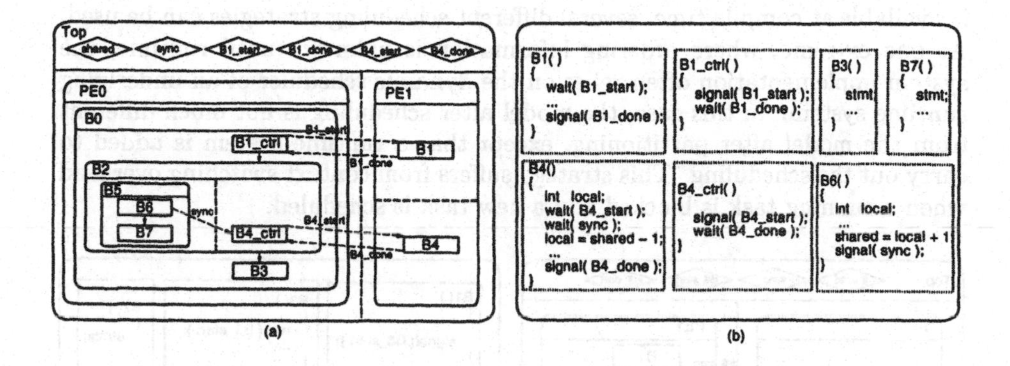

**Figure 1.26** System model after partitioning: (a) system model, (b) atomic behaviors.

The system model after partitioning must reflect the partitioning decision and must be complete in order for us to perform validation. As shown in Figure 1.26 an additional level of hierarchy is inserted to describe the selected partitioning into two processing elements, PE0 and PE1. Additional controlling behaviors are also inserted whenever child behaviors are assigned to different PEs than their parents. For example, in Figure 1.26, behavior B1_ctrl and B4_ctrl are inserted in order to control the execution of B1 and B4, respectively. Furthermore, in order to maintain the functional equivalence between the partitioned model and the original specification, synchronization operations between PEs must be inserted. In Figure 1.26 synchronization variables , *B1_start, B1_done, B4_start, B4_done* are added so that the execution of B1 and B4, which are assigned to PE1, can be controlled by their controlling behaviors B1_ctrl and B4_ctrl through inter-PE synchronization.

However, the model after partitioning is still far from implementation for two reasons: there are concurrent behaviors in each PE that have to be serialized; and different PEs communicate through global variables which have to be localized.

### *1.4.3 Scheduling*

Given a set of behaviors and possibly a set of performance constraints, scheduling determines a total order in invocation time of the behaviors running on the same PE, while respecting the partial order imposed by dependencies in the functionality as well as minimizing the synchronization overhead between the PEs and context switching overhead within the PEs.

Depending upon how much information on the partial order of the behaviors is available at compile time, several different scheduling strategies can be used. At one extreme, where ordering information is unknown until runtime, the system implementation often relies on the dynamic scheduler of an underlying runtime system. In this case, the model after scheduling is not much different from the model after partitioning, except that a runtime system is added to carry out the scheduling. This strategy suffers from context switching overhead when a running task is blocked and a new task is scheduled.

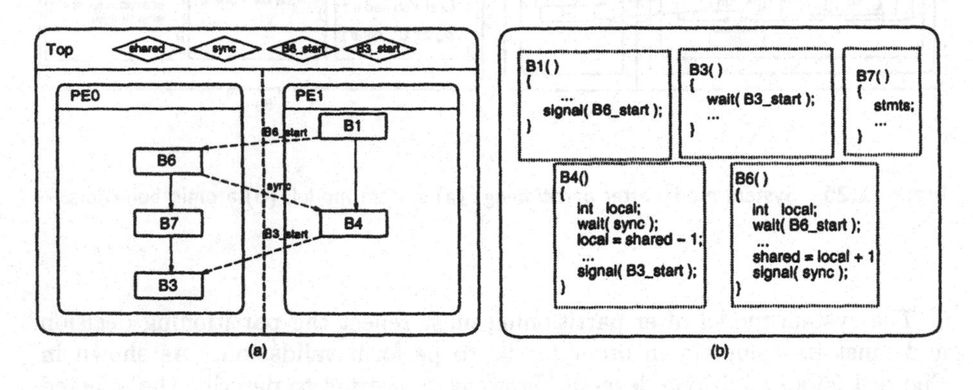


**Figure 1.27** System model after scheduling: (a) scheduling decision, (b) system model, (c) atomic behaviors.

On the other extreme, if the partial order is completely known at compile time, a static scheduling strategy can be taken, provided a good estimation on the execution time of each behavior can be obtained. This strategy eliminates the context switching overhead completely, but may suffer from inter-PE synchronization especially in the case of inaccurate performance estimation. On the other hand, the strategy based on dynamic scheduling does not have this problem because whenever a behavior is blocked for inter-PE synchronization, the scheduler will select another to execute. Therefore the selection of the scheduling strategy should be based on the trade-off between context switching overhead and CPU utilization.

The model generated after static scheduling will remove the concurrency among behaviors inside the same PE. As shown in Figure 1.27, all child behaviors in PE0 are now sequentially ordered. In order to maintain the partial order across the PEs, synchronization between them must be inserted. For example, B6 is synchronized by *B6_start*, which will be asserted by B1 when it finishes.

Note that B1_ctrl and B4_ctrl in the model after partitioning are eliminated by the optimization carried out by static scheduling. It should also be mentioned that in this section we define the tasks, rather than the algorithms of co-design. Good algorithms are free to combine several tasks together. For example, an algorithm can perform the partitioning and static scheduling at the same time, in which case intermediate results, such as B1_ctrl and B4_ctrl, are not generated at all.

### *1.4.4 Communication Synthesis*

Up to this stage, the communication and synchronization between concurrent behaviors are accomplished through shared variable accesses. The task of this stage is to resolve the shared variable accesses into an appropriate inter-PE communication scheme at implementation level. If the shared variable is a shared memory the communication synthesizer will determine the location of such variables and change all accesses to the shared variables in the model into statements that read or write to the corresponding addresses. If the variable is in the local memory of one particular PE all accesses to this shared variable in models of other PEs have to be changed into function calls to message passing primitives such as send and receive. In both cases the synthesizer also has to insert interfaces for the PEs and shared memories to adapt to different protocols on the buses.

The generated model after communication synthesis, as shown in Figure 1.28, differs from previous models in several aspects.

New behaviors for interfaces, shared memories and arbiters are inserted at the highest level of the hierarchy. In Figure 1.28 the added behaviors are *IF0, IF1, IF2, Shared_mem, Arbiter.*

The shared variables from the previous model are all resolved. They either exist in shared memory or in local memory of one or more PEs. The communication channels of different PEs now become the local buses and system buses. In Figure 1.28, we have chosen to put all the global variables in *Shared_mem*, and hence all the global declarations in the top behavior are moved to the behavior *Shared_mem.* New global variables in the top behavior are the buses *lbus0, lbus1, lbus2, sbus.*

If necessary, a communication layer is inserted into the runtime system of each PE. The communication layer is composed of a set of inter-PE communica-

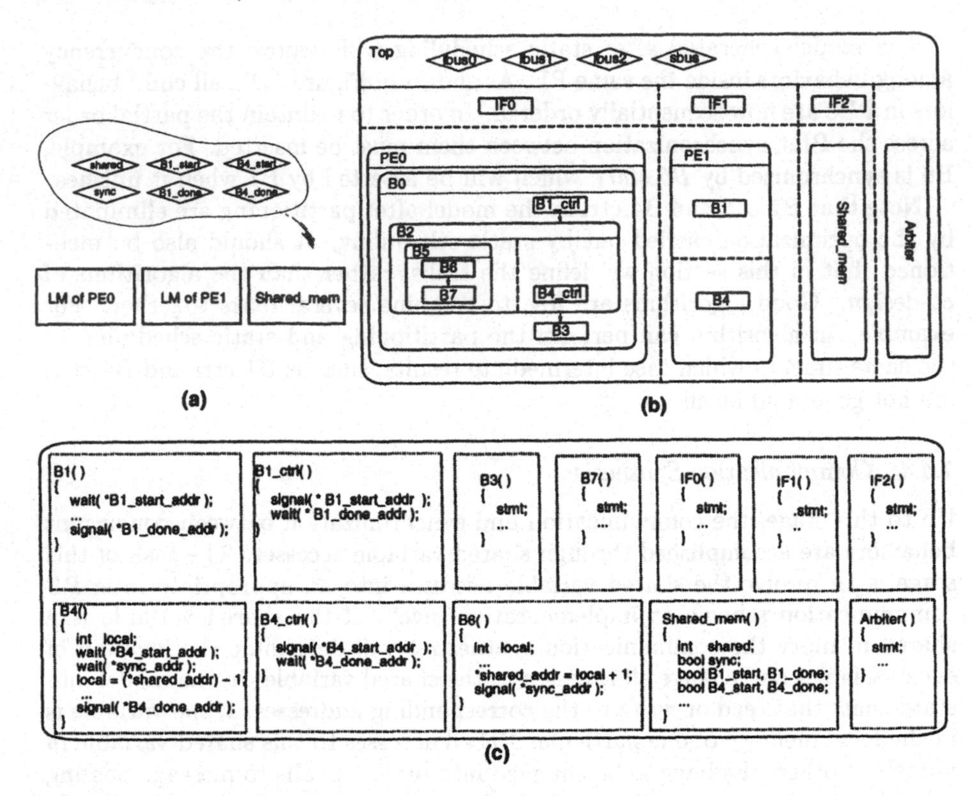


**Figure 1.28** System model after communication synthesis: (a) communication synthesis decision, (b) system model, (c) atomic behaviors.

tion primitives in the form of driver routines or interrupt service routines, each of which contain a stream of I/O instructions, which in turn talk to the corresponding interfaces. The accesses to the shared variables in the previous model are transformed into function calls to these communication primitives. For the simple case of Figure 1.28, the communication synthesizer will determine the addresses for all global variables, for example, ***shared_addr*** for variable ***shared***, and all accesses to the variables are appropriately transformed. The accesses to the variables are exchanged with reading and writing to the corresponding addresses. For example, ***shared = local + 1*** becomes ****shared_addr = local+1***.

### *1.4.5 Analysis and Validation Flow*

Before each design refinement, the input design model must be functionally validated through simulation or formal verification (see Chapter 6). It also needs to be analyzed, either statically, or dynamically with the help of the simulator or estimator, in order to obtain an estimation of the quality metrics, which will be evaluated by the synthesizer to make good design decisions. This motivates the set of tools to be used in the analysis and validation flow of the methodology. Such a tool set typically includes a static analyzer, a simulator, a debugger, a profiler, and a visualizer.

The **static analyzer** associates each behavior with quality metrics such as program size and program performance in case it is to be implemented as software, or metrics of hardware area and hardware performance if it is to be implemented as an ASIC. To achieve a fast estimation with satisfactory accuracy, the analyzer relies on probabilistic techniques and the knowledge of backend tools such as a compiler and high-level synthesizer.

The **simulator** serves the dual purpose of functional validation and dynamic analysis. The simulation model runs on a simulation engine, which in the form of runtime library, provides an implementation for the simulation tasks such as simulation time advance and synchronization among concurrent behaviors.

Simulation can be performed at different levels of accuracy, such as functional, cycle-based, and discrete-event simulation. A functionally accurate simulation compiles and executes the design model directly on a host machine without paying special attention to simulation time. A clock-cycle-accurate simulation executes the design model in a clock-by-clock fashion. A discrete-event simulation incorporates a even more sophisticated timing model of the components, such as gate delay. Obviously there is a trade-off between simulation accuracy and simulator execution time.

It should be noted that, while most design methodologies adopt a fixed accuracy simulation at each design stage, applying a mixed accuracy model is also possible. For example, consider a behavior representing a piece of software that performs some computation and then sends the result to an ASIC. While the part of the software which communicates with the ASIC needs to be simulated at cycle level so that tricky timing problems become visible, it is not necessary to simulate the computation part with the same accuracy.

The **debugger** renders the simulation with break point and single step ability. This makes it possible to examine the state of a behavior dynamically. A **visualizer** can graphically display the hierarchy tree of the design model as well as make dynamic data visible in different views and keep them synchronized at all times. All these efforts are invaluable in quickly locating and fixing the design errors.

The **profiler** is a good complement of a static analyzer for obtaining dynamic information such as branching probability. Traditionally, it is achieved by instrumenting the design description, for example, by inserting a counter at every conditional branch to keep track of the number of branch executions.

### *1.4.6 Backend*

At the stage of the backend, as shown in the lower part of Figure 1.24, the leaf behaviors of the design model will be fed into different tools in order to obtain their implementations. If the behavior is assigned to a standard processor, it will be fed into a compiler for this processor. If the behavior is to be mapped on an ASIC, it will be synthesized by a high-level synthesis tool. If the behavior is an interface, it will be fed into an interface synthesis tool.

A **compiler** translates the design description into machine code for the target processor (see Chapter 5 for more details). The **high-level synthesizer** translates the behavioral design model into a netlist of register-transfer level (RTL) components.

We define an interface as a special type of ASIC which links the PE that it is associated (via its native bus) with other components of the system (via the system bus). Such an interface implements the behavior of a communication channel. An example of such an interface translates a read cycle on a processor bus into a read cycle on the system bus. The communication tasks between different PEs are implemented jointly by the driver routines and interrupt service routines implemented in software and the interface circuitry implemented in hardware. While partitioning the communication task into software and hardware, and model generation for the two parts is the job of communication synthesis, the task of generating an RTL design from the interface model is the job of **interface synthesis**. The synthesized interface must harmonize the hardware protocols of the communicating components.

## 1.5 CONCLUSIONS

Codesign represents the methodology for specification and design of systems that include hardware and software components. A codesign methodology consists of design tasks for refining the design and the models representing the refinements. Hardware/software co-design is a very active research area which is just beginning to be commercialized; most commercial co-design tools are co-simulation engines. The next few years should see advances in both the theory and commercial application of specification languages, architectural exploration tools, algorithms for partitioning, scheduling, and synthesis, and backend tools for custom software and hardware synthesis.

**Acknowledgements**

We would like to acknowledge the support provided by UCI grant #TC20881 from Toshiba Inc. and grant #95-D5-146 and #96-D5-146 from Semiconductor Research Corporation.

We would also like to acknowledge Prentice-Hall Inc., Upper Saddle River, NJ 07458, for the permission to reprint figures from [8] (annotated by †), figures from [2] (annotated by ‡), and the partial use of text appearing in Chapter 2 and Chapter 3 in [8] and Chapter 6 and Chapter 8 in [2].

We would also like to thank Jie Gong, Sanjiv Narayan and Frank Vahid for valuable insights and early discussions about models and languages. Furthermore, we want to acknowledge Jon Kleinsmith, En-shou Chang, Tatsuya Umezaki for contributions in language requirements and model development.

We would also like to acknowledge the book editors Wayne Wolf and Jørgen Staunstrup who helped us to improve this chapter.

# 2 HARDWARE/SOFTWARE CO-SYNTHESIS ALGORITHMS

Wayne Wolf

Department of Electrical Engineering
Princeton University
Princeton, New Jersey USA

## 2.1 INTRODUCTION

This chapter surveys methodologies and algorithms for hardware-software co-synthesis. While much remains to be learned about co-synthesis, researchers in the field have made a great deal of progress in a short period of time. Although it is still premature to declare an authoritative taxonomy of co-synthesis models and methods, we can now see commonalities and contrasts in formal models and algorithms.

Hardware-software co-synthesis *simultaneously* designs the software architecture of an application and the hardware on which that software is executed [9]. The problem specification includes some description of the required **functionality**; it may also include **non-functional requirements** such as performance goals. Co-synthesis creates a **hardware engine** consisting of one or more **processing elements (PEs)** on which the functionality is executed; it also creates



*J. Staunstrup and W. Wolf (eds.), Hardware/Software Co-Design: Principles and Practice*, 47-73.

an **application architecture**, which consists in part of an **allocation** of the functionality onto PEs in the hardware engine. Some of the functions may be allocated to programmable CPUs, while others may be implemented in ASICs which are created as a part of the co-synthesis process. Communication channels to implement the communication implied by the functionality must also be included in the system architecture—in general, this entails both hardware elements and software primitives. In order to complete the implementation, a schedule for execution is required, along with a component mapping and other details.

If either the hardware engine design or the application architecture were given, the problem would not be a co-design problem. Co-synthesis allows trade-offs between the design of the application and the hardware engine on which it executes. While such trade-offs may not be as important in data processing applications, where application programs are regularly changed and the hardware engine is large and immovable, trade-offs are particularly important in embedded computing. Embedded computing applications almost always have cost constraints; they may also have power, weight, and physical size limitations. Being able to, for example, modify the hardware engine to simplify some aspect of the application software is often critical in meeting the requirements of sophisticated embedded systems.

There are several different styles of co-design, depending on what assumptions are made about the specification and the components and what elements of the system are synthesized. The next section sets the background for co-synthesis. Section 2.2 reviews some background material: models for the behavior of the system to be synthesized, models of architectures, and real-time systems. Section 2.4 introduces **hardware/software partitioning**, which maps a behavioral specification onto a hardware architecture consisting of a CPU and multiple ASICs. Section 2.5 describes algorithms for **distributed system co-synthesis**, which synthesizes arbitrary hardware topologies.

There are strong reasons to use embedded CPUs to help implement complex systems. Implementing a system in software on an existing CPU is much simpler than designing logic by hand, so long as other constraints such as speed, power consumption, etc. can be met. Software can also be changed quickly and in many cases much more cheaply than is possible when changing an ASIC.

In this naive view, the implementation choices are between one CPU and an all-custom hardware solution. If we can find a CPU that can execute the code at the required speed and meet the other non-functional constraints, we use the CPU, otherwise we bite the bullet and design custom hardware. But in fact, the choices are more complex and subtle than that—we can design systems which include combinations of CPUs and special-purpose function units which together perform the required function. When we design sophisticated, hetero-

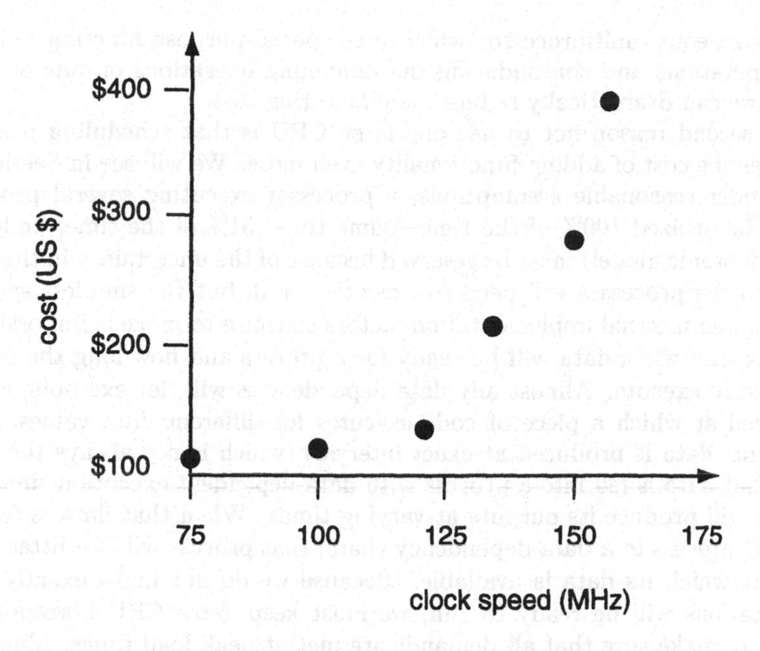


**Figure 2.1** Performance as a function of price.

geneous systems, we need hardware/software co-synthesis to help us evaluate the design alternatives.

Why should we ever use more than one CPU or function unit? Why not use the fastest CPU available to implement a function? There are two reasons. The first is that CPU cost is an exponential function of performance. Figure 2.1 shows the late-1996 retail prices of Pentium Processors (TM) at various clock rates. The chart shows that the processor cost is truly exponential function of its clock rate, a fact that should be no surprise since the yield of chips exponentially decreases with increased speed.

The exponential cost/performance curve of CPUs implies that paying for performance in a uniprocessor is very expensive. If a little extra processing power is needed to add an extra function to the system, the performance required to execute that function will be very expensive. It is often cheaper to use several small processors to implement a function, even when communication overhead is included. And the processors in the multiprocessor need not be CPUs—they may be special-purpose function units as well. Using a special-purpose unit to accelerate a function can result in a much cheaper implementation (as measured in manufacturing cost, not necessarily in design cost) than using even a dedicated CPU for that function. By co-designing a heterogeneous multiprocessor which uses special-purpose function units for some operations and consolidating the remaining operations on one or a few CPUs, we can dramatically reduce manufacturing cost.

The second reason not to use one large CPU is that scheduling overhead increases the cost of adding functionality even more. We will see in Section 2.3 that, under reasonable assumptions, a processor executing several processes cannot be utilized 100% of the time—some time (31% of the time, under the rate-monotonic model) must be reserved because of the uncertainty in the times at which the processes will need to execute. In all but the simplest systems, external and internal implementation factors conspire to make it impossible to know exactly when data will be ready for a process and how long the process will take to execute. Almost any data dependencies will, for example, change the speed at which a piece of code executes for different data values. Even if external data is produced at exact intervals (which is not always the case), when that data is fed into a process with data-dependent execution times, the process will produce its outputs at varying times. When that data is fed into the next process in a data-dependency chain, that process will see jitter in the times at which its data is available. Because we do not know exactly when computations will be ready to run, we must keep extra CPU horsepower in reserve to make sure that all demands are met at peak load times. Since that extra CPU performance comes at exponentially-increasing cost, it often makes sense to divide a set of tasks across multiple CPUs.

The job of hardware/software co-synthesis is to create both a **hardware architecture** and a **software architecture** which implements the required function and meets other **non-functional** goals such as speed, power consumption, etc. The notion of a hardware architecture is familiar—it is described as some sort of block diagram or equivalent. The reader may be less comfortable with a software architecture, but such things do exist. While the division of a program into modules, functions, etc. can be considered its architecture, a more useful view for embedded systems is that the software architecture is defined by the **process structure** of the code. Each process executes sequentially, so the division of functions into processes determines the amount of parallelism which we can exploit in the design of the system. The process structure also influences other costs, such as data transfer requirements. However, dividing a system into too many processes may have negative effects both during and after design: during design, having too many processes may obscure design decisions; as the system executes, context-switching and scheduling overhead may absorb any parallelism gains. As a result, proper divison of a function into a set of software processes is critical for obtaining a cost-effective implementation.

## 2.2 PRELIMINARIES

Some co-synthesis systems which deal with **single-rate** behavior while others can synthesize **multi-rate** behavior. A single-rate system would generally perform a single complex function. A good example of a multi-rate system is an audio/video decoder: the audio and video streams are encoded at different sampling rates, requiring the specification to be written as multiple components, each running at its own rate.

No matter what model is used for behavior, we are often interested in specifying the required performance, this being the most common form of non-functional specification. We generally assume that the behavior model describes one iteration of a behavior which is repeatedly executed. Two performance parameters are commonly used:

- The **rate** at which a behavior is executed is the reciprocal of the maximum interval between two successive initiations of the behavior.
- The **latency** of a behavior is the required maximum time between starting to execute the behavior and finishing it.

Rate and latency are both required in general because of the way in which data may arrive. In simple systems, data will arrive at a fixed time in each sampling period and go through a fixed set of processing. In more complex systems, data may arrive at different points in the processing period; this may be due either to exernal causes or because the data is generated by another part of the system which has data-dependent execution times. Rate specifies the intervals in which data samples may occur, while latency determines how long is allowed to process that data.

The standard model for single-rate systems is the **control/data flow graph (CDFG)**, which is described in more detail in Section 7.5. A CDFG's semantics imply a program counter or system state, which is a natural model for either software or a harwire-controlled datapath. The unified system state, however, makes it difficult to describe multi-rate tasks.

A common multi-rate model is the **task graph**, which was introduced in Section 7.3.5. The most common form of the task graph is shown in Figure 2.2. The task graph has a structure very similar to a data flow graph, but in general the nodes in a task graph represent larger units of functionality than is common in a data flow graph. Edges in the task graph represent data communication. This model describes multiple flows of control because one datum output from a node can be sent to two successor nodes, both of which are activated and executed in some unspecified order. Unfortunately, the terminology used for the task graph varies from author to author. We prefer to refer to the nodes as **processes**, since that is what the nodes most often represent. Some authors,

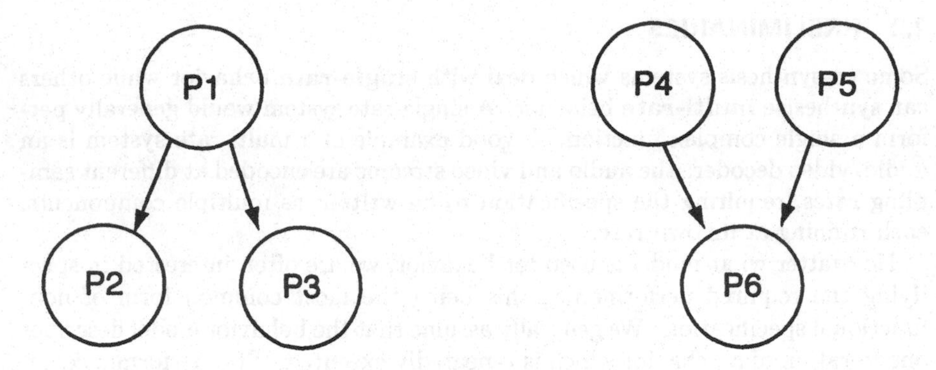


**Figure 2.2** An example task graph.

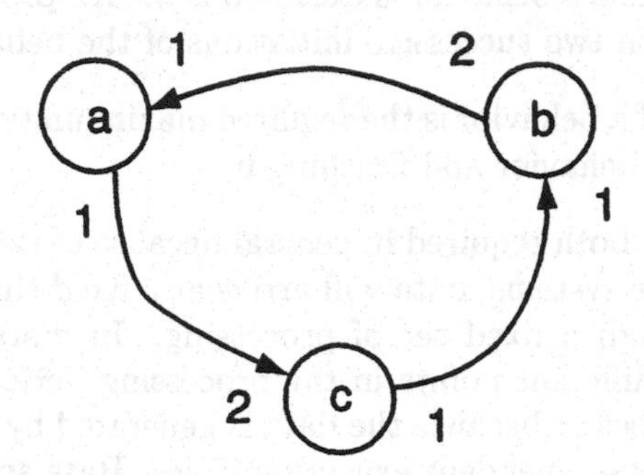


**Figure 2.3** A synchronous data flow graph.

however, refer to each node as a task. The task graph may contain several unconnected components; we prefer to call each component a **subtask**. Subtasks allow the description of multi-rate behavior, since we can assign a separate rate to each subtask.

A multi-rate behavior model particularly well-suited to signal processing systems is the **synchronous data flow graph (SDFG)** [3]. An example is shown in Figure 2.3. As in a data flow graph, nodes represent functions and arcs communication. But unlike most data flow graphs, a valid SDFG may be cyclic. The labels at each end of the arcs denote the number of samples which are produced or consumed per invocation—for example, process *b* emits two samples per invocation while process *a* consumes one per invocation. This implies that *a* runs at twice the rate of *b*. Lee and Messerschmitt developed algorithms to determine whether a synchronous data flow graph is feasible and to construct schedules for the SDFG on either uniprocessors or multiprocessors.

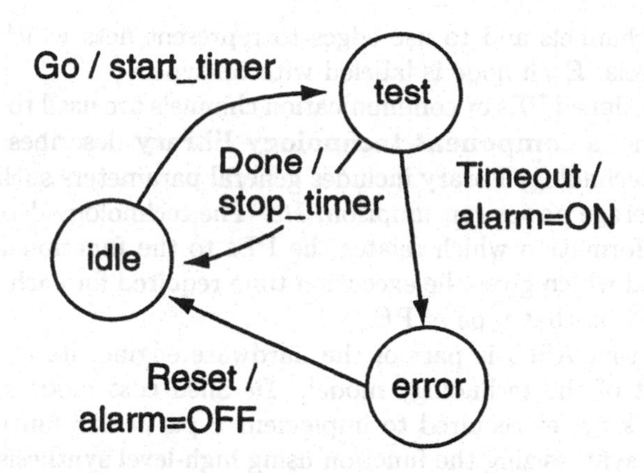


**Figure 2.4** A co-design finite-state machine.

The **co-design finite-state machine** (CFSM) is used as a model of behavior by the POLIS system [10]. A CFSM is an event-driven state machine—transitions between states are triggered by the arrival of events rather than by a periodic clock signal. As a result, the component CFSMs in a network need not all change state at the same time, providing a more abstract model for time in the system. A simple CFSM is illustrated in Figure 2.4. When the machine receives an event, its action depends on its current state and the defined transitions out of those states. Upon taking a transition to a new state, the machine may emit events, which can go either to the outside world or other CFSMs in the system. (Transitions in the figure are annotated with input / output.) A network of CFSMs can be interpreted as a network of non-deterministic FSMs, though that description will generally be less compact than the CFSM network.

## 2.3 ARCHITECTURAL MODELS

We also need architectural models to describe the implementation. The allocation of elements of the functional description to components in the hardware architecture is generally described as a map, so the greatest need is for a description of the hardware engine. Chapter 4 describes CPU architectures in detail; here we concentrate on basic models sufficient for cost estimation during co-synthesis.

The engine itself is generally modeled as a graph. In the simplest form, processing elements are represented by nodes and **communication channels** by edges. However, since an edge can connect only two nodes, busses are hard to accurately model in this way. Therefore, it is often necessary to introduce nodes for the channels and to use edges to represent nets which connect the PEs and channels. Each node is labeled with its type.

When pre-designed PEs or communication channels are used to construct the hardware engine, a **component technology library** describes their characteristics. The technology library includes general parameters such as manufacturing cost, average power consumption, *etc.* The technology description must also include information which relates the PEs to the functional elements—a table is required which gives the execution time required for each function that can be executed on that type of PE.

When a custom ASIC is part of the hardware engine, its clock rate is an important part of the technology model. Detailed cost metrics, such as the number of clock cycles required to implement a particular function, must be determined by synthesizing the function using high-level synthesis or otherwise estimating the properties of that implementation. The synthesis system usually makes assumptions about the communication mechanism used to talk to the rest of the system—shared memory, for example.

Some background in CPU scheduling is useful for co-synthesis since these systems must run several components of the functionality on a CPU. The units of execution on a CPU are variously called **processes, threads**, or **lightweight processes**. Processes on a workstation or mainframe generally have separate address spaces to protect users from each other; in contrast, a lightweight process or thread generally executes in a shared memory space to reduce overhead. We will generally use these terms interchangeably since the details of operating system implementation are not important for basic performance analysis.

The processes on a CPU must be scheduled for execution, since only one may run at any given time. In all but the simplest implementations, some decision-making must be done at run time to take into account variabilities in data arrival times, execution times, *etc.* The simplest scheduling mechanism is to maintain a queue of processes which are ready to run and to execute them in order. A more complex mechanism is to assign **priorities** to each process (either statically or dynamically) and to always execute the highest-priority process which is ready to run. Figure 2.5 illustrates the prioritized execution of two processes *P1* and *P2*, where *P1* has higher priority (presumably because it has a tighter deadline to meet). In the first case, *P2* has become ready to execute and obtains control of the CPU because no higher-priority process is running. When *P1* becomes ready, however, *P2*'s execution is suspended and *P1* runs to completion. When *P1* is running and *P2* becomes ready, however, *P2* has to wait for *P1* to finish. Unlike timesharing systems, embedded and real-time systems generally do not try to time-slice access to the CPU, but allow the highest-priority process to run to completion. Prioritized execution gives significantly higher utilization of the CPU resources—in general, prioritization

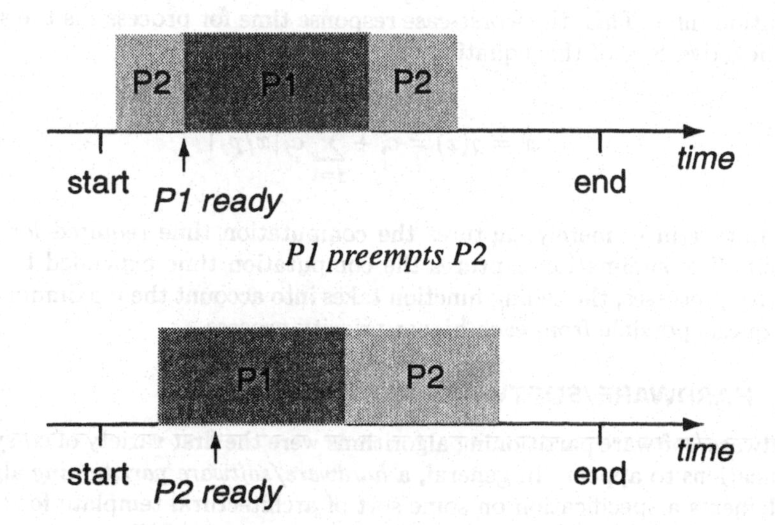


**Figure 2.5** An example of prioritized process execution.

allows a slower CPU to meet the performance requirements of a set of processes than would be required for non-prioritized execution.

The real-time systems community has studied the execution of processes on CPUs to meet **deadlines**. The survey by Stankovic *et al.* [11] gives a good overview of results in real-time scheduling. Two common real-time scheduling algorithms are **rate-monotonic analysis (RMA)** [12], which is a static-priority scheme, and **earliest deadline first (EDF)**, which can be modeled as a dynamic priority scheme. RMA assumes that the deadline for each process is equal to the end of its period. Liu and Layland analyzed the worst-case behavior of a set of processes running on a CPU at different rates. They showed that, under their assumptions, a static prioritization was optimal, with the highest priority going to the process with the shortest period. They also showed that the highest possible utilization of the CPU in this case is 69%—some excess CPU capacity must always be left idle to await possibly late-arriving process activations. EDF does not assume that deadline equals period. This scheduling policy gives highest priority to the active process whose deadline is closest to arriving.

Lehoczky *et al.* [13] formulated the response time (the time from a process becoming ready to the time it completes execution for that period) for a set of prioritized processes running on a single CPU. For ease of notation, assume that the highest-priority process is numbered 1 and that the process under consideration is numbered $i$. Let $p_j$ be the period of process $j$ and $c_j$ be its execution time. Then the worst-case response time for process $i$ is the smallest non-negative root of this equation:

$$x = g(x) = c_i + \sum_{j=1}^{i-1} c_j \lceil x/p_j \rceil \qquad (2.1)$$

The first term $c_i$ merely captures the computation time required for process $i$ itself. The summation captures the computation time expended by higher-priority processes; the ceiling function takes into account the maximum number of requests possible from each higher-priority process.

## 2.4 HARDWARE/SOFTWARE PARTITIONING

Hardware/software partitioning algorithms were the first variety of co-synthesis applications to appear. In general, a ***hardware/software partitioning*** algorithm implements a specification on some sort of architectural template for the multiprocessor, usually a single CPU with one or more ASICs connected to the CPU bus. We use the term ***allocation*** for synthesis methods which design the multiprocessor topology along with the processing elements and software architecture.

In most hardware/software partitioning algorithms, the type of CPU is normally given, but the ASICs must be synthesized. Not only must the functions to be implemented on the ASICs be chosen, but the characteristics of those implementations must be determined during synthesis, making this a challenging problem. Hardware/software partitioning problems are usually single-rate synthesis problems, starting from a CDFG specification. The ASICs in the architecture are used to accelerate core functions, while the CPU performs the less computationally-intensive tasks. This style of architecture is important both in high-performance applications, where there may not be any CPU fast enough to execute the kernel functions at the required rate, and in low-cost functions, where the ASIC accelerator allows a much smaller, cheaper CPU to be used.

One useful classification of hardware/software partitioning algorithms is by the optimization strategy they use. Vulcan and Cosyma, two of the first co-synthesis systems, used opposite strategies: Vulcan started with all functionality in the ASICs and progressively moved more to the CPU to reduce implementation cost; while Cosyma started with all functions in the CPU and moved operations to the ASIC to meet the performance goal. We call these methodologies **primal** and **dual**, respectively, by analogy to mathematical programming.

### 2.4.1 *Architectural Models*

Srivastava and Brodersen [14] developed an early co-synthesis system based on a hierarchy of templates. They developed a multi-level set of templates, ranging from the workstation-level to the board architecture. Each template included both hardware and software components. The relationships between system elements were embodied in busses, ranging from the workstation system bus to the microprocessor bus on a single board. At each stage of abstraction, a mapping tool allocated components of the specification onto the elements of the template. Once the mapping at a level was complete, the hardware and software components were generated separately, with no closed-loop optimization. The components at one level of abstraction mapped onto templates at the next lower level of abstraction. They used their system to design and construct a multiprocessor-based robot arm controller.

Lin *et al.* [15, 16] developed an architectural model for heterogeneous embedded systems. They developed a set of parameterizable libraries and CAD tools to support design in this architectural paradigm. Their specifications are constructed from communicating machines which use a CSP-style rendezvous communication mechanism. Depending on the synchronization regime used within each component, adapters may be required to mediate communications between components. Modules in the specification may be mapped into programmable CPUs or onto ASICs; the mapping is supported in both cases by libraries. The libraries for CPUs include descriptions of various I/O mechanisms (memory mapped, programmed, interrupt, DMA). The communication primitives are mapped onto software libraries. For ASIC components, VHDL libraries are used to describe the communication primitives. Parameterizable components can be constructed to provide higher-level primitives in the architecture.

### 2.4.2 *Performance Estimation*

Hardt and Rosenstiel [17] developed an estimation tool for hardware/software systems. Their architectural model for performance modeling is a co-processor linked to the CPU through a specialized bus. They model communication costs using four types of transfer—global data, parameter transfer, pointer access, and local data—and distinguish between reads and writes. They use a combination of static analysis and run-time tracing to extract performance data. They estimate speed-up based on the ratio of hardware and software execution times and the incremental performance change caused by communication requirements.

Henkel and Ernst [18] developed a clustering algorithm to try to optimize the tradeoff between accuracy and execution time when estimating the performance

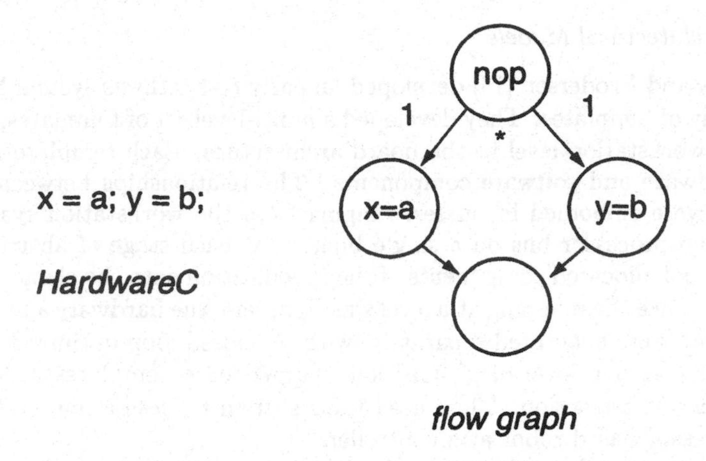


**Figure 2.6** Conjunctive execution in HardwareC.

of a hardware implementation of a data flow graph. They used a heuristic to cluster nodes in the CDFG given as input to Cosyma. Clustering optimized the size of the clusters to minimize the effects of data transfer times and to create clusters which are of moderate size. Experiments showed that clustering improved the accuracy of the estimation of hardware performance obtained by list scheduling the clusters.

### *2.4.3 Vulcan*

Gupta and De Micheli's co-synthesis system uses a primal methodology, by starting with a performance-feasible solution and moving functionality to software to reduce cost. As a result, Vulcan places a great deal of emphasis on the analysis of concurrency—eliminating unnecessary concurrency is the way in which functions are moved to the CPU side of the partition.

The user writes the system function as a HardwareC program [19], which was originally developed for high-level synthesis. HardwareC provides some data types for hardware, adds constructs for the description of timing constraints, and provides serialization and parallelization constructs to aid in the determination of data dependencies. However, the fundamental representation of the system function is as a **flow graph**, which is a variation of a (single-rate) task graph. A flow graph includes a set of nodes representing functions and a set of edges specifying data dependencies. The operations represented by the nodes are typically low-level operations, such as multiplications, in contrast to the process-level functionality often represented by task graphs. Each edge

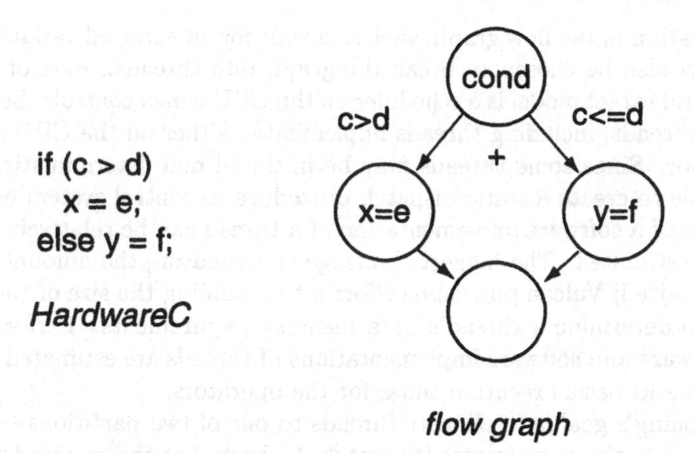


**Figure 2.7** Disjunctive execution in HardwareC.

has a Boolean condition under which the edge is traversed in order to control the amount of parallelism. Figure 2.6 illustrates a flow graph in which two assignments can be executed in parallel; a *nop* node is inserted to allow the flow of control to follow two edges to the two assignments. In this case, each edge's enabling condition is tautologically true. Figure 2.7 illustrates disjunctive execution—the conditional node controls which one of two assignments is created and the edge enabling conditions reflect the condition. The flow of control during the execution of the flow graph can split and merge, depending on the allowed parallelism.

The flow graph is executed repeatedly at some rate. The designer can specify constraints the relative timing of operators and on the rate of execution of an operator. Minimum and maximum timing constraints [20] between two nodes $i$ and $j$ specify minimum time $l$ or on maximum time $u$ between the execution of the operators:

$$t(v_j) \geq t(v_i) + l_{ij}$$
$$t(v_j) \leq t(v_i) + u_{ij}$$

The designer can specify a minimum rate constraint $\rho_i <= R_i$, which specifies that the $j^{th}$ and $j+1^{th}$ executions of operator $i$ should occur at the rate $R_i$. Maximum rate constraints can be similarly specified. Each non-deterministic delay operator must have a maximum rate specified for it so that the total run time of the flow graph can be bounded.

Vulcan divides the flow graph into **threads** and allocates those threads during co-synthesis. A thread boundary is always created by a non-deterministic delay operation in the flow graph, such as a wait for an external variable. Other points may also be chosen to break the graph into threads. Part of Vulcan's architectural target model is a scheduler on the CPU which controls the scheduling of all threads, including threads implemented either on the CPU or on the co-processor. Since some threads may be initiated non-deterministically, it is not possible to create a static dispatch procedure to control system execution.

The size of a software implementation of a thread can be relatively straightforwardly estimated. The biggest challenge is estimating the amount of static storage required; Vulcan puts some effort into bounding the size of the register spill set to determine a thread's data memory requirements. Performance of both hardware and software implementations of threads are estimated from the flow graph and basic execution times for the operators.

Partitioning's goal is to allocate threads to one of two partitions—the CPU (the set $\Phi_H$) or the co-processor (the set $\Phi_S$)—such that the required execution rate is met and total implementation cost is minimized. Vulcan uses this cost function to drive co-synthesis:

$$f(\omega) = c_1 S_h(\Phi_H) - c_2 S_s(\Phi_S) + c_3 \mathcal{B} - c_4 \mathcal{P} + c_5 |m| \tag{2.2}$$

where the $c_i$'s are constants used to weight the components of the cost function. The functions $S_h$ and $S_s$ measure hardware and software size, respectively. $\mathcal{B}$ is the bus utilization, $\mathcal{P}$ is the processor utilization, which must be less than 1 (= 100% utilization), and $m$ is the total number of variables which must be transferred between the CPU and the co-processor.

The first step in co-synthesis is to create an initial partition; all the threads are initially placed in the hardware partition $\Phi_H$. The co-synthesis algorithm then iteratively performs two steps:

- A group of operations is selected to be moved across the partition boundary. The new partition is checked for performance feasibility by computing the worst-case delay through the flow graph given the new thread times. If feasible, the move is selected.

- The new cost function is incrementally updated to reflect the new partition.

Once a thread is moved to the software partition, its immediate successors are placed in the list for evaluation in the next iteration; this is a heuristic to minimize communication between the CPU and co-processor. The algorithm does not backtrack—once a thread is assigned to $\Phi_S$, it stays there.

Experimental results from using Vulcan to co-synthesize several systems show that co-synthesis can produce mixed hardware-software implementations which are considerably faster than all-software implementations but much cheaper than all-hardware designs.

### *2.4.4 Cosyma*

The Cosyma system is described in more detail in Chapter 8, but we summarize its basic algorithms here for easier comparison with other co-synthesis techniques. The Cosyma system [21] uses a dual methodology—it starts out with the complete system running on the CPU and moves basic blocks to the ASIC accelerator to meet performance objectives. The Cosyma design flow is illustrated in Figure 8.1. The system is specified in $C^x$, a superset of the C language with extensions for specifying time constraints, communication, processes, and co-synthesis directives. The $C^x$ compiler translates the source code into an extended syntax graph (ESG), which uses data flow graphs to represent basic blocks and control flow graphs to represent the program structure. The ES graph is partitioned into components to be implemented on the CPU or on the ASIC.

Cosyma allocates functionality to the CPU or ASIC on a basic block level—a basic block is not split between software and custom hardware implementations. Therefore, the major cost analysis is to compare the speed of a basic block as implemented in software running on the CPU or as an ASIC function. When a function is evaluated for reallocation from the CPU to the ASIC, the change in the basic block $b$'s performance (not the total system speedup) can be written as:

$$\Delta c(b) = w(t_{HW}(b) - t_{SW}(b) + t_{com}(Z) - t_{com}(Z \cup b)) \times It(b) \qquad (2.3)$$

where $\Delta c(b)$ is the estimated decrease in execution time, the $t(b)$s are the execution times of hardware and software implementations, $t_{com}(Z)$ is the estimated communication time between CPU and co-processor, given a set $Z$ of basic blocks implemented on the ASIC, and $It(b)$ is the total number of times that $b$ is executed.

The weight $w$ is chosen to drive the estimated system execution time $T_S$ toward the required execution time $T_c$ using a minimum number of basic blocks in the co-processor (which should give a minimal cost implementation of the co-processor). The time components of $\delta c$ are estimated:

- $t_{SW}(b)$ is estimated by examining the object code for the basic block generated by a compiler.
- $t_{HW}(b)$ is estimated by creating a list schedule for the data flow. Assuming one operator per clock cycle, the depth of the schedule gives the number of clock cycles required.
- $t_{com}(Z \cup b)$ is estimated by data flow analysis of the adjacent basic blocks. In a shared memory implementation, the communication cost is proportional to the number of variables which need to be communicated.

Henkel et al. report that limiting the partitioning process to basic blocksbasic block only limited the performance gains that could be achieved—by moving only basic blocks into the co-processor, the speedup gained was typically only 2×. This observation coincides with the observations of compiler designers, who observe a limited amount of intra-basic-block parallelism. As a result of that initial experience, they implemented several control-flow optimizations to increase the number of parallel operations that can be implemented in the co-processor: loop pipelining, speculative execution of branches with multiple branch prediction, and operator pipelining. Cosyma uses the BSS synthesis system to both estimate and implement the co-processor.

Given these optimizations, Cosyma produced CPU-co-processor implementations with speedups ranging from 2.7 to 9.7 times over the all-software implementation. CPU times ranged from 35 to 304 seconds on a typical workstation.

### *2.4.5 Other Partitioning Systems*

Kalavade and Lee [22] developed an iterative algorithm for co-synthesis of mixed hardware-software systems which they call a **global criticality/local phase** algorithm. The two aspects of the algorithm are designed to control heuristic search for performance and implementation cost. The global criticality measure estimates the criticality of a node in the system schedule; this criterion is used to select which nodes in the behavior should be moved into the ASIC side of the partition. The local phase criterion is used to determine a low-cost implementation for a function. Two major criteria are used to select low-cost implementations for some of the functions. First, some functions are much cheaper to implement in one style than in another; the cheapest style is selected early. These heuristics are used to guide a partitioning algorithm which allocates nodes in the behavior to hardware or software implementations. Second, some nodes consume a large amount of resources (area or time) when implemented in one way. Experiments show that this algorithm gives near-optimal solutions when compared to ILP solutions, but in a much smaller amount of CPU time.

Kirovski and Potkonjak [23] developed an synthesis algorithm for low-power real-time systems. Their algorithm uses a template of multiple CPUs communicating via multiple shared bussesbus; the busses also handle input and output data streams. Their algorithm prefers assigning a function to the CPU on which it requires the least power. The algorithm also balances system load across the processors to maximize opportunities for voltage scaling.

## 2.5 DISTRIBUTED SYSTEM CO-SYNTHESIS

Distributed system co-synthesis does not use an architectural template to drive co-synthesis. Instead, it creates a multiprocessor architecture for the hardware engine as part of co-synthesis. The multiprocessor is usually heterogeneous in both its processing elements, communication channels, and topologies. This style of co-synthesis often puts less emphasis on the design of custom ASICs and more on the design of the multiprocessor topology. (The multiprocessor may consist of characterized ASICs as well as CPUs.) Heterogeneous distributed systems are surprisingly common in practice, particularly when one or two large CPUs are used in conjunction with small microcontrollers and ASICs to deliver performance and complex functionality at low cost.

### 2.5.1 *An Integer Linear Programming Model*

Prakash and Parker developed the first co-synthesis method for distributed computing systems [24]. They developed an integer linear programming (ILP) formulation for the single-rate co-synthesis problem and used general ILP solvers to solve the resulting system of equations.

Their methodology started with a single-rate task graph and a technology model for the processing elements, the communication channels, and the processes' execution characteristics on them. Given these inputs, they can create a set of variables and an associated set of constraints.

One set of variables describes the task graph:

- The variables $S_1, \ldots$ denote the processes.
- An input variable $i_{1,x}$ represents the $x^{th}$ input to process $S_1$.
- An output variable $o_{1,y}$ represents the $y^{th}$ output from process $S_1$.
- Input and output variables may also have parameters which specify what fraction of the process can start/finish before the input/output action.
- The volume of data transferred between processes $S_1$ and $S_2$ is written as $V_{1,2}$.

Another set of variables represents the technology model:

- The parameter $D(T, S_1)$ gives the execution time required for process $S_1$ on a processor of type $T$.
- The transfer time for a unit of data within a PE is given as $D_{CL}$. The time required to transfer a unit of data across a communication channel is $D_{CR}$.

The model allows for both local data transfers and multi-hop data transfers.

A set of auxiliary variables and help define the implementation:

- Process execution timing variables: start time, end time.
- Data transfer timing variables: transfer start time, transfer end time.
- Process-to-PE mapping variable: $\delta_{d,1} = 1$ denotes that process $S_1$ will be allocated to PE $d$.
- Data transfer type variables: local or multi-hop (non-local).

And finally, a set of constraints define the structure of the system:

- Processor-selection constraints: exactly one of the $\delta_{d,1}$ must be equal to 1.
- Data-transfer type constraints: makes sure that a data transfer is either local or multi-hop, but not both and not neither.
- Input-availability constraints: make sure that data is not used by the sink process until after it is produced by the source process.
- Output-availability constraints: make sure that the data obeys the fractional output generation parameters.
- Process execution start/end constraints: uses the processor mapping to determine the amount of time from the start of process execution to process completion.
- Data-transfer start/end constraints: similar to process execution constraints, but using data transfer times $D()$.
- Processor-usage-exclusion: makes sure that processes allocated to the same PE do not execute simultaneously.
- Communication-usage-exclusion: similarly, makes sure that multiple communications are not scheduled on the same link simultaneously.

Given that they used general ILP solvers, they could solve only relatively small problems—their largest task graph included nine processes. Solving that problem took over 6000 CPU minutes on a processor of unspecified type. However, they did achieve optimal solutions on those examples which they could solve. Their examples provide useful benchmarks against which we can compare heuristic co-synthesis algorithms.

### 2.5.2 Performance Analysis

During co-synthesis, a proposed allocation must be tested for feasibility by determining if the processes can be feasibly scheduled to meet their deadlines. One way to do this is by simulation—running a set of input vectors through the process set and measuring the longest time required for any input vector. Simulation has been used by D'Ambrosio and Hu [25, 26] and Adams and Thomas [27], among others, to determine the feasibility of a system configuration. However, if not all input combinations are tested, the longest running time measured may not be the actual worst case; exhaustive simulation is not feasible for practical examples. In fact, simulation fails to give worst-case running times on even examples of only moderate size. The only way to be sure that a given system architecture is guaranteed to meet its deadlines is to use an algorithm which bounds the worst-case execution time. Multiprocessor scheduling is NP-complete, so obtaining the tightest bound is impractical. However, it is possible to get tight bounds in a reasonable amount of CPU time.

Yen and Wolf [28] developed an algorithm which computes tight bounds on the worst-case schedule for a multi-rate task graph on a distributed system. The algorithm is given the task graph, the allocation of processes to processors, the processor graph, and the worst-case execution times of the processes. The algorithm uses a combination of linear constraint solving and a variant of the McMillan-Dill max constraint algorithm to bound the execution time. Linear constraints are induced by data dependencies in the task graph—the start time of a sink node is bounded below by the completion time of the source node. Max constraints are introduced when processes share a PE—the start time of a node depends on the max of the completion times of its predecessors.

The goal of the algorithm is to find **initiation** and **completion** times for each process in the task graph. The times computed as $[min, max]$ pairs to bound the initiation and completion times. The algorithm iteratively applies linear-constraint and max-constraint solution phases to the task graph. At each step, the initiation and completion time bounds are tightened. The algorithm terminates when no more improvement occurs in a step.

The linear-constraint procedure is a modified longest path algorithm. It computes the latest request time and latest finish time for each process by tracing through the task graph from the source nodes. A simple longest path algorithm would add worst-case completion times for each process along the path. To obtain more accurate estimates, this procedure also estimates the **phases** of the processes. If one process is known to be delayed with respect to another, we can reduce the maximum amount of interference that one process can create for another based on the phase difference between their initiation times.

The max constraint procedure computes a request time and a finish time for each process, as measured relative to the start time for the task graph. Two runs of the algorithm are used to produce both upper and lower bounds on request and finish times. The max constraints induced by co-allocation of processes model the preemption of one process by another. McMillan and Dill showed that, although solving constraint systems including both min and max constraints is NP-complete, max-only constraint systems can be solved efficiently. Because multiprocessor scheduling is NP-complete, this procedure does not guarantee the minimum-length schedule, but it does produce tight bounds in practice in a very small number of iterations (typically $< 5$).

Balarin and Sangiovanni-Vincentelli [29] developed an efficient algorithm for the validation of schedules for reactive real-time systems. They model a system as a directed acyclic graph of tasks, each with a known priority and execution time; all tasks run on a single processor. Events are initiated by events (either external or internal) and the validation problem is cast in terms of the times of events in the system—namely, for every critical event transmitted from task $i$ to task $j$, the minimum time between two executions of $i$ must be larger than the maximum time between executions of $i$ and $j$. They compute ***partial loads*** at various points in the schedule to determine whether any deadlines are not met.

Memory is an important cost in any system, but is particularly important for single-chip systems. Li and Wolf [30] developed a performance model for hierarchical memory systems in embedded multiprocessors. Caches must often be used with high-performance CPUs to keep the CPU execution unit busy. However, caches introduce very large variations between possible behaviors. Naive worst-case assumptions about performance—namely, assuming that all memory accesses are cache misses—are unnecessarily pessimistic. Li and Wolf model instruction memory behavior at the process level, which allows context switching to be efficiently modeled during the early phases of co-synthesis. Process-level modeling generates much tighter bounds on execution time which take into account the actual behavior of the cache. These estimates are also conservative—they do not overestimate performance. Agrawal and Gupta [31] developed a system-level partitioning algorithm to minimize required buffer memory. They use compiler techniques to analyze variable lifetimes in a multi-rate system and use the results to find a partitioning which minimizes buffering requirements.

### *2.5.3 Heuristic Algorithms*

Wolf [32] developed a heuristic algorithm to solve Prakash and Parker's single-rate co-synthesis problem. Given a single-rate task graph, this algorithm:

allocates a set of heterogeneous processing elements and the communication channels between them; allocates processes in the task graph to processing elements; and schedules the processes. The input to the algorithm includes the task graph and a technology description of the PEs and communication links. The major objective of co-synthesis is to meet the specified rate for execution, and the minor objective is to minimize the total cost of PEs, I/O devices, and communication channels. The algorithm is primal since it constructs a performance-feasible solution which it then cost-minimizes.

Co-synthesis starts by allocating PEs and ignoring communication costs; it later fills in the communication and devices required. Co-synthesis proceeds through five steps:

1. Create an initial feasible solution by assigning each process to a separate PE. Perform an initial scheduling on this initial feasible solution.

2. Reallocate processes to PEs to minimize total PE cost, possibly eliminating PEs from the initial feasible solution.

3. Reallocate processes again to minimize the amount of communication required between PEs.

4. Allocate communication channels.

5. Allocate I/O devices, either by using devices internally implemented on the PEs (counters on microcontrollers, for example) or by adding external devices.

The initial reallocation to minimize PE cost is the most important step in the problem, since processing elements form the dominant hardware cost component. This step consists of three phases: **PE cost reduction, pairwise merge,** and **load balancing.** The PE cost reduction step tries to move several processes at a time to achieve a good solution in a small number of moves. The procedure scans the PEs, starting with the least-utilized PE. It first tries to reallocate that PE's processes to other existing PEs. If all the processes can be reallocated, it eliminates the PE. If some processes cannot be moved to existing PEs, it replaces the PE with another, lower-cost PE which is sufficient to execute the remaining processes. Once all PEs have been cost-reduced, the algorithm tries to merge a pair of PEs into a single PE; the success of this step depends on the availability of a PE type in the component library which is powerful enough to run the processes of two existing PEs. After pairwise merging, the load is rebalanced. These three steps are repeatedly applied to the design until no improvement is obtained.

Experiments showed that this heuristic could find optimal results to most of Prakash and Parker's examples and near-optimal results for the remaining

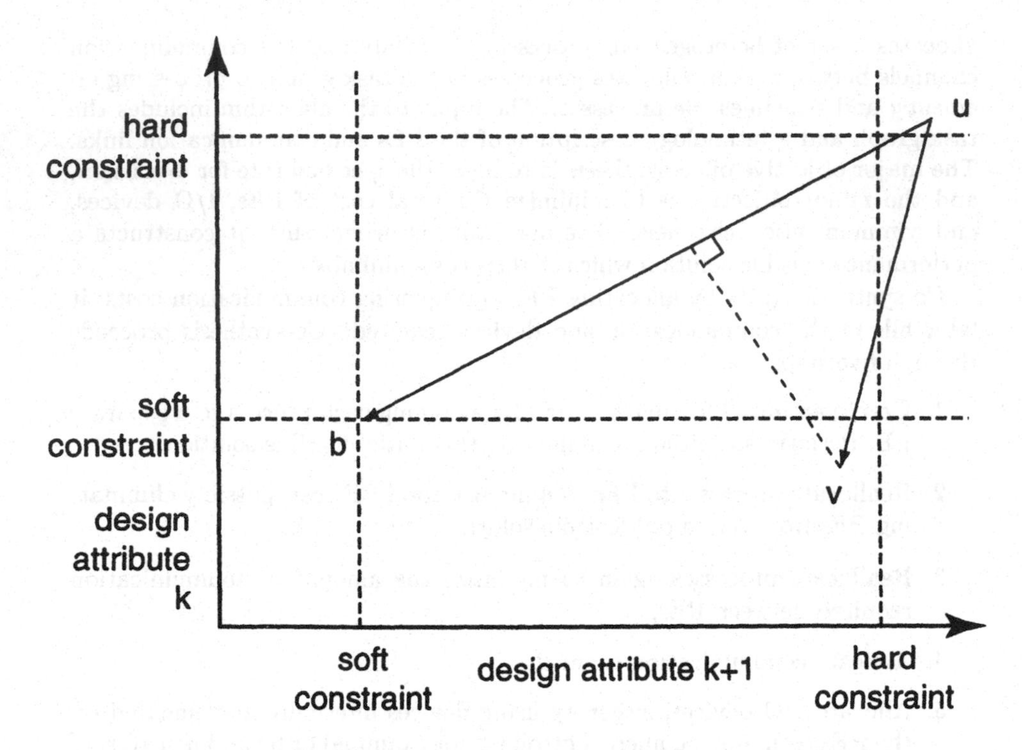


**Figure 2.8** Components of sensitivity.

examples. It also showed good results on larger examples. The algorithm requires very little run time thanks to the multiple-move strategy used during PE cost minimization.

Yen and Wolf [33] developed a co-synthesis algorithm for multi-rate systems specified as task graphs. Each task can execute at its own rate. Each subtask is given up to three requirements by the designer: a **period**, or the time between two executions of the task; a **hard deadline**, which is the maximum allowed time between the initiation and completion of the task; and a **soft deadline**, which is a desired time between the task's initiation and completion. Period and deadlines can be different because the task may not start execution at the beginning of its period. A hard deadline must be satisfied; a soft deadline may be broken, but a cost is incurred for doing so. The problem specification also includes a technology description, which gives the execution times for the processes on the available processing elements and the data rates of the communication links.

Sensitivity-driven co-synthesis makes single-process moves of processes during co-synthesis. A move consists of reallocating a process to a different processing element; a move may require either adding or deleting a processing element and its associated communication links. The goodness of a move is estimated as shown in Figure 2.8. The desired solution is at point *b*, where both the hard and soft deadlines are met. The current solution is at point *u* and the proposed move changes the solution to point *v*. The quality of the proposed move is proportional to the projection of $\rightarrow uv$ onto $\rightarrow ub$, which captures the extent to which this single move advances the implementation toward the minimum-cost solution.

An initial feasible solution is constructed by assigning each process to a separate PE. While this is wasteful, it is difficult to construct a smaller feasible hardware engine without entering full-blown co-synthesis. It is relatively easy to eliminate unnecessary PEs during co-synthesis using straightforward heuristics.

The main optimization loop tests the single-move options. It first computes the sensitivity for each possible single-process move as illustrated in Figure 2.8, evaluating each process moved onto each possible PE. It chooses the move with the highest gain; if no move is feasible, it adds a PE to create a feasible move. It implements the chosen move by reallocating the process and recomputes the schedule for the next step. Co-synthesis executes this optimization loop twice, in two phases. The first phase tries to consolidate the implementation into as few PEs as possible so that idle PEs can be eliminated. The second phase tries to balance the system load. Not only does load balancing improve the quality of the implementation by minimizing the risk that approximations made during co-synthesis will cause some node to become overloaded in the implementation; it also allows for some further improvements in the allocation to try to meet the soft deadline.

Experiments with sensitivity-driven co-synthesis show that it can achieve optimal or near-optimal results for the small examples for which optimal results are known. It gives good results on larger task graphs for which computing the optimal result is too expensive. CPU times are higher than for Wolf's co-synthesis algorithm but much better than for pure ILP solutions.

The TOSCA system [34, 35] provides the designer with a set of design transformations for design space exploration. TOSCA's design representation and transformations target control-dominated systems which can be described as systems of communicating machines. Optimizations include: unfolding entry and exit actions; unfolding FSM enabling conditions; flattening hierarchies; replacing timers with counters; and collapsing a set of communicating processes into a single process.

### *2.5.4 System Partitioning*

Some co-synthesis methodologies are adapted from graph partitioning algorithms but do not make strong assumptions about the hardware architecture. An example is the system of Adams and Thomas [27]. They used partitioning algorithms to divide a unified behavioral description, equivalent to a single task, into hardware and software components. Their synthesis process uses operator-level partitioning, but explicitly compares hardware and software implementations for clusters. They implemented their methodology in a tool called Co-SAW (Co-Design System Architect's Workbench).

The first-step in synthesis, preprocessing, includes two parallel phases:

- One phase simulates the complete system model to obtain execution traces for system-level performance analysis.
- Another phase first clusters operators in the behavioral description to form tasks; it then estimates the costs of hardware and software implementations of the clusters.

The data gathered in the preprocessing step is used to drive synthesis. The synthesis algorithm modifies the initial partitioning (*code motion*) to try to expose useful parallelism. They use non-deterministically generated moves to select code motion operations both within and between processes. They showed that they could effectively explore the design space using Co-SAW.

Hou and Wolf [36] developed a partitioning algorithm used to increase the granularity of a task graph before co-synthesis. Partitioning is driven by a cost function which estimates the execution time for subgraphs of a task without explicit scheduling. Their algorithm restricts partitioning moves to create clusters in which data never flows out of the cluster and then back in. This restriction improves the estimation of the total execution time of a cluster, since a new cluster can never be blocked by waiting for data to be returned from an external process. They use a non-greedy clustering algorithm which selects clusters which meet the criteria and determine whether combining the processes into a single process will unnecessarily increase the execution time of the task. They showed that their algorithm could reduce the number of processes in tasks, greatly reducing the time required for sensitivity-driven co-synthesis, without significantly changing the quality of the co-synthesized implementation.

Shin and Choi [37] developed a partitioning and scheduling algorithm for embedded code. Their algorithm partitions a CDFG into a set of threads and then generates a combined static/dynamic scheduler for the system. They partition the CDFG based on accesses to hardware. After generating a large number of small threads, they combine threads in deadlock-free combinations to generate a reasonable number of threads. They then generate a scheduler which statically schedules threads whose initiation times are known and dynamically schedules threads which have data-dependent initiation times.

### *2.5.5 Reactive System Co-Synthesis*

Control-dominated systems are often A **reactive system** reacts to external events; such systems usually have control-rich specifications. Several research efforts have targeted reactive real-time systems. both at the process-abstraction and the detailed design levels. Rowson and Sangiovanni-Vincentelli [38] describe an interface-based design methodology for reactive real-time systems and a simulator to support that methodology.

POLIS [10] is a co-design system including both synthesis and simulation tools for reactive real-time embedded systems. As described in Section 2.2, POLIS uses the CFSM model to represent behavior. Central to the POLIS synthesis flow is the **zero-delay hypothesis**—events move between communicating CFSMs in zero time. (This hypothesis is also used by Esterel and other reactive system modeling methodologies.) The communication within the system can be analyzed by collapsing the component CFSMs into a single reactive block; the effects of non-zero response time can be taken into account by analyzing and modifying this block. After analysis, a partitioning step allocates functions to hardware or software implementations. Hardware and software components can be synthesized from the state transition graphs of the resulting CFSM models [39, 40]. Elements of the system can be mapped into microcontroller peripherals by modeling them with library CFSMs [41]. Chiodo et al. [42] describe a design case study with the POLIS system.

The Chinook system [43] is another co-synthesis system for reactive real-time systems. The main steps in the Chinook design flow are:

- hardware-software partitioning, which chooses hardware or software implementations and allocates to processors;
- device driver synthesis and low-level scheduling, which synthesizes an appropriate driver for a peripheral based on the results of allocation;
- I/O port allocation and interface synthesis, which assigns I/O operations to physical ports on the available microcontrollers;
- system-level scheduling, which serializes operations based on the more accurate timing estimates available at this point in synthesis; and
- code generation.

Ortega and Borriello [44] developed an algorithm for synthesizing both inter-CPU and intra-CPU communication as part of the Chinook system. Their system can insert queues, synthesize bus protocol implementations, and related operations to implement inter-CPU communication; it uses shared memory to implement intra-CPU communication.

### *2.5.6 Communication Modeling and Co-Synthesis*

Communication is easy to neglect during design but is often a critical resource in embedded systems. Communication links can be a significant cost of the total system implementation cost. As a result, bandwidth is often at a premium. For example, the automotive industry recently adopted an optical bus standard, but since the optical link is plastic for low cost and ruggedness, its bandwidth is limited. Limited bandwidth makes it important that required communications are properly scheduled and allocated to ensure that the system is both feasible and low-cost.

Daveau *et al.* [45] first formulated communication co-synthesis as an allocation problem in the Cosmos system [46]. Like TOSCA, Cosmos is targeted to communication-rich systems and provides a set of transformations for design exploration; however, Cosmos can map to a wider range of hardware architectures. The givens for the problem are: a set of processes and communication actions between them; and a library of communication modules (busses, etc.). The objective is to allocate the communication actions to modules such that all the communications are implemented and feasibly scheduled; some cost objective may also be considered. Cosmos is described in more detail in Chapter 10.

Solar models a communication module by its protocol, its average transfer rate, its peak transfer rate, and its implementation cost. To allocate modules to communication actions, they build a decision tree to map out possible allocations. Nodes in the decision tree (except for the leaf nodes) correspond to the logical channel units which must be implemented. A path through the decision tree therefore enumerates a combination of logical channel units. An edge enumerates an allocation decision for the edge's source node. Each leaf node is labeled with the cost of the path which it terminates. The algorithm performs a depth-first search of the node to evaluate allocation combinations and choose a minimum-cost allocation. Results show that the algorithm can quickly identify a minimum-cost allocation of communication modules for the system.

The CoWare system [47] is a design environment for heterogeneous systems which require significant effort in building interfaces between components. CoWare is based on a process model, with processes communicating via ports using protocols. SHOCK is a synthesis subsystem in CoWare which generates both software (I/O drivers) and hardware (interface logic) interfaces to implement required communication. The interfaces are optimized to meet the requirements of the specification, taking into account the characteristics of the target components.

Yen and Wolf [48] developed a model for communication channels in embedded computing systems and a co-synthesis algorithm which makes use of that model. This model assumes that communication on a bus will be prioritized, with some communications requiring higher priority than others. Communication can be modeled by dummy processes which represent send and receive operations; these communication processes are allocated to the interface between the PE which executes the process doing the sending/receiving and the communication channel on which the communication is performed. The total delay incurred by a communication action has two components: the intrinsic delay of the communication and the wait required to obtain the bus from competing processes. These components are similar to the case for the delay of processes on a CPU. However, communication is treated differently because busses cannot usually be preempted in the middle of a transaction. As a result, this work assumes that a communication is **non-preemptable**—a lower-priority communication will finish once it has started, even if another higher-priority communication request arrives after the lower-priority request has started. Non-preemptability complicates the calculation of the worst-case delay for a communication, but that delay can still be solved for by numerical techniques.

## 2.6 CONCLUSIONS

Our understanding of co-synthesis has been hugely increased over the past several years. We now have a much better understanding of the types of co-synthesis problems which are of interest, a set of models for those problems, and some strong algorithms for both analysis and co-synthesis.

Of course, there remains a great deal of work to be done. Outstanding challenges range from the development of more accurate models for both hardware and software components to the coupling of co-synthesis algorithms with software and system design methodologies. Beyond this, larger, more realistic, and more examples would be a great benefit in driving co-synthesis research in realistic directions. However, given the enormous amount of progress over the past few years, we can expect the state of the art in co-synthesis to be well advanced over the next five to ten years.

### Acknowledgments

Work on this chapter was supported in part by the National Science Foundation under grant MIP-9424410.

# 3 PROTOTYPING AND EMULATION

Wolfgang Rosenstiel

Technische Informatik
Universität Tübingen
Tübingen, Germany

## 3.1 INTRODUCTION

ASICs will continue to grow increasingly complex. Errors of specification, design and implementation are unavoidable. Consequently, designers need validation methods and tools to ensure a perfect design before the production is started. Errors caught after fabrication incur not only added production costs, but also delay the product, which is an increasingly serious detriment in today's fast-paced international markets. "First-time-right silicon" is, therefore, one of the most important goals of chip-design projects.

Figure 3.1 shows loss due to relatively late marketing. In order to get "first-time-right silicon," a variety of approaches is needed. Figure 3.2 describes the necessary steps of synthesis and validation for the design of integrated circuits. There are four methods to reach the goal of "first-time-right silicon":

- specification on high levels of abstraction followed by automatic synthesis;
- simulation on various levels;



*J. Staunstrup and W. Wolf (eds.), Hardware/Software Co-Design: Principles and Practice*, 75-112.

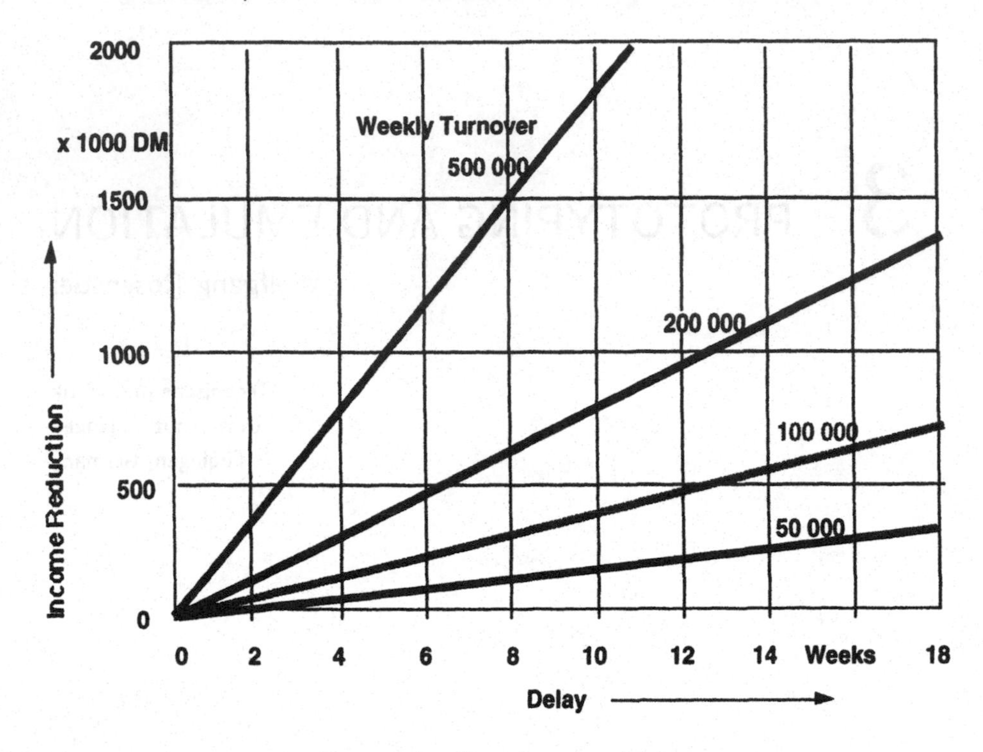


**Figure 3.1** Cost of product delays.

- formal verification;
- prototyping and emulation.

In the last few years the growing significance of synthesis has become apparent, that is synthesis in a general sense and—more specifically—synthesis on higher levels of abstraction, such as RT level synthesis and the so-called high-level synthesis, i.e. synthesis from behavioral descriptions. The increasing number of available commercial tools for RT and high-level synthesis indicates that the abstraction level of the design entry will increase to higher levels in the near future. Especially with respect to hardware/software co-design high level synthesis gathers momentum. Only by integrating high-level synthesis in the hardware software co-design cycle real hardware/software co-design will be possible. The investigation of various possibilities with different hardware/software trade-offs is only sensible if the different hardware partitions can be implemented as fast as software can be compiled into machine

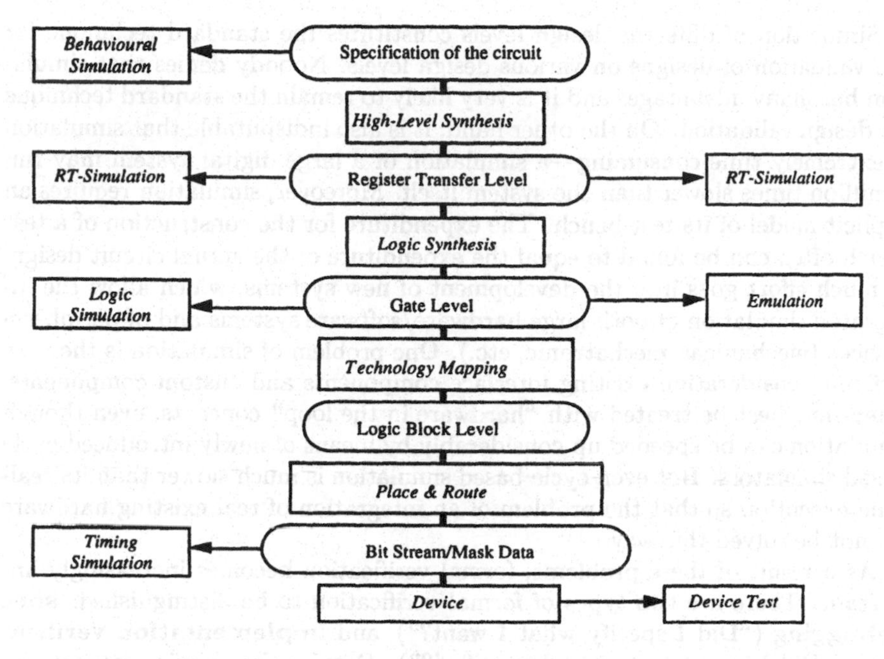


**Figure 3.2** Design flow.

code. High-level synthesis is, therefore, the enabling technology for hardware software co-design.

Another advantage of high-level synthesis is its high level of abstraction and as a consequence of it a shorter formal specification time of the circuit to be implemented. This more compact description also results in a much faster simulation. High-level synthesis combined with hardware/software co-design also supports the investigation of many different design alternatives resulting in higher quality designs. Design exploration compensates for the reduced efficiency and low optimality of synthesized designs when compared to manually optimized solutions, which may only be done for very few design alternatives.

The CAD vendor Synopsys reports from user experience that, through the application of behavioral compilation, design effort may be reduced by a factor of 10 and that, despite this shorter design time, area and delay could be reduced by 10-20% compared to manual designs. The higher quality of the implementation is mainly due to the investigation of further design alternatives and more sophisticated pipelining schemas, which automatically optimize clock frequency and latency.

Simulation of different design levels constitutes the standard technique for the validation of designs on various design levels. Nobody denies that simulation has many advantages and it is very likely to remain the standard technique for design validation. On the other hand, it is also indisputable that simulation is extremely time-consuming—a simulation of a large digital system may run a million times slower than the system itself. Moreover, simulation requires an explicit model of its test bench. The expenditure for the construction of a test bench often can be found to equal the expenditure of the actual circuit design.

much effort goes into the development of new systems, which allow the integrated simulation of both large hardware/software systems and of peripheral devices (mechanical, mechatronic, etc.). One problem of simulation is the very difficult consideration existing interface components and custom components. They may best be treated with "hardware in the loop" concepts, even though simulation can be speeded up considerably by means of newly introduced cycle based simulators. But even cycle based simulation is much slower than its real-time execution so that the problem of an integration of real existing hardware cannot be solved this way.

As a result of these problems, formal verification becomes increasingly important. There are two types of formal verification to be distinguished: **spec debugging** ("Did I specify what I want?") and **implementation verification** ("Did I implement what I specified?"). Despite the growing importance of formal verification and despite the introduction of some early verification tools, it is limited in many ways. One of its major restrictions is the high complexity of verification procedures, which limit an application to comparably simple design modules. It is difficult to include time-related constraints into the verification process. The difficulty of including an environment also occurs for formal verification. Another important technique for the "first -time-right silicon" goal is therefore an increasing use of emulation and prototyping. This contribution is devoted to this issue. Emulation and prototyping support hardware/software co-design by means of simple coupling with real hardware— in particular with standard microprocessors which run parts of a total system in software—and are closely related to other contributions in this book

## 3.2 PROTOTYPING AND EMULATION TECHNIQUES

Validation in hardware/software co-design has to solve two major problems. First, appropriate means for software and hardware validation are needed; second, these means must be combined for integrated system validation. Since validation methods for software are well known, the main effort is spent on hardware validation methods and the integration of both techniques. Today, the main approaches for hardware validation are simulation, emulation [49],

[50], [51] and formal verification [52]. Simulation provides all abstraction levels from the layout level up to the behavioral level. Formal verification of behavioral circuit specifications has up to now only limited relevance, since the specifications which can be processed are only small and can only use a very simple subset of VHDL. All these validation and verification are harder to implement during hardware/software co-design: low-level validation methods are not really useful, since hardware/software co-design aims at automatic hardware and software design starting from behavioral specifications. This would be like writing programs in C++ code and debugging the generated assembly code. This even holds for debugging at the RT level , since HW/SW co-design usually implies high-level synthesis for the hardware parts. The RT description of the hardware is produced automatically and may therefore be very difficult to interpret for a designer. Though simulation can operate at the algorithmic level, there are some problems with it, too. First, simulation is very time-consuming and thus it may be impossible to simulate bigger systems in the same way as systems consisting of hardware and software parts. Second, simulation is too slow to bring simulated components to work with other existing components. Thus, it is necessary to model the outside world within the simulator, which requires at least an enormous effort.

One great advantage of emulation and prototyping in contrast to simulation is higher speed. Emulation, for instance, is only 100 times slower than real time. Furthermore, the integration of the environment is much easier. Likewise high expenditure for the generation of test benches can be avoided.

Today, emulation is the standard technique CPU design and is increasingly used for ASIC design. Emulation can be expected to become an essential part of co-design methodologies since—compared with simulation—the speed advantage of emulation with respect to co-emulation is even more apparent. One of the main disadvantages of emulation in contrast to simulation is that timing errors are hard or impossible to detect. Therefore, emulation mostly serves the testing of functional correctness. There are further disadvantages: slow compilation once the circuit changes, different design flows for implementation and emulation, and the high expenditure, which may be up to one Dollar per emulated gate. However, it is possible to overcome many of these disadvantages by not restricting emulation to the gate level as is normally done, but by aiming at emulation on higher abstraction levels, which then may be more viable in real time.

First steps in this direction have been taken by Quickturn with the HDL-ICE system, which allows to relate the probed signals to a register-transfer specification. Synopsys has introduced the Arkos system which also performs a cycle based emulation of an RT-level design. A similar approach is taken by Quickturn in the CoBalt system. Potkonjak et. al. have described a method to do design for debugging of an RT level circuit based on emulation which multiplexes output pins of a design and register contents to the same pins, whenever they are not used by the design itself. This approach concentrates rather on a minimization of output pins needed for debugging then on source code based debugging. Although solutions for the RT level start to evolve on the market there is a big gap between the algorithmic level and the RT level where emulation is available today. In this contribution we will therefore first present classical gate level emulation techniques based on FPGA structures. Later we will explain how these classical gate level emulation systems are currently evolving to support the RT level. Still in the research state is emulation support at higher levels together with high-level synthesis and hardware/software co-design. This outlook on current state of the art in emulation research will then be described by explaining our own approach which we call **source-level emulation** (SLE).

As mentioned before, emulation nowadays forms part of usual validation techniques in microprocessor design, but also is applied to ASICs. SUN Microsystems, for instance, reports at length on the use of emulation in connection with the design of the UltraSparc-architecture. With a speed of 500,000 cycles per second it was possible to test the boot sequence of Solaris or to emulate with the same speed the start and use of window systems, such as Openwin, etc. Problems mentioned in this context are related to the integration of memory but also to the simulation of pre-charging and other dynamic effects of circuit design. An interesting side effect which was reported, is that the return on investment costs for emulation sytems, which are still very expensive, are distributed by 50% each on return on investment before first silicon and 50% after first silicon. Thus, the availability of a complete emulation system allows also the improvement of fault diagnosis in the case that there are still errors in the implemented chip.

In addition, it is possible to test further developments of this chip with respect to what-if design alternatives. Bug fixes are conceivable as well. Tandem has also given a detailed account of the results derived from the use of emulation. They pointed out that without emulation it would have been impossible to make complete operating system tests, to simulate input/output behavior and to integrate finished ASICs step by step into the total system. The emulated components are gradually substituted by real chips. This allows a complete validation at any point of the development time. Yet, various users of emulation systems record unanimously that emulation is generally only used after thorough simulations so that there is only a low expectation of design errors anyway. Simulation mainly aims at the reduction of the number of functional and timing errors, whereas emulation in general serves realistic loads in demonstrating functional correctness.

Some of the problems which have been mentioned in connection with emulation systems are caused by the different design flows required for implementation and emulation. Figure 3.3 demonstrates these varying design flows. Alternative 1 is today still very common. Cycle simulation is passed for so

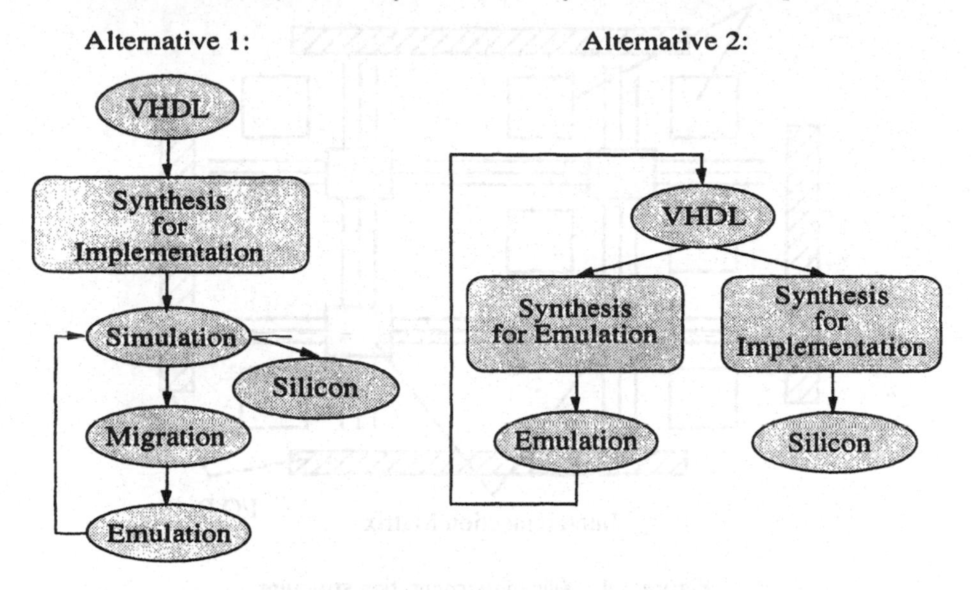


**Figure 3.3** Different design flow integrations for emulation.

long until no errors occur anymore. This is followed by a translation into the final ASIC implementation. The disadvantage of this alternative is that the migration which is required by emulation results in an implementation which is different from the subsequent translation into an ASIC. The ensuing validation problem is to be solved by an increased application of alternative 2. Here, a synthesis for emulation is implemented, for example on the basis of a gate level or a future RT level VHDL description. Accordingly, errors bring on changes in the VHDL source. This identical VHDL source file is then also converted by synthesis into a corresponding ASIC. Postulating an identical VHDL subset with identical semantics enables a sign-off on the basis of the VHDL source code. In the future, alternative 2 will support a RT level sign-off by means of a more abstract description on RT level.

Today's existing emulation systems are distinguished into FPGA based and custom based solutions. Additionally, the FPGA based approaches may be categorized as off-the-shelf FPGAs and FPGAs which have been developed especially for emulation. Essentially, off-the- shelf FPGAs go back to the FPGAs of Xilinx. Currently, XC3000 and XC4000 families are the predominant FPGAs for this application. Their superiority comes from their static RAMs-based implementation, which allows unlimited reprogrammability. Figures 3.4, 3.5 and

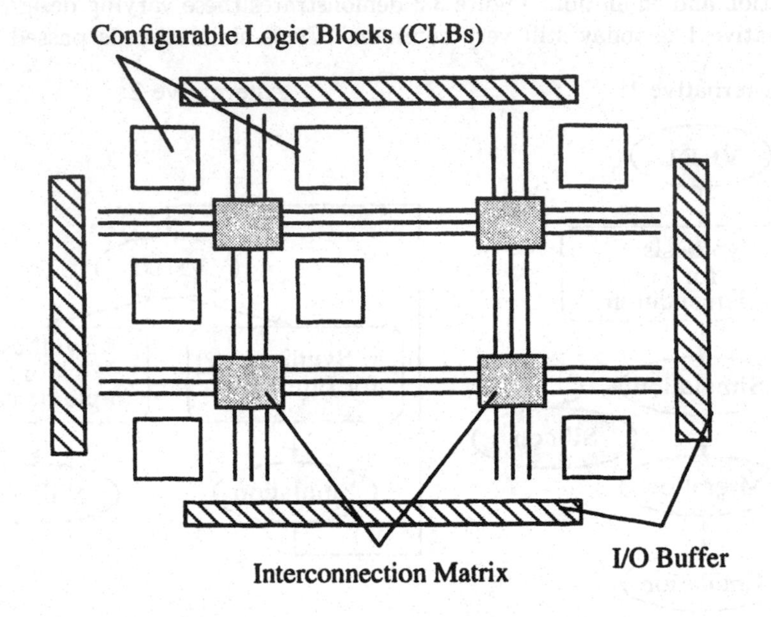


**Figure 3.4** Xilinx interconnection structure.

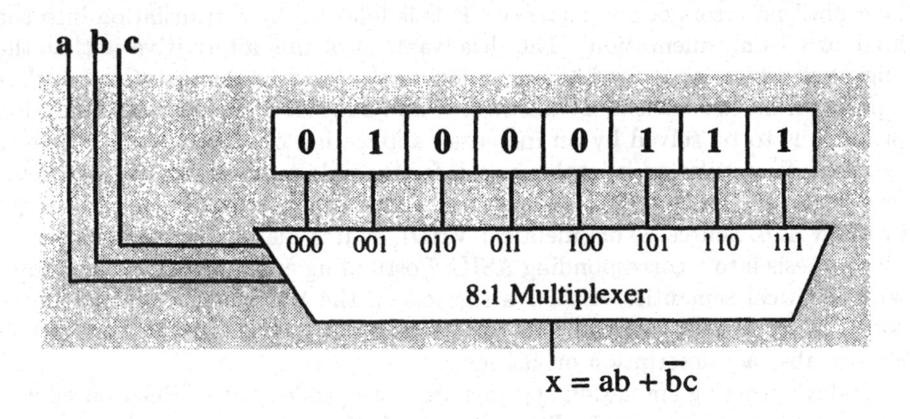


**Figure 3.5** CLB principle.

3.6 show respectively the basic set-up of the Xilinx device architecture, the

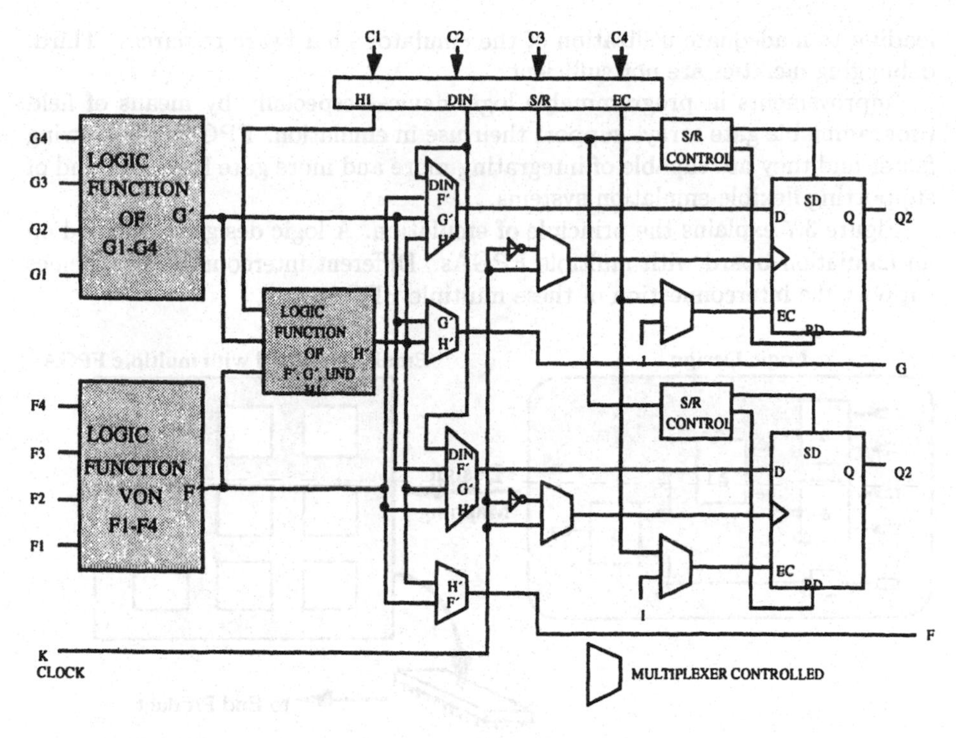


**Figure 3.6** CLB structure.

general function of the look-up tables as well as the internal architecture of a CLB of the Xilinx XC4000 series.

## 3.3 PROTOTYPING AND EMULATION ENVIRONMENTS

This section describes the functionality and characteristics of today's existing prototyping environments. There are many different approaches for prototyping architectures. Most of them address the prototyping of hardware but not of embedded systems. Even if their gate complexity is sufficient for total systems, they display various disadvantages with respect to hardware/software co-design and prototyping of whole systems. First, until now there is no integration of microprocessors for the execution of the software parts supported. Integration facilities are not flexible enough to adequately support prototyping in hardware/software co-design; in particular, it is very difficult for the emulator's synthesis software to route buses which are usually introduced by the integration of microprocessors. Second, hardware partitioning is too inefficient, leading to inadequate utilization of the emulator's hardware resources. Third, debugging facilities are not sufficient.

Improvements in programmable logic devices especially by means of field programmable gate arrays support their use in emulation. FPGAs are growing faster and they are capable of integrating more and more gate functions and of supporting flexible emulation systems.

Figure 3.7 explains the principle of emulation. A logic design is mapped on an emulation board with multiple FPGAs. Different interconnection schemes support the interconnection of these multiple FPGAs.

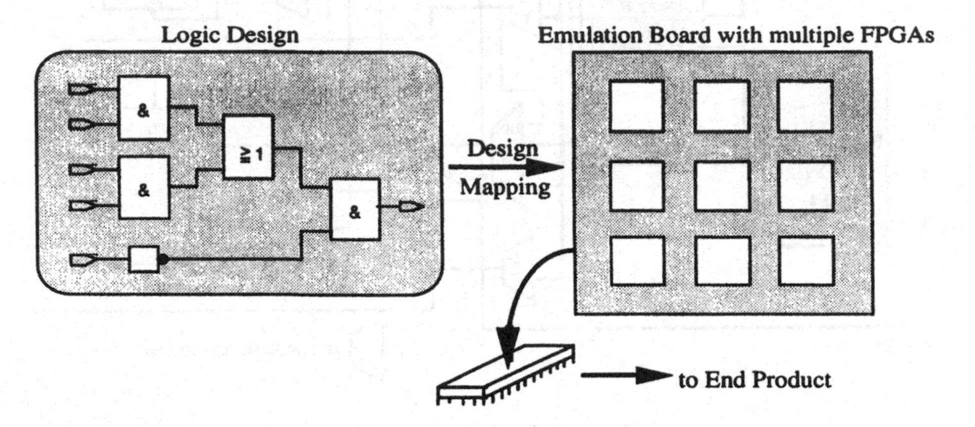


**Figure 3.7** FPGA mapping.

Figure 3.8 explains the general structure of such an emulation system. Just as individual emulation systems differ with respect to their use of FPGAs, there are are differences between emulation systems on the market concerning their interconnection structure. The systems which are available on FPGAs may be arranged in three classes. Below concrete examples for each class are provided:

Some systems also use programmable gate arrays for their interconnection call structure, others take recourse to special interconnect switches (ex. Aptix), and there are some which apply especially developed custom interconnect chips for the connection of individual FPGAs.

Until now, we has not suggested any distinctions between prototyping and emulation. In fact, the dividing-line between the two terms is blurred. Many manufacturers of emulation systems claim that these may be used in particular for prototyping. In my opinion, it is actually impossible to separate these two terms clearly. "Prototype" tends to refer to a special architecture which is cut to fit a specific application. This special architecture is often based on FPGAs and describes the simulation in the programmable logic of the system which is

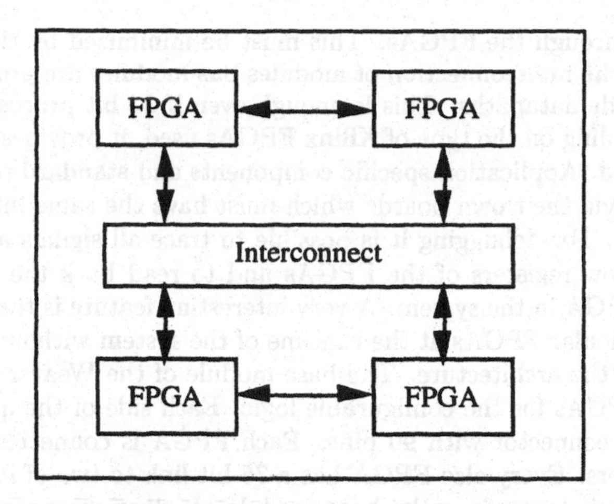


**Figure 3.8** Principal structure of an FPGA-based emulation.

to be designed. However, interconnection structures as well as gate capacities of the prototype are often tuned in to the evolving system. One's own prototype has the advantage of achieving a higher clock rate due to specific cutting to fit for an application. The best elements for the corresponding application can be selected, and thus, RAMs, microcontroller, and special peripheral components etc. can be used. Its weak point is, however, that new expenditure for the development of such a special prototype occurs for every new prototype. Therefore, it is appropriate to ask to what extent this disadvantage overthrows the above mentioned advantages. Some kind of compromise between general purpose emulation systems, which will be elaborated later on, and special prototypes may be seen in the Weaver system. It was developed by the University of Tübingen and Siemens Munich under the auspices of the Basic Research ESPRIT project COBRA (Co-Design Basic Research Action).

### *3.3.1 The Weaver Prototyping Environment*

The Weaver prototyping environment is designed especially for prototyping in the domain of hardware/software co-design. It is a modular and extensible system with high gate complexity. Since it fulfills the requirements on a conventional prototyping system it is also useful for hardware prototyping. This section describes the hardware and supporting software for this environment.

Weaver uses a hardwired regular interconnection scheme. This way fewer signals have to be routed through programmable devices, which results in a better performance. Nevertheless, it will not always be possible to avoid routing of signal through the FPGAs. This must be minimized by the supporting software. For the interconnection of modules bus modules are provided. These offer 90-bit-wide datapaths. This is enough even if 32 bit processors are integrated. Depending on the type of Xilinx FPGAs used, it provides 20K to 100K gates per board. Application specific components and standard processors can be integrated via their own boards which must have the same interface as the other modules. For debugging it is possible to trace all signals at runtime, to read the shadow registers of the FPGAs and to read back the configuration data of any FPGA in the system. A very interesting feature is the ability to reconfigure particular FPGAs at the runtime of the system without affecting the other parts of the architecture. The base module of the Weaver board carries four Xilinx FPGAs for the configurable logic. Each side of the quadratic base module has a connector with 90 pins. Each FPGA is connected with one of these connectors. Every also FPGA has a 75 bit link to two of its neighbours. A control unit is located on the base module. It does the programming and readback of the particular FPGAs. A separate bus comes to the control unit of each base module in the system. The programming data comes via this bus serially. This data is associated with address information for the base module and the FPGA on the base module. This way the control unit can find out if the programming data on the bus is relevant for its own base module and if so if it can forward the programming data to the FPGA for which it is intended. The readback of configuration data and shadow registers is done in the same way. The ROM is used to store the configuration data for each FPGA of a base module. While the startup of the control unit reads out the ROM and programs each FPGA with its configuration data. Figure 3.9 shows the base module equipped with Xilinx 4025 FPGAs.

An I/O module provides a connection to a host. Several interfaces are available. Examples are serial and parallel standard interfaces as well as processor interfaces to PowerPC and Hyperstone. With such a module the host works as I/O preprocessor which offers an interface for interaction to the user. A RAM module with 4 MB static RAM can be plugged in for the storage of global or local data. It can be plugged in a bus module so that several modules have access to it in the same way, or it may be connected directly to a base module. Then this module has exclusive access to the memory. Requests of other modules must be routed through the FPGAs of the directly connected base module. This is possible but time-consuming. The bus module allows the plugging together of modules in a bus-oriented way. Given the option of having a bus on each side of the base module, this architecture can be used to build multiprocessing systems with an arbitrary structure. For the integration of standard processors they must be located on their own modules which have to meet the connection conventions of the other modules. A standard processor module for

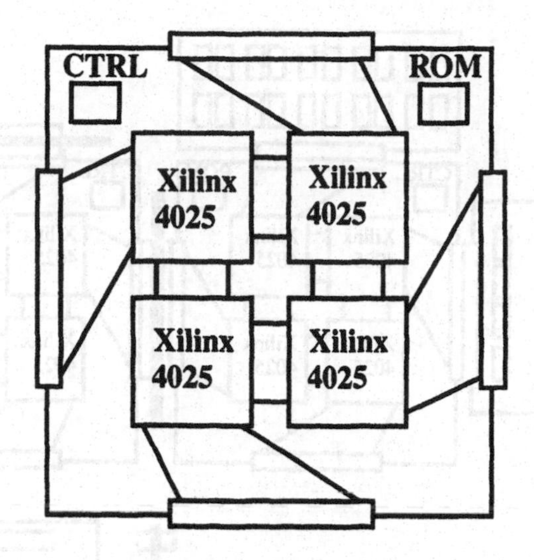


Figure 3.9 The base module.

the Hyperstone 32 bit processor is already included in the set of module types. Even though this list of module types is sufficient for many applications, there may be even more applications which require other types of modules like networking modules or DSP modules. One may think of any other type of module, and certainly the list of existing module types will grow in the future. With these modules arbitrary structures can be built. The structure depends on the application which is prototyped. Thus a running system contains all modules needed but not one more. It is important to reduce overhead and to keep the price low. An example of a more complex structure is depicted in Figure 3.10.

This picture shows an architecture in top view and in side view. As can be seen, the architecture is built in three dimensions and consists of four basic modules, a RAM module and an I/O module. On the right side is a tower of three basic modules which are connected via three bus modules. On the left side is another basic module with a local RAM module and an I/O module. The gross amount of gates in this example is about 400K. Thus it is an example which would be sufficient for many applications. One may assume that the left basis carries an application specific software executing processor, which works with the RAM on the RAM module and which communicates with software on a host for providing a user interface via the I/O module. The tower on the right side may carry some ASICs.

After the hardware/software partitioning the resulting behavioral hardware description has to be synthesized via existing high-level synthesis tools. The

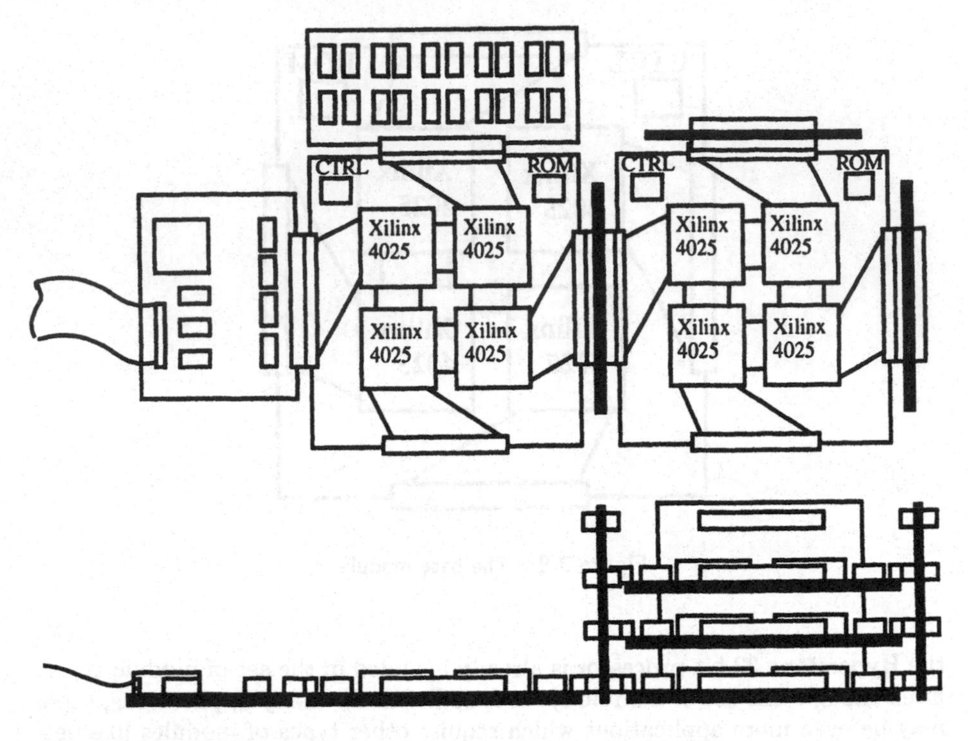


**Figure 3.10** A more complex structure.

hardware debugging software provides a user interface for the readback of configuration data, for signal tracing and for shadow register reading. A special partitioning tool was developed to partition design on several FPGAs [53].

This example of a prototyping system is used as experimental environment for the source-level emulation which will be elaborated lateron. In the following the five most important emulation systems are outlined.

These five emulation systems are manufactured by the Quickturn, Mentor, Zycad, Aptix, and Synopsys. The new simulation systems from Synopsys (Arkos) and Quickturn (CoBalt) are elaborated at the end of this contribution since they have just been brought on the market and since they utilizes very different emulation architectures. While the other systems from Quickturn, Mentor, Zycad and Aptix are based on off-the-shelf or in-house-developed FPGAs and since they essentially depend on a rather complicated and lengthly configuration process, such as depicted in Figure 3.11, the Arkos and CoBalt systems are based on special processors, on which an existing logic or RT level design is mapped by a compiler.

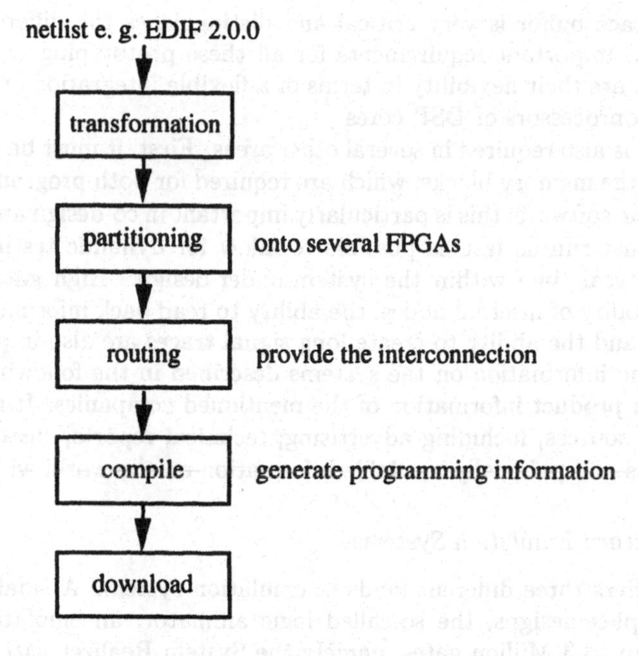


**Figure 3.11** The FPGA configuration process.

All these emulation systems can operate in two operation modes, shown in Figure 3.12. By means of a host system individual stimuli are applied to an emulator in a test vector mode (a). The resulting test answers are evaluated this way as well.

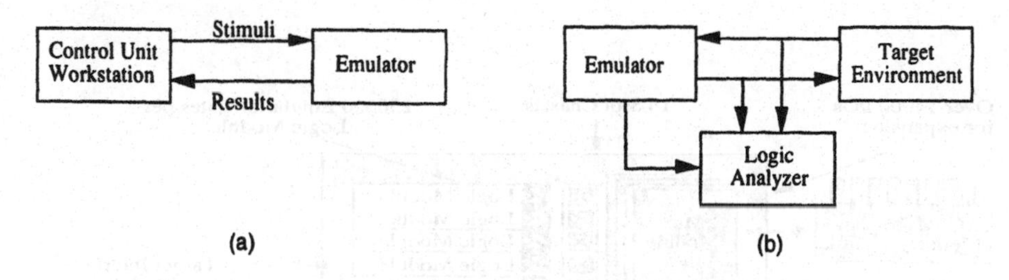


**Figure 3.12** Operation modes of emulation systems.

The emulated system in the target environment is operated in a dynamic mode (b). A logic analyzer is controlled by the emulator and connected to the communication signals between the emulator and the target environment. The logic emulator traces the signals and allows the later analysis of this trace. The size of the trace buffer is very critical and distinguishes the different emulation systems. Important requirements for all these prototyping or emulation environments are their flexibility in terms of a flexible integration of microcontrollers, microprocessors or DSP cores.

Flexibility is also required in several other areas. First, it must be possible to implement large memory blocks, which are required for both program and data elements of the software; this is particularly important in co-design applications. The clock must run as fast as possible to allow for dynamic testing and for validation in real time within the system-under-design. High gate capacity, high observability of internal nodes, the ability to read back information out of the FPGAs, and the ability to create long signal traces are also important.

Most of the information on the systems described in the following sections is taken from product information of the mentioned companies. It is collected from various sources, including advertising, technical reports, customer experience reports and publically available information on the world wide web.

### *3.3.2 Quickturn Emulation Systems*

Quickturn offers three different kinds of emulation system. A smaller version for less complex designs, the so-called logic animator, an emulation system for designs up to 3 Million gates, namely the System Realizer with the HDL-ICE software support and quite recent accelerator architecture CoBalt which is similar to the Synopsys Arkos system. Table 3.1 gives an overview of the Logic Animator, which is not described here in much detail. Tab. 3.2 describes two versions of the System Realizer : the basic M250 and the more sophisticated M3000. Figure 3.13 gives an overview of the hardware architecture in the M3000.

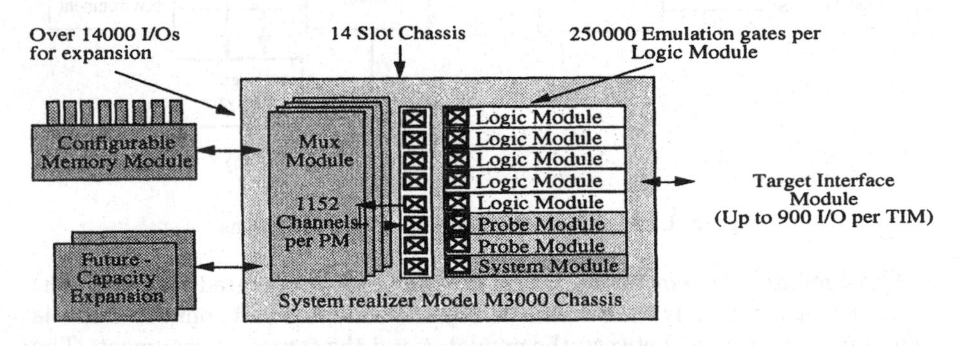


**Figure 3.13** Realizer hardware architecture.

| Model | Animator Model 550 |
|---|---|
| Architecture | Single board, reprogrammable system |
| Capacity | 50K gates typical |
| Memory support | Memory compiler for synchronous RAM (1 port and 2 port) and ROM |
| Prototyping speed | 8 - 16 MHz typical |
| I/O connections | 448 bi-directional pins (for cores, logic analysis and target interface) |
| Software | Single-pass automatic compilation, interactive speed optimization, incremental probe changes |
| Design entry | EDIF, TDL, NDL, Verilog netlist formats, over 40 ASIC libraries available |
| EDA compatibility | Verilog, ViewSim, or VSS simulation environments |
| Debug | Connects to commercial logic analyzers (HP 1650 and 16500 families, Tektronix) |

(Source: Quickturn Product Description)

**Table 3.1** Logic Animator overview.

| **Features and Specifications** | **Model M3000** | **Model M250** |
|---|---|---|
| Architecture | Hierarchical multiplexing architecture<br>Custom interconnect chip<br>Xilinx 4013 FPGA | Hierarchical multiplexing architecture<br>Custom interconnect chip<br>Xilinx 4013 FPGA |
| Design entry<br><br><br><br><br>Emulation | Netlist formats:<br>Verilog structural<br>EDIF, TDL, NDL<br>Over 55 ASIC libraries available<br>Single-pass automatic compilation | Netlist formats:<br>Verilog structural<br>EDIF TDL NDL<br>Over 55 ASIC libraries available<br>Single-pass automatic compilation |
| Debug Software | Functional test<br>Vector debug<br>Interactive readback | Functional test<br>Vector debug<br>Interactive readback |
| Capacity<br><br>Memory support | 250,000 - 3,000,000 emulation gates<br>Memory compiler for RAM (single or multi-port) and ROM<br>Configurable Memory Module up to 14MB each single port up to 875k, 4 write ports, 16 read port | 250,000 emulation gates<br><br>Memory compiler for RAM (single or multi-port) and ROM |
| Emulation speed | 1-4MHz typical | 4-8MHz typical |
| I/O connections | Up to 7200 signal pins<br>4,000 expansion pins (for CMM or capacity expansion) | 600 signal pins |
| Debug | Integrated logic analyzer<br>1152 probes (thousands possible through multiple probe modules)<br>128k memory depth<br>8-event trigger and acquire<br>16MHz operation | Integrated logic analyzer<br>1152 probes<br><br>128k memory depth<br>8 events, complex trigger and acquire<br>16MHz operation |

(Source: Quickturn Product Description)

The System Realizer allows a maximum capacity of 3 million gates. The architecture allows a modular system of up to twelve 250,000 emulation gate logic modules with probe modules through a programmable backplane. Stand alone 250,000 emulation gates integrated system for mid-range designs (basic model of the System Realizer: M250) is also possible. The software support includes partitioning and mapping on the different FPGAs. There is also an integrated instrumentation available to capture and process real-time data for design debugging. The Quickturn System Realizer model especially supports a very flexible memory emulation. Small scattered memories can be compiled into the FPGAs internal RAMs to emulate single, dual, and triple-port memories. For larger memories configurable memory modules (CMMs) are offered. They are specificly designed for memory applications, such as large multiport cache arrays, multi-port register files, asynchronous FIFOs, microcode ROMs and large on-chip RAM arrays. In addition, special target interface modules (TIM) interface the emulator with the target system. Standard adapters route system I/O signals to signals in the emulated design. Furthermore, complex ASIC functions including microcontrollers, CPUs, peripheral controllers etc. are integrated into the emulation environment by plugging real chips into this TIM. The System Realizer currently supports 3 and 5 Volt designs. Like all other emulation systems it allows debug in the test vector mode and in the dynamic mode. In the test vector mode test vectors are applied and results are stored. This vector debug mode supports breakpoints, query signals, read-back of internal nodes, single step and continue. A special functional test mode supports regression testing where 128 K of vectors can be applied at up to 4 MHz in an IC tester-like functional validation environment. An automatic vector compare logic for quick go/no go testing is built-in. In the dynamic mode a logic analyzer can be connected to the emulation environment. This logic analyzer includes a state machine based trigger and acquire capability with up to 8 states, 8 event trigger support, 128K channel depth and 1152 channels. Moreover, further 1152 channel capacity modules may be added in place of emulation modules. The trigger-and-acquire conditions are very similar to advanced commercial logic analyzers.

### *3.3.3 Mentor SimExpress Emulation System*

Like the Quickturn system, the Mentor emulation system also consists of FPGAs. The difference is that the Mentor system is based on a special purpose full custom FPGA architecture developed by the French company Meta Systems, which recently has been taken over by Mentor. Based on Mentor product information this special purpose FPGA emulation architecture is 100 times faster than off-the-shelf FPGA based emulators. The special purpose FPGAs especially support on-chip logic analyzers with access to every net in the design. It also in particular supports high speed fault simulation. The special purpose FPGA also allows higher emulation clock speed up to a maximum of 20 MHz. Typical values are 500 KHz to 4 MHz. In a full version the capacity allows up to 1.5 million user gates plus memory mapping. The same as in the Quickturn system the system configuration consists of a basic cabinet with 23 universal slots. One, two, four and six cabinet configurations are available. For these universal slots different card types are available. All these card types can be plugged into these universal slots. There are actually four different card types available. Namely logic card (10K-12K user gates), I/O card (336 pins), memory card (64 Mbytes) and stimulus card (12 Mbytes).

The custom FPGA architecture which is used in the Mentor emulation system also uses reprogrammable logic blocks (comparable to Xilinx). These programmable logic blocks contain flexible four-input/one-output modules together with a configurable latch or flip-flop. The logic cards themselves allow 24 programmable 32 Kbyte RAM chips. Furthermore, as mentioned before, special dedicated 64 Mbyte memory cards are available. The custom FPGA chips moreover support an automatic mapping of clock distribution trees to avoid timing races and hazards. A special purpose custom interconnection chip supports a scalable network architecture and improves the efficiency of the compilation process and the design access.

As all other systems a test vector and a dynamic mode are supported. The test vector mode supports an interactive debugging, taking advantage of the logic analyzer build into the custom FPGA chip. This custom FPGA chip also supports accesses to all design modes without recompiling the design. Additionally, a signal name cross-reference supports debugging in the context of the signal names and design hierarchy of the original netlist. In the dynamic mode traces and triggers are used. There are four system wide trigger conditions and a 62-state sequencer to generate breakpoints and to perform complex event analysis.

### *3.3.4 Zycad Paradigm RP and XP*

Zycad offers two different systems. The so-called Paradigm XP for the acceleration of gate- level logic simulation and fault simulation, and the Paradigm RP for rapid prototyping. The Paradigm RP was based on Xilinx FPGAs and will in the future be based on the new fine grain FPGAs from Gatefield, the GF100K family. The Paradigm XP can be seen as a family of application specific supercomputers, optimized for logic and fault simulation. It supports accelerated logic simulation with 10 million events per second, which corresponds to 3000 SPECints. It also supports accelerated concurrent fault simulation and mixed-level simulation. The Paradigm XP supports also a logic simulation capacity of up to 16 million gates and a capacity of up to 4 million gates for fault simulation. The back annotation is supported via the standard delay format (SDF). Paradigm XP boards are modular with each having a simulation capacity of 256,000 logic and 64,000 fault primitives. A single Paradigm XP board consists of six 391-pin PGAs connected via a 500-bit wide, high speed parallel bus. 128 Mbytes of pipelinedpipelining memory are possible. The concurrent fault algorithm allows a maximum of 16 faults per simulation pass which is up to 16 times faster than a serial algorithm running on an accelerator. The so-called SuperFault software supports thousands of faults per pass and fault simulation of large circuit designs and complex systems but requires more memory.

Figure 3.14 from the Zycad homepage compares the compile time, simulation time and post process time for logic simulation running on Paradigm XP with a pure software simulation.

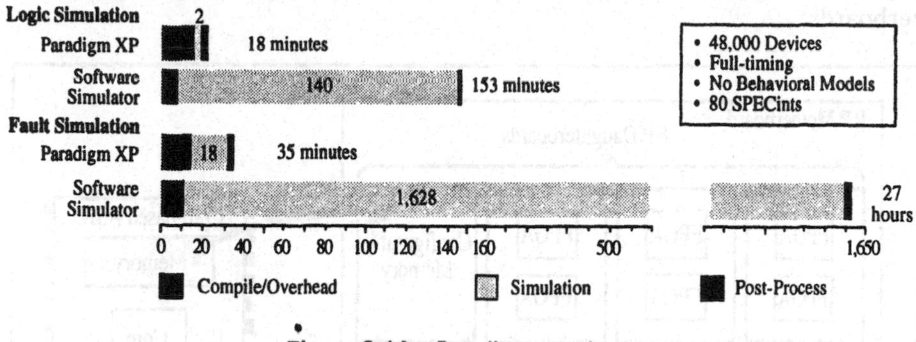


**Figure 3.14** Paradigm speed-up.

The Paradigm RP and the corresponding software called Concept Silicon support prototyping and emulation. It consists of a so-called motherboard and daughterboard architecture based on a modular cabinet structure, the so-called RP2000 system. This RP2000 system supports up to four Paradigm RP motherboard modules. The Paradigm RP uses a backplane with Xilinx 3090 FPGAs to create a crossbar interconnection. It is proposed that these Xilinx 3090 FPGAs are just used for a programmable interconnection. The concept of this backplane guarantees that no further FPGAs are needed to exchange signals between the daughterboards. The backplane also includes a removable prototyping board (proto-board) for target system interconnect and for breadboarding on the prototype itself. This proto-board also contains the connectors for the logic analyzer interface. Each daughterboard which can be plugged into the motherboard and the Xilinx 3090 backplane contains 2 Xilinx 4013 FPGAs. Other daughterboards are also available for example daughterboard containing 2 Altera Flex 81500 FPGAs connected by a 55 bit local bus to take advantage of their additional I/0. Further daughterboards support configurable memory of up to four independent memory instances of up to 4 KB depth as well as single and dual port memories with widths of up to 32 bits and 64 bits.

Special features of the Paradigm RP2000 and the Concept Silicon Software are incremental compilation, memory compilation and partitioning enhancements. The incremental compilation especially supports incremental global routing and incremental FPGA compilation. Only changed nets and changed FPGAs are recompiled.

Figure 3.15 shows the mapping of a core-cased ASIC design to an RP2000. The memory of a core-based ASIC is compiled into the configurable memory daughterboards. The core is compiled into a "daughtercore" assuming that the used core processor is also available as a daughtercore card and the standard logic of this core-based ASIC is compiled into the Xilinx 4013 FPGA daughterboards.

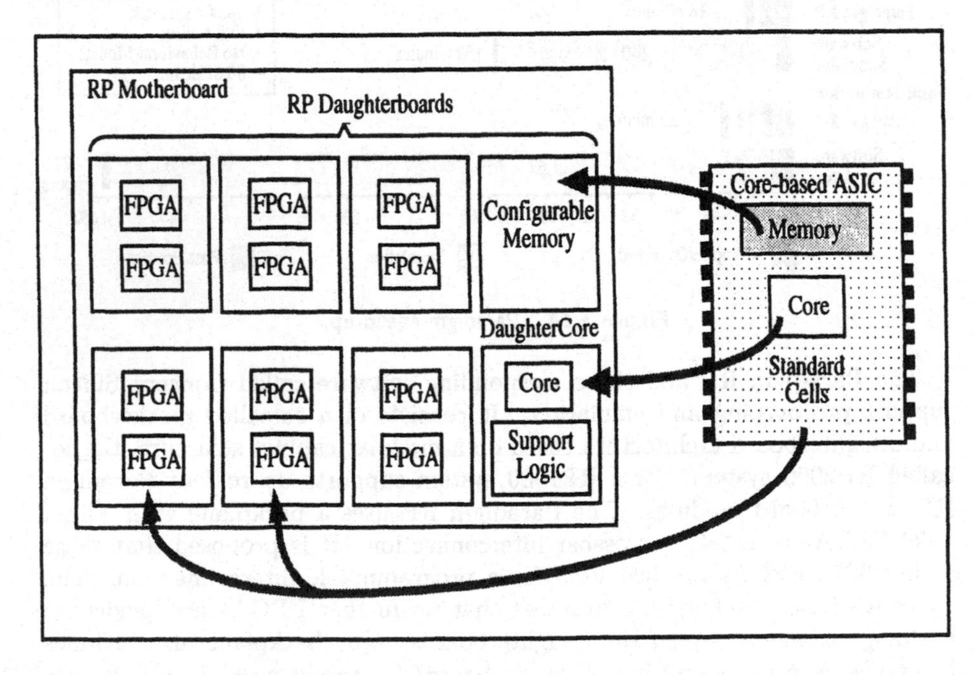


**Figure 3.15** Mapping of core-based ASICs.

Figure 3.16 shows the interconnection of the daughterboards via the Xilinx 3090 crossbar. An important concept in this context are the so-called *daughter-*

*cores.* The idea is that for core- based ASICs the corresponding daughtercores are available and support hardware in the loop for the used processor. The idea of Zycad is that such daughtercores are developed together with customers using the Zycad system. Actually daughtercores are available with a Thompson DSP core and a LSI logic multiplier core. The RP2000 system does not contain special logic analyzer modules, but allows to interface the system with an IEEE-488 compatible logic analyzer. An overview on the complete RP2000 structure is given in Figure 3.17.

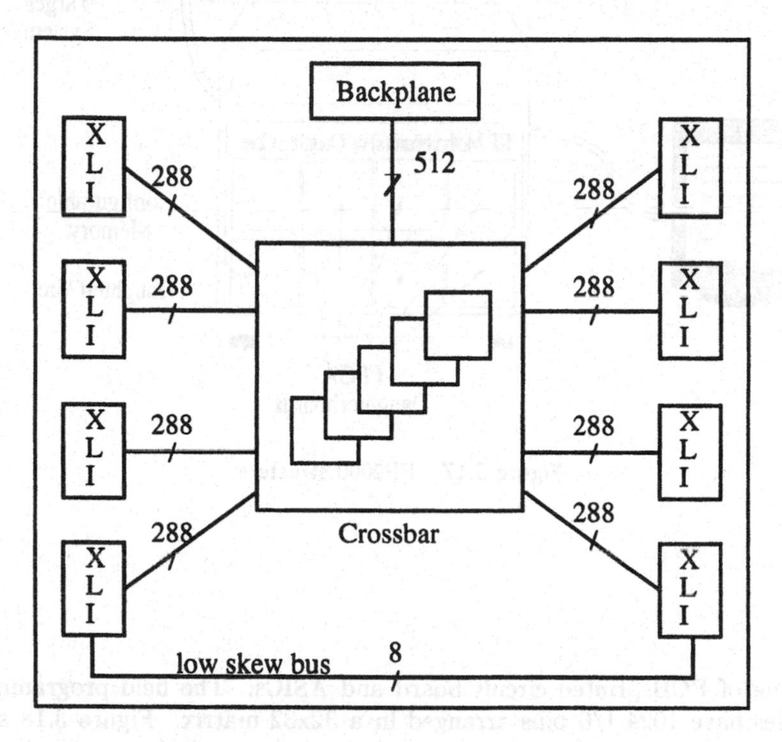


**Figure 3.16** Daughterboard interconnection.

### *3.3.5 Aptix Prototyping System*

Aptix prototyping environment is based on the Aptix switches, which are called **field programmable interconnect components** (FPICs). They are special switching elements which can be programmed and which support the routing of signals between the chips of boards. By means of these field programmable switches a fast breadboarding of system designs is possible, allowing fast pro-

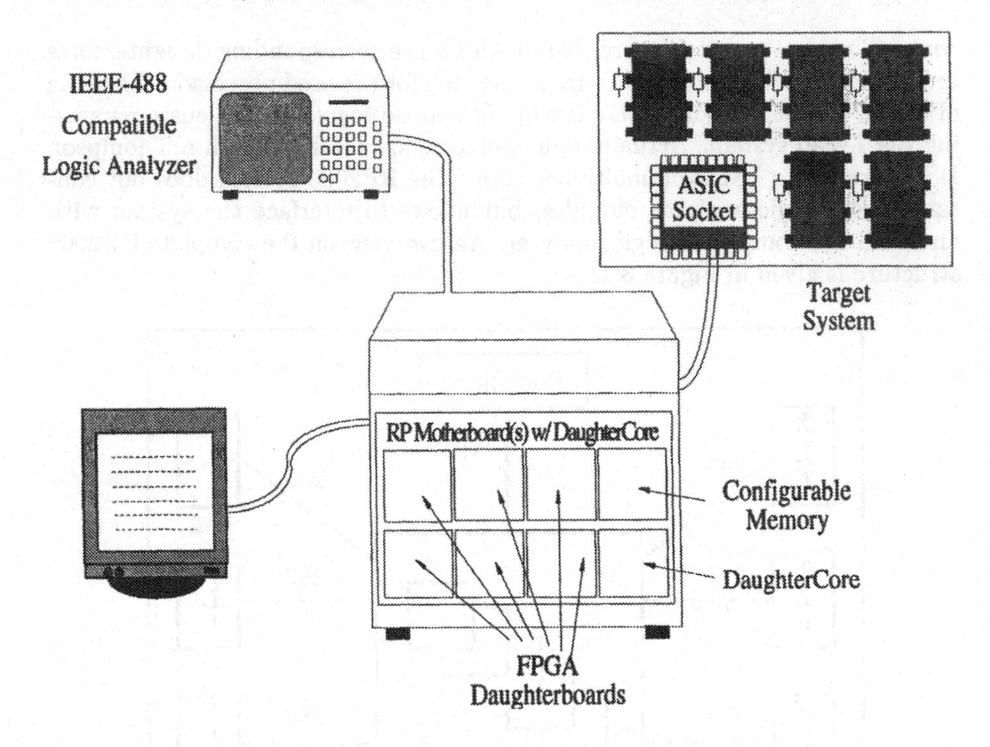


**Figure 3.17** RP2000 structure.

totyping of PCBsprinted circuit board and ASICs. The field programmable switches have 1024 I/0 pins arranged in a 32x32 matrix. Figure 3.18 shows how the different board components are interconnected to these programmable switches. Figure 3.19 shows such a field programmable circuit board.

New versions of the Aptix prototyping system (MP4 system explorer, MP4Pro Explorer) support the integration of the shelf components and FPGAs of any type. This environment also supports test vector and dynamic mode. In the test vector mode up to 544 probe points for debugging are available. An additional stimulus response interface module is available with up to 288 pattern generation channels. The host system compares the uploaded results with original simulation results. Tab. 3.3 gives an overview on the MP4 and the MP4Pro architectures.

| Feature | MP4 | MP4Pro |
|---|---|---|
| Architecture | Open architecture<br>Accepts all components | Open architecture<br>Accept all components |
| FPGAs | Support for all popular families | Support for all popular families |
| Interconnection | Fully reprogrammable 4 FPIC with 1024 pins each<br>4ns propagation delay through FPICs | Fully reprogrammable 4 FPIC with 1024 pins each<br>4ns propagation delay through FPICs |
| Emulation Speed | 20 - 35 MHz typical<br>50MHz high speed electronic bus | 10 - 20 MHz typical |
| Debug | 256 probes interface to HP logic analyzers and pattern generators (HP 1660 and 16500 families supported) | 256 probes interface to HP logic analyzers and pattern generators (HP 1660 and 16500 families supported), Stimulus Response Interface with 544 probes and 288 pattern generation channels |
| Pattern generation | All popular simulators plus ASCII | All popular simulators plus ASCII |
| vector formats | interface | interface |
| Capacity | 4 FPICs 2,880 free holes<br>Up to 20<br>FPGAs (208 QFP) | 4 FPICs 2,880 free holes<br>Up to 20<br>FPGAs (208 QFP) |
| Design input | EDIF, Xilinx XNF | EDIF, Xilinx XNF |
| System programming | Ethernet access through the Host | Ethernet access through the Host |
| Interface Module | On-board flash memory for standalone operation<br><br>Controlled by on-board microcontroller | Interface Module On-board flash memory for standalone operation<br>Controlled by on-board microcontroller |
| Partitioning | Manual partitioning | Automatic partitioning software |
| I/O connections | 448 I/O plus 160 bus pins | 448 I/O plus 160 bus pins |
| Software platforms | Sun SPARC, HP 700, PC | Sun SPARC, HP 700, PC |

**Table 3.3** MP4 architecture overview.

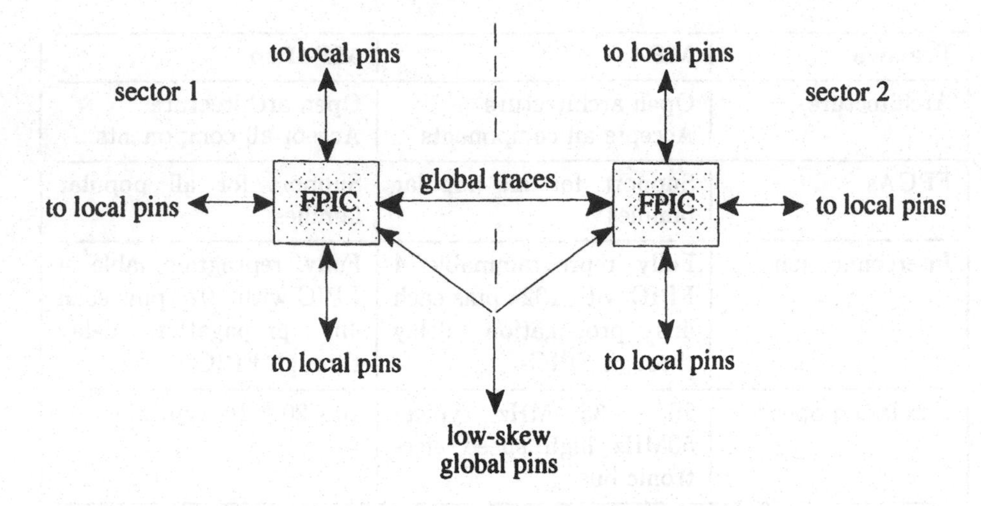


**Figure 3.18** Programmable interconnection.

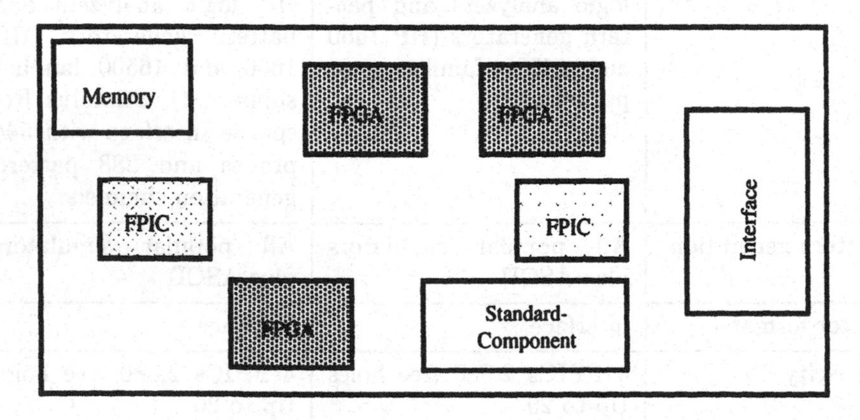


**Figure 3.19** Programmable circuit board.

### *3.3.6 Arkos (Synopsys) and CoBalt (Quickturn) Emulation Systems*

These systems are completely different from all other FPGA based emulation and prototyping environments. They are not FPGA- but processor-based. They contain processors specialized for logic computation and use a special interconnection architecture. In contrast to the FPGA based emulation and prototyping environments for the Arkos and CoBalt systems the design is compiled into a program that is loaded and executed on these logic processors. Compiling the hardware description into code for these specialized logic processors especially supports debugging. A compilation is of course also much faster than mapping on FPGAs. In addition, special custom logic processors are fast enough for the emulation speed to be compared to FPGA based emulators. Arkos and CoBalt support emulation speed in the range of 1 MHz.

Quickturn and Synopsys each announced their new generation of custom processor-based emulation systems in late 1996. These new generations of emulation systems offer certain unique capabilities in design capacity, compile time and debug capability. Although the Synopsys Arkos emulator and the Quickturn CoBalt (Concurrent Broadcast Array Logic Technology) System have different architectures, they provide similar benefits to the user including fast compile times, enhanced runtime and debug performance, and improved design visibility.

Just recently Quickturn and IBM entered into a multi-year agreement to co-develop and market the CoBalt compiled-code emulation system based on a 0.25-micron custom processor and an innovative broadcast interconnect architecture developed at IBM, which is coupled with Quickturn's Quest II emulation software. CoBalt provides verification capabilities for designs between 500,000 and 8,000,000 logic gates. The Synopsys Arkos emulation system is also based on a custom processor and was originally introduced by Arkos Design Systems in 1994 and was re-engineered after Synopsys acquired that company in 1995. With this new agreement, both companies are reaffirming efforts in their respective areas of expertise and are developing a close cooperation.

## 3.4 FUTURE DEVELOPMENTS IN EMULATION AND PROTOTYPING

Advances in design automation enable synthesis of designs from higher levels of abstraction. Advances in synthesis have natural consequences for emulation, which may lead to problems, in particular in connection with debugging. Interactive debugging by means of a designer requires that in mapping the description on the emulation environment the designer is enabled to reconstruct the transformations which have been implemented by the synthesis system. In the case of implementation of emulation starting off from a description on RT level, the resulting necessity of keeping up the pace may mostly be reconstructed by cross reference lists. However, it is considerably harder to keep up the interrelation between the description done by the designer and the final emulated circuit in emulation from a design created by high-level synthesis. A particular complication ensues from the fact that the time response of the circuit is determined or optimized only through high-level synthesis.

As a solution to this problem we have suggested the so-called source-level emulationsource-level emulation (SLE), which aims at closing the gap between description, or simulation on behavioral level and hardware emulation. The basic idea of SLE is the retention of a relation between hardware elements and the source program (which may be given as behavioral VHDL) during the execution of the application on an emulation hardware. In analogy to a software debugger it aims at a potential debugging on the basis of the source code. Consequently, it should be possible to set up breakpoints based on the VHDL source text, to select the active variable values and to continue the program. This kind of source emulation is of particular relevance in connection with hardware/software co-design and co-emulation so that in the end debugging implementation will be possible independently of the implementation in hardware or software—debugging on the the same level and with the same user interface.

Our approach supports symbolic debugging of running hardware analogous to software debugging, including examining variables, setting breakpoints, and single step execution. All this is possible with the application running as real hardware implementation on a hardware emulator. Back annotating the values read from the circuit debugging can be done at the source code level. For certain applications it is not even necessary to capture the environment of the application in test vectors. It is possible to connect the emulator to the environment of the application. There is no need to develop simulation environments in VHDL. Avoidance of the test bench development is another advanced side effect of this approach, since these simulation environments are at least as error-prone as the application itself. Figure 3.20 shows the steps of HLS together with the data that these produce. Both actions, the sharing of components and the sharing of registers add multiplexers to the circuit.

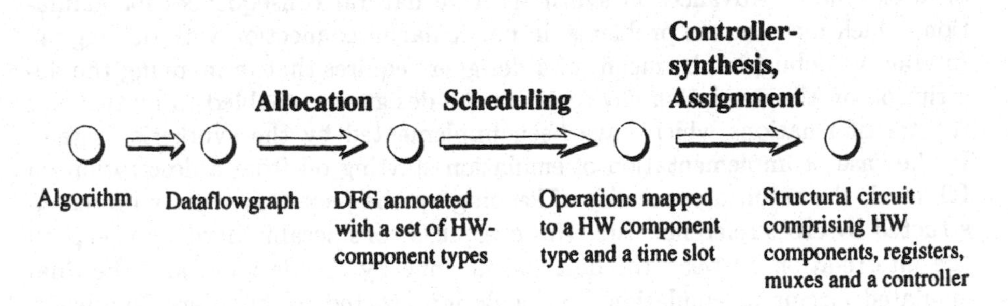


**Figure 3.20** Steps in high-level synthesis.

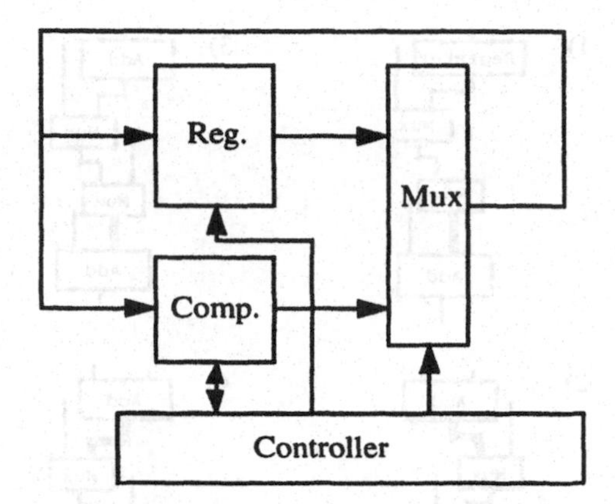


**Figure 3.21** Target architecture.

### *3.4.1 Target Architecture*

To get a better view on what SLE is, it is necessary to take a closer look on the relation between the behavioral VHDL source code and the generated RT level circuit. Figure 3.22 shows different possible implementations of variables by nets at the RT level. In case 1) the requested variable (bold line) is an input net of the component and connected to register via several other components.

Only the register has to be read for retrieving the value of this net. In case 2) it is backtracked from the component which is connected to a register via some copying components. The corresponding value may be retrievable by reading the register in the next clock cycle. In case 3) the input net is connected to another computing component. It is impossible to retrieve the value of the net directly. In this case the input nets of all other components have to be recursively backtracked. Until registers are reached at all inputs of the resulting combinational circuit, the value of the requested net must be computed from the values of these registers by the debugging software. Therefore, the debugging software maintains a model of the circuit and the components. In case 4), where the requested output net is connected to another computing component, the input nets of the component of which the output is requested must be backtracked as in case 3).

The time overhead for the software computation does not affect the running hardware since such computations occur only at breakpoints when the hardware stops. It only requires a library where a behavioral model of each component is stored.

High-level synthesis maps operations to components in the circuit. Thereby it can happen that an operation corresponds to several components as well as

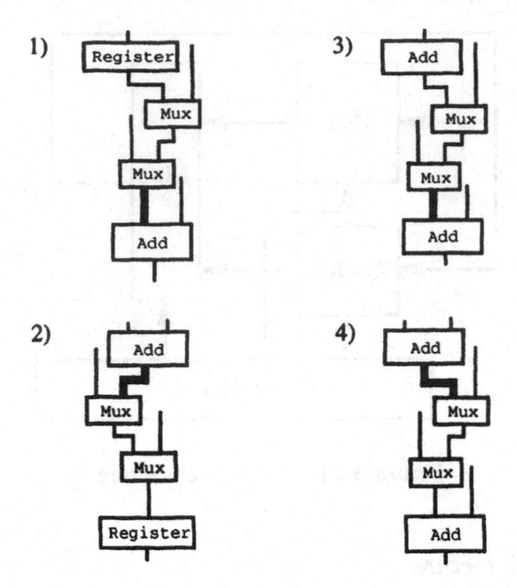


**Figure 3.22** Possible net connections.

several operations corresponding to the same component. For SLE it is necessary to keep track of the correspondence between operations and components after high-level synthesis, and to resolve such dependencies. This correlation is used to set breakpoints at particular steps in the algorithm. Therefore the corresponding components are identified together with the state of the controller at which this components execute the requested operation. The state of the controller is known from scheduling and the component is known from the assignment.

When the correspondence between the behavioral specification and the synthesized RT level is known, a special synthesis step is needed, which does not destroy the structure for debugging through retiming optimization etc. Also some additional hardware is required for the retrieval of register values from outside and for setting breakpoints and which is to interrupt the hardware when a breakpoint is reached. A special debugging controller has been developed to provide these additional debugging support operations. A special target hardware system has been developed as a back-end environment for this source level emulation system (Figure 3.21).

This hardware is based on the Weaver environment presented in section 3.3.1 and allows the access of the requested debugging structure inside the circuit. As mentioned before, the programming and the controlling of breakpoints is done by the debugging controller. This component allows the host to control the circuit operation. Figure 3.23 shows the scheme of this debugging controller.

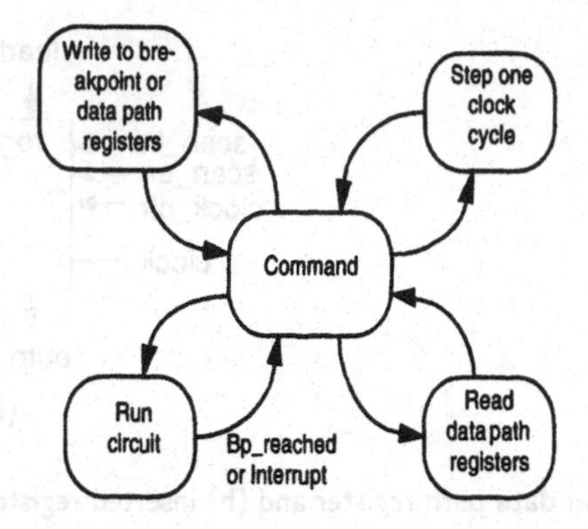


**Figure 3.23** Simplified state diagram of the debugging controller.

The command state is the initial state of the controller. In this state the controller waits for a command and the circuit stops. The following commands are implemented:

- Write a breakpoint.
- Write values to the data path registers.
- Read the contents of the data path registers.
- Step one clock cycle.
- Run the circuit.

Each command is decoded and the controller switches to the corresponding macro state, which controls the required action. In each macro state, the controller returns to the command state after the action is finished. Only the ***run circuit*** state is kept until a breakpoint is reached or a interrupt command is issued by the host. The circuit is interrupted by setting a clock control signal to 0, which then prevents the data path registers and the circuit controller from operation.

Each register in the data path is exchanged according to Figure 3.24 by a register which supports the required operation control. The inputs added to the original data path register are used for programming or to read back of the registers via a scan path and for enabling normal data path operation. These additional inputs are controlled by the debugging controller. The debugging controller itself is not application dependent. It is implemented together with the application on the Weaver system for reasons of simplicity, but it can also be implemented outside the programmable logic.

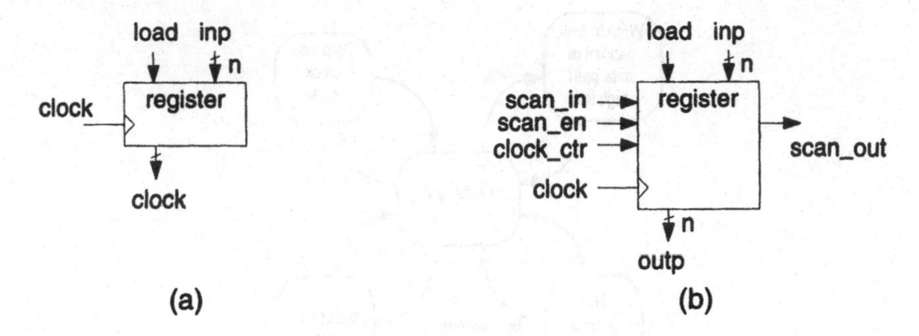


**Figure 3.24** (a) Original data path register and (b) inserted register for debug operations.

## 3.5 EXAMPLE

As example we have chosen Euclid's gcd algorithm. The VHDL code is shown in Figure 3.25.

```
entity gcd is
  port (I1, I2: in Bit_Vector (15 downto 0);
        Ou: out Bit_Vector (15 downto 0));
end gcd;

architecture beh of gcd is
begin
process
  variable X1, X2: Bit_Vector(15 downto 0);
begin
  while (true) loop
    wait for t1;
    read(X1,I1);
    read(X2,I2);
    while X1 /= X2 loop
      if X1 < X2 then
        X2 := X2 - X1;
      else
        X1 := X1 - X2;
      end if;
      wait for t2;
    end loop;
    wait for t3;
    write(OU,X1);
  end loop;
 end process;
 end beh;
```

**Figure 3.25** The example VHDL code.

It is a very simple algorithm but it shows the principles of the approach. Figure 3.26 shows the intermediate flowgraph, the generated circuit and the controller graph of the gcd algorithm.

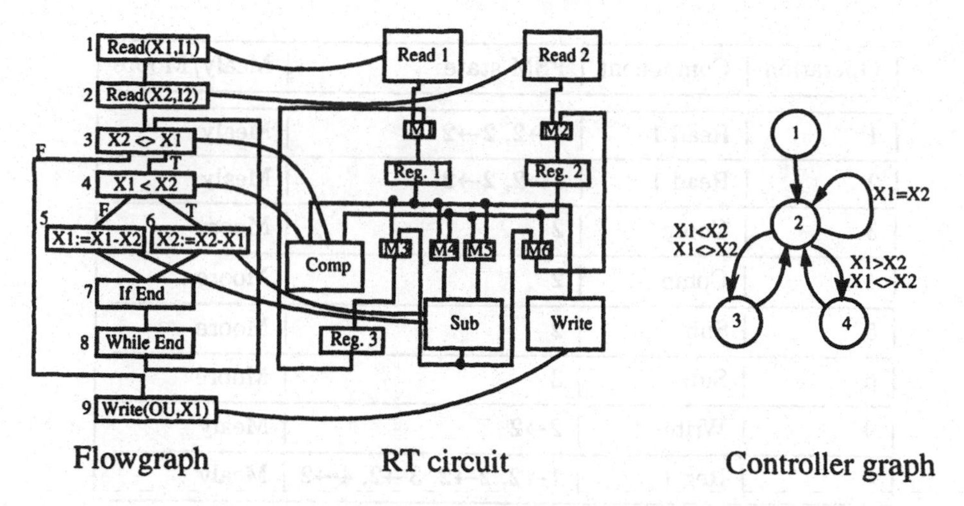


**Figure 3.26** FG, circuit, and controller of the gcd algorithm.

The relationship between a flow graph and VHDL code is described in more detail in Chapter 7. The lines between the flow graph and circuit show the relationships between operations and components. Both subtractions are executed by the same component, and both comparisons are performed by the same comparator. Thus the while-condition and the if-condition of the VHDL source are evaluated in parallel at the same time. The conditional transitions in the controller graph are marked with their conditions. Table 3.4 shows the full relationship between operations, components and controller states. The operations and controller states are numbered according to their numbers in Figure 3.26, and the components are also named according to Figure 3.26. The column Mealy/Moore is used to describe the behavior of the controller output which controls the component. For synchronous components this is of type Mealy, and for others it is of type Moore. If it is Mealy, then the component belongs to a state transition rather than to a state. In this case the outputs of the comparator determine together with the controller state whether the component is executed or not.

For instance, operation No. 1 is executed by the component Read 1 in the controller states 1 and 2. It is controlled by a Mealy output of the FSM and therefore also the comparator outputs are relevant to decide whether the component is actually executed or not. In state 1 it is executed anyway, since there is no relevant condition. But in state 2 it is only executed if the following state is also 2, that is if the condition X1=X2 is true.

| Operation | Component | FSM state | Mealy/Moore |
|---|---|---|---|
| 1 | Read 1 | 1→2, 2→2 | Mealy |
| 2 | Read 1 | 1→2, 2→2 | Mealy |
| 3 | Comp | 2 | Moore |
| 4 | Comp | 2 | Moore |
| 5 | Sub | 2 | Moore |
| 6 | Sub | 3 | Moore |
| 9 | Write | 2→2 | Mealy |
| - | Reg 1 | 1→2, 2→2, 3→2, 4→2 | Mealy |
| - | Reg 2 | 1→2, 2→2, 3→2, 4→2 | Mealy |
| - | Reg 3 | 2→3, 2→4 | Mealy |

**Table 3.4** Relation operation-component-FSM state.

| Mux | State 1 | State 2 | State 3 | State 4 | Type |
|---|---|---|---|---|---|
| M1 | 0, 1→2 | 0, 2→2 | 1, 3→2 | 1, 4→2 | Mealy |
| M2 | 0, 1→2 | 0, 2→2 | 1, 3→2 | 1, 4→2 | Mealy |
| M3 | - | 0, 2→3<br>1, 2→4 | 0, 3→2 | 1, 4→2 | Mealy |
| M4 | - | 0 | 1 | - | Moore |
| M5 | - | 0 | 1 | - | Moore |
| M6 | - | 0, 2→3<br>1, 2→4 | 0, 3→2 | 1, 4→2 | Mealy |

**Table 3.5** Multiplexer controls.

Table 3.5 shows how the multiplexers are controlled in each controller state. Some of the multiplexers are controlled by controller outputs with Mealy behavior, others with Moore behavior. Those with Mealy behavior have a state transition after the controlling value in the table. Also these may have different values for different transitions in the same state. For instance, in state 2 the multiplexer M3 puts its first input through (control = 0) if the following state is state 3, and the second input is put through (control = 1) if the following state is state 4. In our example we request the input variable X1 of the operation No. 5. Table 3.4 shows that this operation is executed by the component sub and that the clock must be interrupted in controller state No. 2. An evaluation of the multiplexer controls given in table 3.5 shows that the requested input of this component is connected directly via a put-through component to register Reg 1. Thus, this register holds the requested value. Doing the same for Reg 2 that at the same time Reg 2 holds the value of X2 as the other input variable of operation No. 5. The values correspond also to the inputs of the operations 3 and 4. The value of Reg 3 cannot be assigned to a variable since the history of the circuit is not known.

## 3.6 RESULTS

The described technique is implemented within our synthesis tool named CADDY [54][55][56]. All changes to the circuit controller and to the data path are applied automatically during the generation of the VHDL code by CADDY. There is no manual work by the user required for the generation of the debug model. The debugging controller communicates with a SUN workstation via the parallel port. The applications listed below for the results were all synthesized, downloaded and run on our Weaver [57] prototyping board. The following applications were done:

- GCD: A standard example for high-level synthesis.
- SIRDG: A circuit which computes single image random dot stereograms. It receives a source image with the height information via a parallel interface and writes the generated image to a host via a parallel interface.
- DCT: A circuit which computes a two-dimensional 8x8-DCT. It is implemented as a coprocessor for the Hyperstone processor [58]. The communication with the processor is done via the external processor bus.

In table 3.6, the area values of the different circuits are listed. All values are given in CLBs for the Xilinx XC4000 series. Please note, that the constant overhead for the debugging controller (appr. 44 CLBs) virtually can be subtracted from the overhead numbers for the examples.

- Orig. refers to the original circuit without debug overhead.

| Circuit | Orig. | Dbg 1 | Dbg 2 | Mealy | Moore | #Bp | #States |
|---|---|---|---|---|---|---|---|
| GCD | 60 | 128/84 | 141/97 | 3 | 3 | 6 | 4 |
| SIRGD | 99 | 190/146 | 213/179 | 2 | 5 | 11 | 21 |
| DCT | 387 | 474/430 | 476/432 | 1 | 6 | 28 | 33 |

**Table 3.6** Area values of the different circuits.

| Circuit | Orig. | Dbg 1 | Dbg 2 |
|---|---|---|---|
| GCD | 5.7 Mhz | 2.0 Mhz | 2.0 Mhz |
| SIRDG | 7.7 Mhz | 4.8 Mhz | 4.9 Mhz |
| DCT | 3.5 Mhz | 3.5 Mhz | 3.5 Mhz |

**Table 3.7** Delay.

- Dbg 1 includes the overhead for debugging without the possibility of data dependent breakpoints. It shows the value with/without debugging controller.
- Dbg 2 represents the full version with data dependent breakpoints.
- Mealy denotes the number of bits needed for the Mealy identifiers, Moore denotes the number of bits of the Moore identifiers. The sum of Mealy and Moore is added to the output of the circuit controller.
- #Bp denotes the number of different detectable breakpoints in the specification.
- #States lists the number of controller states of the circuit controller.

Table 3.7 shows the delay that is added by the additional logic for debugging. The values are obtained by the Xilinx tool *xdelay*. This tool provides a quite pessimistic estimation. All designs run at a significantly higher clock speed on our Weaver board. Nevertheless, it shows the relation between the different implementations. There is not much difference between the two versions with debugging overhead, since the evaluation of conditional breakpoints does not slow down the circuit additionally. Our approach mainly adds controller delay. This is the reason, why the GCD circuit is so much slower. The delay of this circuit is already dominated by the controller delay since its data path is very simple. Delay in the DCT example is mainly determined by the combinatorial multiplier. To this path, we do no add any delay. Therefore, the debug versions can be clocked with equal frequency.

## 3.7 CONCLUSIONS

This contribution has given a survey on the state of technique in the fields of prototyping and emulation. Both existing commercial prototyping and emulation systems have been presented as well as actual research activities. The existing commercial systems are featured by different starting-points. Most of the systems take user programmable gate arrays like off-the-shelf FPGAs, especially those from Xilinx. Mentor uses FPGA structures which are especially optimized for emulation. These systems differ regarding the interconnection structure among these numerous FPGAs where a logic design is mapped onto. Some systems use again programmable gate arrays as interconnection structure. Others take special user programmable switches for interconnect. Another alternative would be to use especially developed VLSI custom switch circuits. A new starting-point has been developed by Synopsys and Quickturn. The Arkos and CoBalt emulation systems of Synopsys and Quickturn do not use FPGAs but closely coupled special processors with an instruction set whose instructions are optimized for execution of logic operators. A circuit design will then be translated into Boolean equations which on their part are compiled into machine code for these special logic processors.

Future work aims especially to support emulation at higher levels of abstraction. These approaches will take into account that automatic synthesis systems will be used more intensively, as these are necessary for the increase of design productivity. Increased design productivity and short design times are necessary for the "first-time-right silicon" concept. Emulation and prototyping are an important contribution to reach this aim. Source-level emulation, which is presented in the last chapter, does moreover help to support hardware/software co-design by an improved hardware/software co-simulation, respectively hardware/software co-emulation.

### Acknowledgments

This work has been partially supported by grants from the European Commission (Project EC-US-045) and Deutsche Forschungsgemeinschaft (DFG Research Programme: Rapid Prototyping of Embedded Systems with Hard Time Constraints). I thank Prof. D. Gajski for his hospitality, since parts of this chapter were written during my stay at UC Irvine. I would also like to thank my research assistants Julia Wunner and Gernot Koch for helping me to put this chapter together.

# 4 TARGET ARCHITECTURES

Rolf Ernst

Technische Universität Braunschweig
Institut für Datenverarbeitungsanlagen
Braunschweig, Germany

## 4.1 INTRODUCTION

This chapter surveys the architectures of CPUs frequently used in modern embedded computing systems. It is legitimate to ask why target architectures should be a co-design problem when computer architecture and VLSI design have been taught at universities for decades and are, therefore, widely known in the CAD and system design communities. Throughout this chapter, however, we will see a large variety of co-design target architectures which substantially differ from the general-purpose RISC architectures that are familiar to students today, but are of the same economic importance. We will define a set of orthogonal specialization techniques as a framework to explain the design process which leads to the observed variety. A number of examples from different application areas will support our findings. While in this chapter the set of specialization techniques is primarily used to understand and to teach target architecture design, it shall also serve as a basis for future work in target system synthesis.



*J. Staunstrup and W. Wolf (eds.), Hardware/Software Co-Design: Principles and Practice*, 113-148.

The main application for hardware/software co-design is embedded system design. Embedded systems realize a well defined set of system tasks. In contrast to general-purpose computer design, the application is defined a priori which allows **target system specialization**. On the other hand, a certain level of adaptability is still required to allow for late changes of the design specification or to increase the reusability of a component. *Designing specialized systems which provide adequate flexibility is a central issue in embedded system design.* Secondly, embedded system design is subject to many **non functional constraints** with strong influence on design objectives and architectures.

The most important non-functional constraints are:

- **strict cost margins** Often, the design objective is not only to minimize cost but to reach a certain market segment consisting of a price range and an expected functionality. If cost is above the intended price margin, functionality and performance may have to be redefined. This is different from typical general-purpose microprocessor design for PC and workstation applications, where performance plays a dominant role.

- **time-to-market and predictable design time** This constraint also applies to general-purpose processor design except that in embedded system design, the specification includes all software functions. Furthermore, development can lead to a family of similar products or include other embedded products, which might not yet be completely defined.

- **(hard) time constraints** Many embedded systems are subject to hard real-time constraints. By definition, when hard real-time constraints [59] are violated, a major system malfunction will occur. This is often the case in safety critical systems. For such systems, timing predictability is of primary importance. Modern RISC processors with long pipelines and superscalar architectures, however, support dynamic out of order execution which is not fully predictable at compile time. So, the obtained speedup is to a large part useless in the context of hard real-time requirements. This is not an issue in a PC environment.

- **power dissipation** is certainly also a problem in PC processor design. In portable systems, such as PDAs and mobile phones, however, battery life time is also an important quality factor.

- **safety** Many embedded applications such as automotive electronics or avionics have safety requirements. There is a host of work in self checking and fault tolerant electronic systems design [60, 61] but, in embedded systems, mechanical parts must be regarded as well. Safety requirements can have very subtle effects on system design when we take EMI (electromagnetic integrity) into account [62]. As an example, the clock frequency may be constrained since (differential mode) radiation of a digital system increases roughly with the square of the clock frequency.

- **physical constraints**, such as size, weight, etc...

We may assume that designers will only accept computer-aided design space exploration and system optimization if the resulting target architectures show a quality comparable to manual designs. We need a thorough understanding of these architectures and their underlying design techniques to be able to match their quality in computer aided co-design. This not only holds for an automated design process, but even more for design space exploration where the designers would expect to have the same design space available as in manual designs. Moreover, target architecture knowledge is a prerequisite for efficient and accurate estimation of performance, cost, power dissipation and safety.

While current co-synthesis tools are limited to a small set of architectures and optimization techniques, future CAD tools for 100 million transistor IC designs will have to cover a wide range of architectures for efficient exploration of a huge design space. Co-design and co-synthesis will have to support multiple-process systems which are mapped to multithreaded, heterogeneous architectures with many different communication and execution models and an inhomogeneous design space which is constrained by predefined components such as standard processors or peripheral devices. A comparison with current high-level synthesis tools [63, 64] which focus on single threaded systems based on a homogeneous architecture and design space shows the tremendous challenge of system co-synthesis.

To explore the inhomogeneous co-design space, we need design transformations which allows us to map a system onto different types of target architectures, such as a parallel DSP, a microcontroller or an ASIC or an application specific processor. However, we will show that the wide variety of different target architectures and components can be explained with a small set of specialization techniques and communication principles which might be exploited in co-design spacc exploration.

The next section introduces the specialization techniques which will then be applied to explain the different classes of target architectures in use today. We will use small systems which will be sufficient to show the main target architecture characteristics. We will focus on the main principles and will exclude other aspects such as test, fault tolerance, or production. The examples are selected to demonstrate the concepts and shall not imply any judgment on design quality. We will assume a working knowledge in general-purpose computer architecture, such as represented e.g. by [65].

## 4.2 ARCHITECTURE SPECIALIZATION TECHNIQUES

We can classify architecture specialization in five independent **specialization techniques** for components and additional techniques to support load distribution and to organize data and control flow in a system of specialized components. **Components** shall be defined to have a ***single control thread*** while **systems** may have ***multiple control threads***. The reference system used for comparison is always a general-purpose programmable architecture. We start bottom up with specialization techniques for individual components and will then derive specialization techniques for systems.

### *4.2.1 Component specialization techniques*

The first technique is **instruction set specialization**. It defines the operations which can be controlled in parallel. It can be adapted to operations which can frequently be scheduled concurrently in the application or which can frequently be chained. An example for the first case is the next data address computation in digital signal processors (DSPs) which can mostly be done in parallel to arithmetic operations on these data, and an example for the second case are multiply-accumulate or vector operations which can also be found in signal processing applications.

The potential impact of instruction set specialization is the reduction of instruction word length and code size (chaining) compared to a general-purpose instruction set for the same architecture which in turn reduces the required memory size and bandwidth as well as the power consumption.

The second technique is **function unit and data path specialization**. This includes word length adaptation and the implementation of application specific hardware functions, such as for string manipulation or string matching, for pixel operations, or for multiplication-accumulation. Word length adaptation has a large impact on cost and power consumption while application specific data paths can increase performance and reduce cost. As a caveat, specialized complex data paths may lead and to a larger clock cycle time which potentially reduces performance. This problem, however, is well known from general-purpose computer architecture. In embedded systems, however, a slower clock at the same performance level might be highly desirable to reduce board and interconnect cost or electromagnetic radiation.

The third technique is **memory specialization**. It addresses the number and size of memory banks, the number and size of access ports (and their word length) and the supported access patterns such as page mode, interleaved access, FIFO, random access, etc. This technique is very important for high performance systems where it allows to increase the potential parallelism or reduce the cost compared to a general-purpose system with the same performance. The use of several smaller memory blocks can also reduce power dissipation.

The fourth technique is **interconnect specialization**. Here, interconnect includes the interconnect of function modules, as well as the protocol used for communication between interconnected components. Examples are a reduced bus system compared to general purpose processors to save cost or power consumption or to increase performance due to a minimized bus load. Interconnect minimization also helps to reduce control costs and instruction word length and, therefore, increases the concurrency which is "manageable" in a processor component. One can also add connections to specialized registers (memory specialization) which increases the potential parallelism. Examples include accumulator registers and data address registers or loop registers in digital signal processors.

The fifth technique is **control specialization**. Here, the control model and control flow are adapted to the application. As an example, microcontrollers often use a central controller without pipelining. Pipelining is employed in high performance processors while distributed control can be found in complex DSP processors as we will see later. The individual controller can be a microprogrammed unit or a finite-state machine.

### *4.2.2 System Specialization*

System specialization has two aspects, **load distribution** and **component interaction.**

Unlike load distribution in parallel systems, embedded systems are typically heterogeneously structured as a result of component specialization. Unlike load distribution in real-time systems (which may be heterogeneous as well), the target system is not fixed. In hardware/software co-design, the distribution of system functions to system components determines how well a system can be specialized and optimized. ***So, hardware/software co-design adds an extra dimension to the cost function in load distribution, which is the potential specialization benefit.***

Two load distribution objectives seem to be particularly important to embedded system design, control decomposition and data decomposition.

**Control decomposition** (which could also be called **control clustering**) partitions system functions into locally controlled components *with the objective to specialize component control.* As a result, some of the components are fully programmable processors, some are hard coded or hard wired functions without a controller (e.g. filters and peripherals) and some components are implemented with a set of selectable, predefined control sequences, possibly with chaining. An example for the latter case are floating point co-processors, DMA co-processors or graphics co-processors. The components control each other which requires a global control flow. Again, the global control flow is specialized to the system and is heterogeneous in general. We will explain the global control flow techniques in the next section.

**Data decomposition** (or **data clustering** partitions the system functions and their data with the objective to specialize data storage and data flow between components. This leads to different, specialized memory structures within and between the components, as we have seen in the context of component memory specialization.

### *4.2.3 System Specialization Techniques*

Control specialization which is restricted to individual components results in a system of communicating but otherwise *independently controlled components.* Control specialization in a system can go further and minimize component control such that it becomes dependent on other components. Such asymmetric control relationships require global control flow. Again, there is a large variety of global control mechanisms, but they adhere to few basic principles. We will distinguish independently controlled components, dependent co-processors, incrementally controlled co-processors and partially dependent co-processors.

The first class are **independently controlled components**. If the components are programmable processors, we would classify the system as MIMD [66]. There is no global control flow required between such components.

On the other extreme are components which completely depend on the controller of another component. They shall be called **dependent co-processors**. The controlling component shall be the *master component*, usually a processor, with the *central controller*, usually a processor control unit. Figure 4.1 shows the typical implementation of the global control flow required for dependent co-processors. The *master component* sends a control message to the *co-processor* which it derives from its own state (in case of a processor: instruction). As a reaction, the *co-processor* executes a single thread of control and then stops to wait for the next *control message.* So, the *co-processor* behavior can be modeled with a **wait** and a **case** statement as shown in Figure 4.1.

If needed, additional control parameters are passed in a *control parameter set*, typically via *shared memory* communication. This alleviates the *master component* from control data transmission. Examples are small peripheral units such as UARTs (serial I/O units) or PIOs (parallel I/O units). From an instruction set point of view, we can consider the co-processor functions as complex, multi-cycle microcoded processor functions. Master component execution can continue in parallel to the co-processor (multiple system control threads) or mutually exclusive to the co-processor (single system control

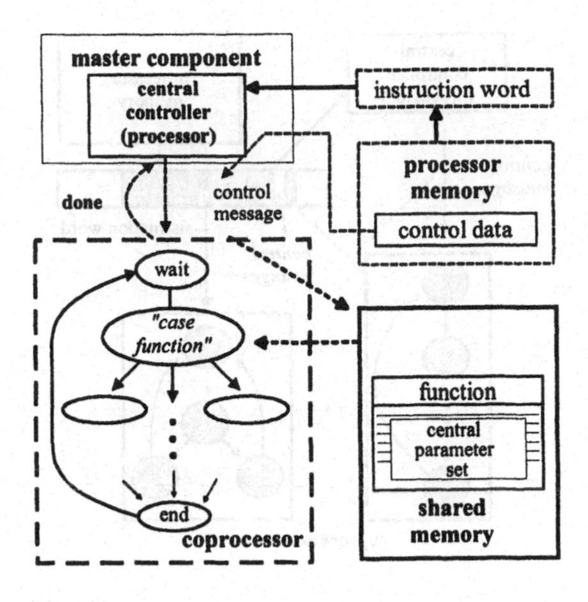


**Figure 4.1** A dependent co-processor.

thread). Dependent co-processors can also be found in many general-purpose architectures, such as floating point co-processors, or graphics co-processors [67]. It is also the co-processor model of many co-synthesis systems, such as Cosyma (see Chapter 8).

The second class shall be called **incrementally controlled co-processors**. Incrementally controlled co-processors can autonomously iterate a control sequence. A central controller (the master component frame is omitted for simplicity) only initiates changes in the control flow and in the control data of the *co-processor(s)* (Figure 4.2). This technique reduces the amount of control information for locally regular control sequences. In case of such regularity, the same instruction word can be shared among the main compoment (e.g. a *processor*) and several co-processors such that the control "effort" is shifted between the units in subsequent instructions using different instruction formats. A good illustration of this approach is found in some digital signal processors with incrementally controlled concurrent address units which exploit highly regular address sequences (see ADSP21060 [68]).

Both dependent co-processors and incrementally controlled co-processors are a part of a control hierarchy. Even if components are independent with their own control memory, there can still be a control hierarchy where a master

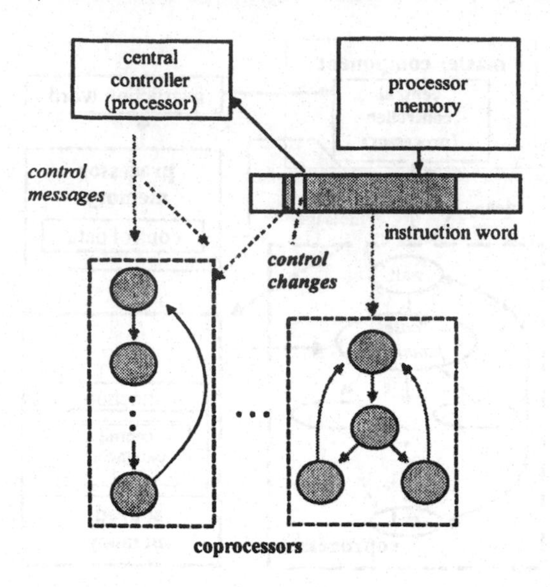


**Figure 4.2** Incrementally controlled co-processors.

component defines the program to be executed by a co-processor. Such co-processors shall be called **partially dependent co-processors**. Partially dependent co-processor implementation is strikingly similar in very different architectures, such as digital signal processors and complex peripheral controllers. The central controller typically downloads a linked set of function blocks each consisting of a function designator and a parameter block. The links can form a circle for iterative execution. The mechanism only allows very simple round robin linear sequences of functions but each function may correspond to a complex (micro)program with data dependent control flow. This is outlined in Figure 4.3. Obviously, these independent co-processors are well suited for shared memory architectures.

The control hierarchy can be extended to several levels as we will see in some of the more complex examples.

### *4.2.4 Memory Architectures*

Memory architecture is highly amenable to application-specific specialization. The application data set can be decomposed and mapped to several specialized memories to optimize cost, access time or power consumption. A priori knowl-

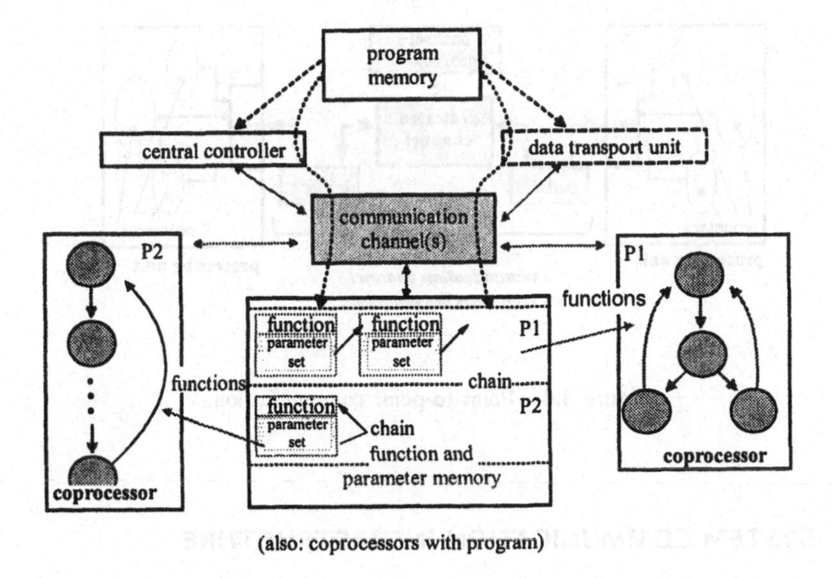


**Figure 4.3** Partially dependent co-processors.

edge of application data structures and data flow can be used in several ways. Besides extending the component memory specialization techniques to systems, **system memory specialization** exploits the data flow within a system. A priori knowledge of data flow is used to insert and optimize buffers, assign memories for local variables and distribute global data such as to minimize access time and cost. This is known from DSP design, such as the TMS320C80 [69], which will be discussed as an example, or the VSP II [70], but has recently also drawn some attention in general-purpose processor design, where part of the memory architecture is specialized to multimedia application, e.g. using stream caches in addition to standard caches. There are even formal optimization techniques, in particular for static data flow systems [3]. One can also exploit knowledge of data access sequences either to reduce memory access conflicts with memory interleaving or e.g., stream caches to reduce address generation costs by taking advantage of linear or modulo address sequences typical for many signal processing or control algorithms. Again, there are optimization approaches which can be used [71, 72]. Reducing memory conflicts, however, can lead to less regular address sequences. So, regular address sequences and memory access optimization are potentially conflicting goals.

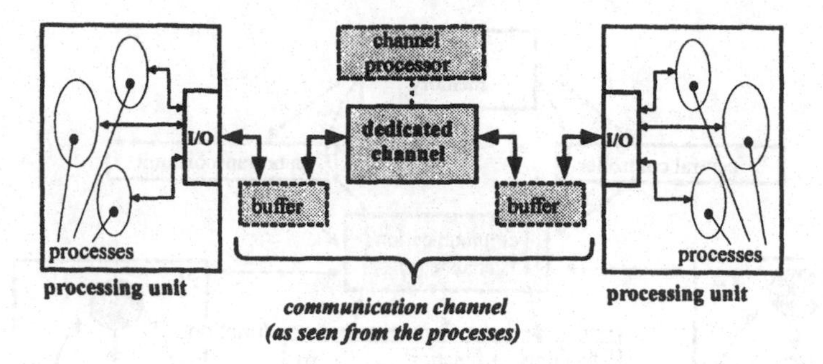


**Figure 4.4** Point-to-point communication.

## 4.3 SYSTEM COMMUNICATION INFRASTRUCTURE

The system specialization techniques are based on component communication. So, the communication infrastructure plays an important role in system specialization and is specialized itself. Even though the various communication mechanisms are largely known to the co-design community, this variety is often neglected or at least hardly exploited in co-design, where most approaches are constrained to a small subset of these mechanisms, as described in Chapters 2 and 10. When summarizing the principles of system communication in the next section, we will, therefore, put emphasis on the impact on the co-design space, in particular process and communication scheduling.

System communication is based on physical communication channels. The most simple medium is the **point-to-point channel** connecting two components (Figure 4.4). Both components can execute several processes which share the component using some scheduling approach. Communication between processes shall use abstract logical communication channels, which in the following shall represent any kind of process communication mechanism, even including shared memory as we will see.

In Figure 4.4, the physical channel is dedicated to one logical channel at a time. If there is no buffering, communication requires a rendezvous mechanism where sender and receiver are active and participate in the communication at the same time. Other logical channels are blocked as long as the channel is occupied. Rendezvous and exclusive access imply potential deadlocks for parallel processes. Typical examples for this simple communication mechanism are communicating controllers in hardware, as e.g. used in the TOSCA [73] or the Polis systems [74] or in the LOTOS based system of UP Madrid [75].

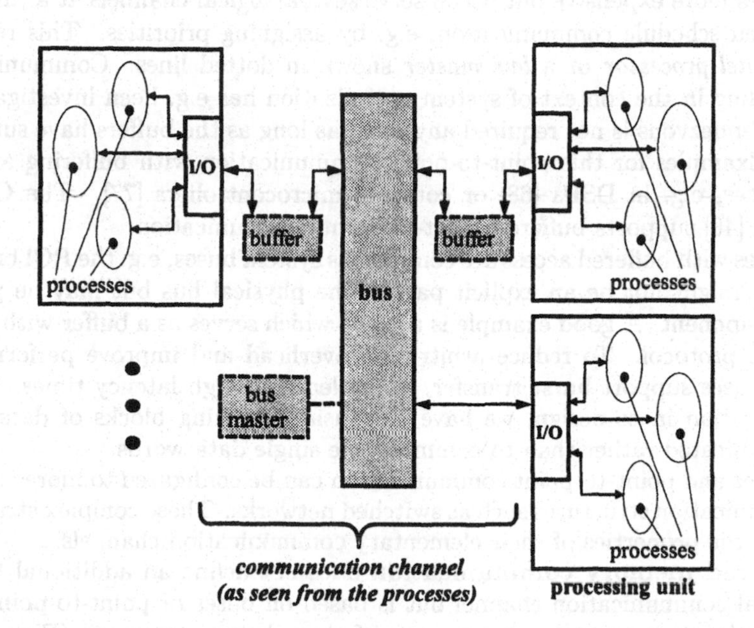


Figure 4.5 Bus communication.

**Buses** connect several components and are used to share the physical communication channel for different point-to-point communications (Figure 4.5). In addition, buses also allow multicast communication on the physical channel. Resource sharing requires communication scheduling, either explicit [76, 48] or implicit by hardware arbitration. If there is no buffering, the same constraints hold as for point-to-point communication, i.e. exclusive access and rendezvous. Most of the microprocessor address and data buses are unbuffered.

If **buffers** are included (dotted boxes in Figure 4.4, 4.5), the physical channel becomes more expensive but it can serve several logical channels at a time and can even schedule communication, e.g. by assigning priorities. This requires a *channel processor* or a *bus master* shown in dotted lines. Communication scheduling in the context of system optimization has e.g. been investigated in [48]. Rendezvous is not required any more as long as the buffers have sufficient size. Examples for this point-to-point communication with buffering are link interfaces, e.g. in DSPs [68] or complex microcontrollers [77]. The Cosmos system [46] supports buffered point-to-point communication.

Buses with buffered access are common as system buses, e.g. the PCI bus [78]. Buffers might not be an explicit part of the physical bus but may be part of the component. A good example is a cache which serves as a buffer with a very specific protocol. To reduce arbitration overhead and improve performance, these buses support burst transfer, which leads to high latency times. So, for optimization in co-design, we have to consider forming blocks of data to be communicated rather than to communicate single data words.

Buses and point-to-point communication can be configured to more complex communication structures such as switched networks. These complex structures inherit the properties of their elementary communication channels.

**Shared memory communication** does not define an additional type of physical communication channel but is based on buses or point-to-point communication between components, one of them being a memory (Figure 4.6). Shared memories avoid copying and coherent control if global data shall be shared. Several memory blocks can be multiplexed to implement buffers without access conflicts (see [68, 69]). Since it is based on several physical communication channels, however, shared memory communication is not zero delay and access conflicts can slow down communication. While in software development, memory access is an atomic operation which can hardly be influenced, communication channels, arbitration and access to shared memory are co-design parameters which must be taken into account for optimization.

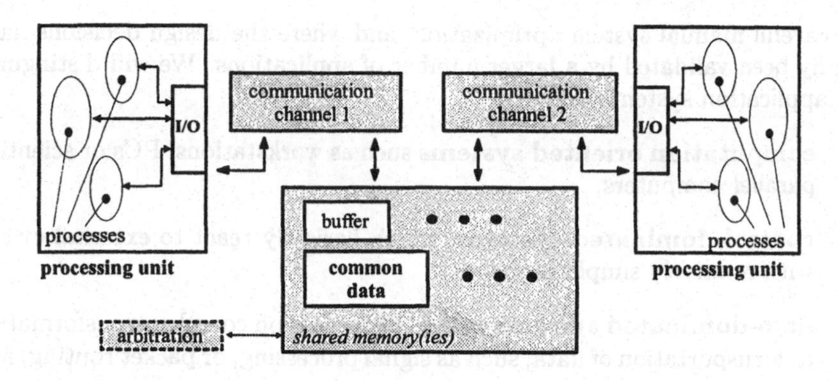


**Figure 4.6** Shared memory communication.

## 4.4 TARGET ARCHITECTURES AND APPLICATION SYSTEM CLASSES

The main difference between general-purpose, highest volume microprocessors and embedded systems is specialization. We have seen a variety of specialization techniques on the component and system levels to optimize an embedded system for a given target application or a class of target applications. Nevertheless, efficient specialization requires suitable system properties. So, given the specialization techniques, a designer should identify the specialization potential of a target application with respect to the specialization techniques, before a decision on the architecture is made. Moreover, specialization must not compromise flexibility, i.e. it must possibly cover a whole class of applications (DSP) and include some flexibility for late changes. This also limits specialization. What is needed is a target application analysis which *identifies application properties* that can be used for specialization, *quantifies* the individual specialization effects, and *takes flexibility requirements* into account.

Unfortunately, there are no formal methods today which would support the three tasks, except for quantitative techniques from general-purpose computer architecture [65], which may help in the second issue. Therefore, system specialization and design is currently for the most part based on user experience and intuition.

To give an impression on specialized architectures are designed and chosen in practice, we will discuss target architectures for several important but very different application system classes. We selected high volume products which have been used for a few years, where we may assume that a large effort went into careful manual system optimization, and where the design decisions have already been validated by a larger number of applications. We will distinguish four application system classes,

- **computation oriented systems** such as workstations, PCs or scientific parallel computers,
- **control-dominated systems**, which basically react to external events with relatively simple processes,
- **data-dominated systems** where the focus is on complex transformation or transportation of data, such as signal processing, or packet routing, and
- **mixed systems** such as mobile phones or motor control which contain extensive operations on data as well as reaction to external events. Here, we will only look at the latter three classes.

## 4.5 ARCHITECTURES FOR CONTROL-DOMINATED SYSTEMS

Control-dominated systems are reactive systems showing a behavior which is driven by external events. The typical underlying semantics of the system description, the input model of computation, are coupled FSMs or Petri nets. Examples are StateCharts or Esterel processes ([46]). Time constraints are given as maximum or minimum times. These shall be mapped to communicating hardware and software components possibly after a transformation which decomposes and/or merges part of the FSMs (of the Petri net).

The resulting co-design problems are the execution of concurrent FSMs (or Petri net transitions) reacting to asynchronous input events, which have an input data dependent control flow and are subject to time constraints. Very often, control applications are safety critical (automotive, avionics). So, correctness (function and timing), FSM transition synchronization and event scheduling are the major problems while computation requirements are typically very low.

Figure 4.7 shows the example of a minimal software implementation of a set of coupled FSMs on a **microcontroller**. Microcontrollers are small systems-on-a-chip used in many low-cost applications from simple functions in automotive electronics to dish washer control. Parallel input and output units (*PIO*) which can be found in any microcontroller latch the FSM input and output signals. FSM transitions must be serialized on the processor which is a resource sharing problem. After transformation, each FSM is mapped to a process. These processes consist of a small variable set containing the FSM state and the FSM transition functions encoded as short program segments in chains of branches or in jump tables. Program size grows with the number of states and transitions. In microcontrollers, program memories often require more space than

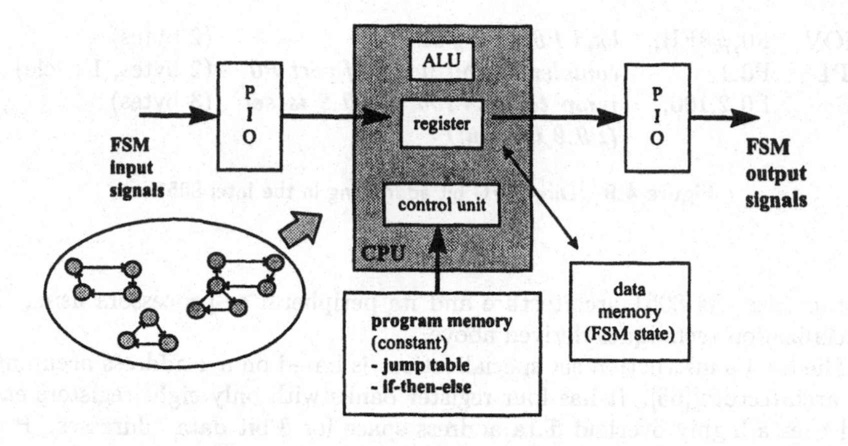


**Figure 4.7** Microcontroller FSM implementation.

the CPU. So, code size is an important issue. In microcontroller architectures, FSM communication uses shared memory. Since each FSM transition typically requires a **context switch**, context switching times are critical.

This short application overview reveals a number of specialization opportunities:

- most of the operations are Boolean operations or bit manipulation,
- there are few operations per state transition,
- the data set containing the FSM states is typically small.

We will see in an example that this observation is used to apply all component specialization techniques plus control decomposition with different types of co-processors.

### 4.5.1 *8051— An 8-bit Microcontroller Architecture*

The 8051 (e.g. [79, 80]) is a quasi standard 8-bit architecture which is used as a processor core (CPU) in numerous microcontroller implementations by many companies. It is probably the highest volume processor architecture overall. Its main application are control-dominated systems with low computing requirements. Typically, it is combined with a number of peripheral units most of them having developed into standard co-processor architectures. It has 8-bit registers and a complex instruction set architecture (CISC). We will

| | | | |
|---|---|---|---|
| MOV | P0,#8FH; | *load P0 with #8FH* | (2 bytes) |
| CPL | P0_1; | *complement bit no. 1 of port P0* | (2 bytes, 1 cycle) |
| JB | P0_2,100; | *jump to PC+100 if P0_2 is set (P0_2 is input)* | (3 bytes) |

**Figure 4.8** Direct I/O bit addressing in the Intel 8051.

characterize the 8051 architecture and its peripheral co-processors using the specialization techniques derived above.

The 8051's instruction set specialization is based on a 1 address accumulator architecture [65]. It has four register banks with only eight registers each and uses a highly overlaid data address space for 8-bit data addresses. Both features together enable very dense instruction coding of 1 to 3 bytes per instruction. This corresponds to the need for dense encoding and makes use of the observation that the variable set is typically small. To avoid a performance bottleneck and to allow short programs, the single address architecture and address overlay are complemented with a rich variety of data addressing modes (direct, indirect, immediate for memory-memory, memory-register, and register-register moves) which make it a CISC architecture [65]. Unlike other CISC architectures, however, there are no complex pipelined multi-cycle operations making interrupt control easy and fast. Also, instruction execution times are data independent making running time analysis easier. It is rather slow with 6 to 12 clock cycles per instruction cycle and clock frequencies up to currently 30 MHz. Among the more application specific operations are registered I/O with direct I/O bit addressing for compact transition functions and branching relative to a data pointer as a support of fast and compact branch table implementation.

Figure 4.8 shows three examples with timing data taken from [79]. The first instruction is an immediate store to a registered 8-bit output: The output port P0 is loaded with the hex value 8F. Execution time is 1 instruction cycle. The second operation is a registered I/O operation to a single output bit, where a value of an output bit is complemented. Again, the execution time is a single instruction cycle only. And finally, in the third example, a conditional jump is executed on a single bit of an input port, executed in 2 instruction cycles. Especially this last instruction shows how the I/O signals are directly included in the control flow, very much like a carry bit of an accumulator. Such an integration of I/O events into the instruction set reflects the FSM view of the processor illustrated in Figure 4.7.

The 8051's datapath and function units are also specialized. The data path is only 8-bit wide. This is completely sufficient for Boolean operations and

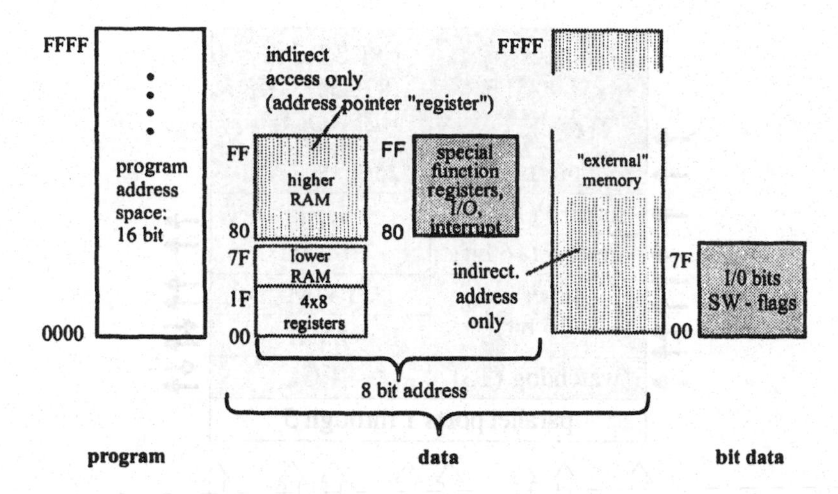


**Figure 4.9** The Intel 8051 address space.

bit manipulation required for FSM implementation and simple I/O protocols. Multiply and division functions are software implemented. Some of the typical peripheral co-processors have specialized data paths.

The 8051's memory architecture specialization is illustrated in Figure 4.9. The lower 8-bit *address space* is overlaid with 5 different memory spaces. Which address space is selected depends on the operation and the address mode. The CPL operation e.g. always addresses one of the 128 *I/O bits* and *SW flags*, while MOV operations to the lower 128 addresses access the data RAM and register banks and the higher addresses either access the I/O ports, interrupt registers or extended data RAM, interestingly with higher priority to I/O. The register set consists of four banks of 8-bit registers. Bank selection is part of the program status word, so switching between banks is done with interrupt or subroutine calls in a single cycle. Complementing the short instruction execution times, this allows an extremely fast context switch for up to four processes. To give an example, even with slow instruction cycle times [79] still lead to context switching times of less than a microsecond. All processes use a single memory space and share the same local memory such that communication between software FSMs can easily be implemented in shared memory.

Except for small 8-bit word length buses, interconnect is not particularly specialized but is based on a single bus due to the low data transportation requirements in such a system.

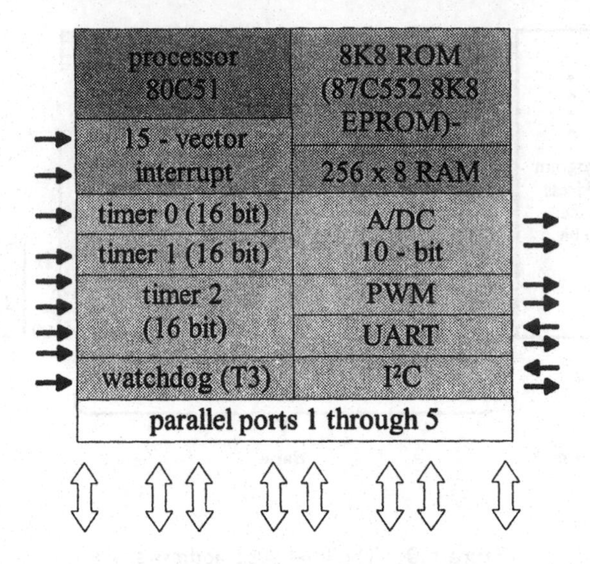


**Figure 4.10** The Philips 80C552.

Component control is specialized to low power consumption and fast reaction to external events. Registered input signals can directly be linked to vectored interrupts which supports fast FSM process invocation on external events. Several power down modes which disable part of the microcontroller function can be applied to reduce power consumption. Figure 4.10 shows a complete 8051 microcontroller [80]. Two memories are on chip, an 8 kByte *EPROM* and a small 256 byte data *RAM*. The small size of both memories reflects the extreme cost sensitivity of high volume control-dominated applications. The 5 × 8 bit *parallel I/O ports* and the universal asynchronous receive transmitter (*UART*) can be considered as dependent co-processors. 3 *timers* can be used to capture the time between events or to count events, or they can generate interrupts for process scheduling. A pulse width modulation (*PWM*) unit is an FSM with a counter and a comparator which can generate pulses of a programmable length and period. Both can be classified as *incrementally controlled co-processors*. PWM is a popular technique to control electrical motors. Combined with a simple external analog low pass filter, it can even be used to shape arbitrary waveforms where the processor repeatedly changes the comparator values while the PWM unit is running. There is also an $I^2C$ unit to connect microcontrollers via the $I^2C$ bus. A peripheral *watchdog* is a timer co-processor which releases an interrupt signal when the processor does not write a given pattern to a register in a given time frame. It can be classified as an independently controlled component. Such watchdogs are used to detect whether the program is still executed as expected which is a helpful feature for safety critical systems. Finally, there is an analog-digital converter (*ADC*) on the chip (a dependent co-processor).

There are two types of system communication mechanisms in this microcontroller. The processor uses shared memory communication by reading and writing from or to memory locations or registers. Only in few cases, point-to-point communication is inserted between co-processors, e.g. between the interrupt and timer co-processors, i.e. almost all of the system communication relies on processor operations. This is flexible since the system can easily be extended but it limits performance.

Since several software processes can access a single co-processor, there are potential resource conflicts requiring process synchronization and mutual exclusion. Peripheral resource utilization is an optimization problem and is, e.g. addressed in [81].

### *4.5.2 Architectures for High-Performance Control*

Applications which contain signal processing or control algorithms are less suited for 8-bit processors. There are basically two approaches to increase computation performance without compromising short latency times and control concurrency. One is to use a higher performance core processor and off load most of the control tasks to more sophisticated programmable co-processors and the other one is to adapt interrupt handling in higher performance processors to support combined execution of FSM processes and signal processing applications.

The ***Motorola MC68332*** [77] is an example for the first approach. The processor core is the CPU 32, which is a 68000 processor enhanced by most of the 68030 features. Other than the 8051, it uses pipelining and a standard register set such that the context switch becomes more expensive. It supports virtual memory and distinguishes user and supervisor modes. So, it aims at the use of operating systems. The processor itself is little specialized except for specific table lookup instructions with built-in interpolation for compressed tables. Tables are commonly used in control engineering e.g. for fuel injection in combustion engine control. For such applications, table interpolation allows higher data density and reduces cost. The *CPU 32* can be combined with different co-processors over an in-house standard ***inter module bus (IMB)*** (Figure 4.11). One of those co-processors is the ***time processing unit***, *TPU* [82].

The **TPU**, shown in Figure 4.12, is a partially dependent co-processor with complex control functions. It consists of a set of 16 concurrent ***timer channels***

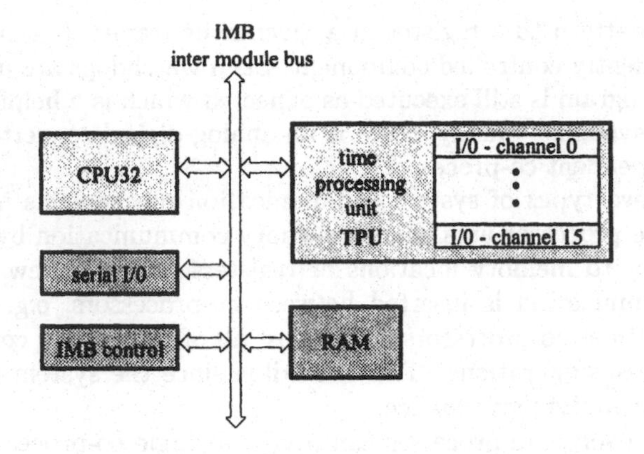


**Figure 4.11** The Motorola MC68332.

which communicate with a microcoded processor, called *microengine*. The microengine can execute predefined microcoded control programs which are described as a FSM and which use some of the timer channels. Examples are pulse-width modulation (PWM), stepper motor motion control with linear acceleration/deceleration and chains of events combining many timer channels such as for accurate pulse sequences for ignition and fuel injection in combustion engines [82]. Specialized functions such as a "missing tooth" function to detect fly wheel sensor faults allow to handle error situations. This is an example of a highly specialized instruction set. For more flexibility, user defined microprograms can be executed from the shared system RAM memory, but at the cost of RAM access conflicts between the core processor (CPU 32) and the TPU with the consequence of reduced performance.

Microprogram execution is based on an interplay of *timer channels* and microengine. Timer channels are programmed by the microengine which waits for the channel to signal the event it was programmed for. According to our classification and depending on the programmed function, timer channels are dependent or incrementally controlled (and extremely specialized) co-processors of the microengine. Since the TPU is a co-processor itself, we have a 2 level control hierarchy. The 16 channels can execute a variety of timing functions such as detecting signal transitions, measuring periods, or forming pulses.

The *microengine* can execute several control programs with different channels in parallel, but only one at a time. Therefore, a hardware *scheduler* serializes the execution admitting context switches whenever a microprogram waits for a channel event. It implements a round robin schedule with microprogram

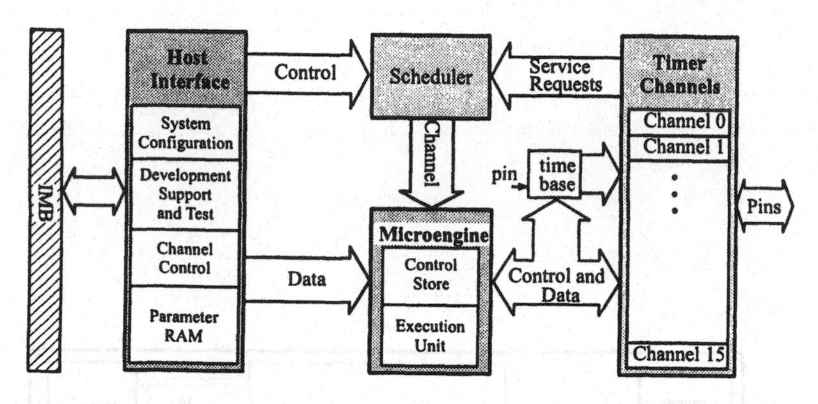


Figure 4.12 The Motorola TPU.

priorities. Since channel events are inherently asynchronous, event queues are installed in the scheduler.

The control parameter sets (see Figure 4.3) are communicated in a shared memory block which is part of the host interface. This avoids access conflicts with the main processor which can even reprogram the TPU while running using a double buffer mechanism.

So, in summary, we see specialized memories with data decomposition (event queues, timer registers, function block shared memory, main RAM memory, etc. ), specialized interconnect (microengine-timers, channels-scheduler-loop), specialized control and control decomposition (2 levels of control flow hierarchy, hardware scheduler, microcode), specialized instruction sets, and specialized function units and data paths.

The architecture bottleneck is the single microengine which can only be interruptedinterrupt in wait states. The result are worst case latencies for high priority channel events of 270 clock cycles (17 $\mu s$ at 16 MHz clock rate) and even 2180 clock cycles for low priority channel events (136 $\mu s$). So, the TPU emphasis is on parallel complex reactions to asynchronous events rather than on very short latency times.

Since the TPU only handles control tasks, the MC68332 is equipped with a second co-processor, the **QSM**, to off-load the CPU from data transfers.

A completely different approach to the same application area is the **Siemens 80C166** (x167 for automotive control) [83] (see Figure 4.13). This architecture shows less specialization than the MC68332. Rather than off-loading the CPU from interrupts, the designers have tried to improve the context switch of a 16-bit processor. The processor provides several register banks with bank

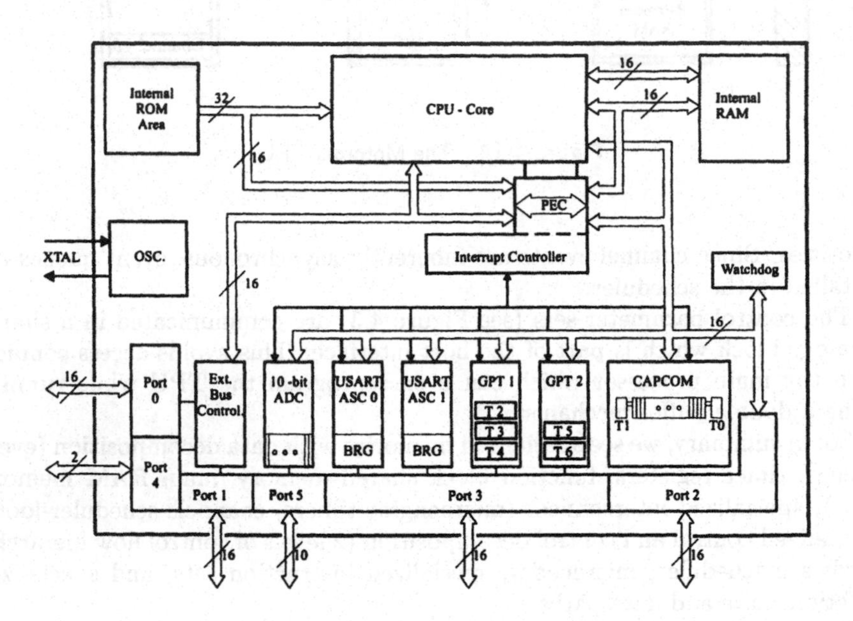


Figure 4.13 The Siemens 80C166.

switching similar to the 8051 and, alternatively, a register windowing technique for local registers as e.g. known from the SPARC architecture which can also support fast context switch. Instruction execution is interruptable. Together, this leads to a maximum context switching time of 10 clock cycles, i.e. 250ns at 40 MHz clock rate. A *PEC unit* inserts single cycle DMA transfer steps on the processor bus thereby delaying the CPU execution one cycle at a time. So, even in case of concurrent DMA transfers, CPU latency is hardly longer. The architecture uses a set of "capture&compare" units as incrementally controlled co-processor which can communicate to jointly implement more complex waveforms and timing functions (*CAPCOM*). Nevertheless, we may expect a higher frequency of interrupts than in the case of an MC68332, but these interrupts meet a CPU which is optimized to handle higher interrupt frequencies.

There is no obvious advantage of one of the architectures, and in fact, both architectures are even used in the same motor control application of different automotive companies. These architectures are a good example that there is no obvious global optimum and that a higher degree of specialization does not guarantee much higher performance for a complete class of applications.

## 4.6 ARCHITECTURES FOR DATA-DOMINATED SYSTEMS

The term **data-dominated systems** can be used for systems where data transport requirements are dominant for the design process (e.g. in telecommunication) or where data processing is the dominant requirement (e.g. in signal processing). In this overview, we want to focus on static data flow systems [3], i.e. systems with periodic behavior of all system parts. Often, such systems are described in flow graph languages, such as in signal flow graphs which are common in digital signal processing or control engineering (see e.g. [84]). Similar to data flow graphs, signal flow graphs represent data flow and functions to be executed on these data.

Such data-dominated applications typically have a demand for high computation performance with an emphasis on high throughput rather than short latency deadline. Large data variables, such as video frames in image processing, require memory architecture optimization. Function and operation scheduling as well as buffer sizing are interdependent optimization problems [3].

The flow graph input model of computation offers many opportunities for specialization:

- The periodic execution often corresponds to a (mostly) input data independent system data flow.

- The input data is generated synchronously with a fixed period, the sample rate.

Such systems can be mapped to a static schedule thereby applying transformations for high concurrency, such as loop unrolling. The resulting static control and data flow allows to specialize control, interconnect function units and data path while, at most, a limited support of interrupts and process context switch is required. Furthermore, static scheduling leads to regular and a priori known address sequences and operations which gives the opportunity to specialize the memory architecture and address units.

**Digital signal processors (DSP)** are the most prominent examples of target architectures for this class of applications. The various signal processing applications have led to a large variety of DSPs, with word lengths of 16 to 64-bit, fixed or floating point arithmetic, with different memory and bus architectures targeting high performance, low cost or low power.

We have selected 2 complex and very different examples which include most of the typical DSP architectural elements, the ADSP21060 [68] and the TMS320C80 [69].

### *4.6.1 ADSP21060 SHARC*

The **ADSP21060** is in many aspects similar to an earlier architecture, the TMS320C40. Figure 4.14 illustrates the global processor architecture. The IC consists of a *core* DSP *processor*, two large *dual-ported SRAM banks* with 2 Mbit capacity each, a versatile *I/O processor* and a host interface. The core processor consists of a single-cycle 40 bit fixed point/floating point computation unit with three concurrent function units, a *multiplier*, a *barrel shifter* and an *ALU*. This combination of units is typical for DSPs and is due to the frequent sequences of multiply-add-shift operations in DSP applications. There are specialized operations needed for DSP applications, such as saturation and clipping, absolute value and shift/rotate-by-n. The three function units and register moves are controlled concurrently in a single field of the instruction word. A 16 entry data register file holds variables and intermediate results. The ADSP is a load/store architecture, i.e. all operations are based on registers, like in a RISC processor. Up to 2 load/store operations can be executed at a time using the 40 bit *DMD bus* and the 48-bit *PMD bus*. Except for the special DSP operations, the computation unit is less specialized than other SHARC components and can be used for standard data processing as well. The processor clock frequency is 40MHz [68].

A higher degree of specialization can be found in the address and control units. The addresses for load/store operations are generated by 2 **data address generation units** (*DAGs*). The DAGs are specialized to generate linear address sequences of the type $adr = (a + b*i) \bmod c$. 4 registers, modify (offset), base (a), index (b), and length (c) are used for this purpose. Such linear

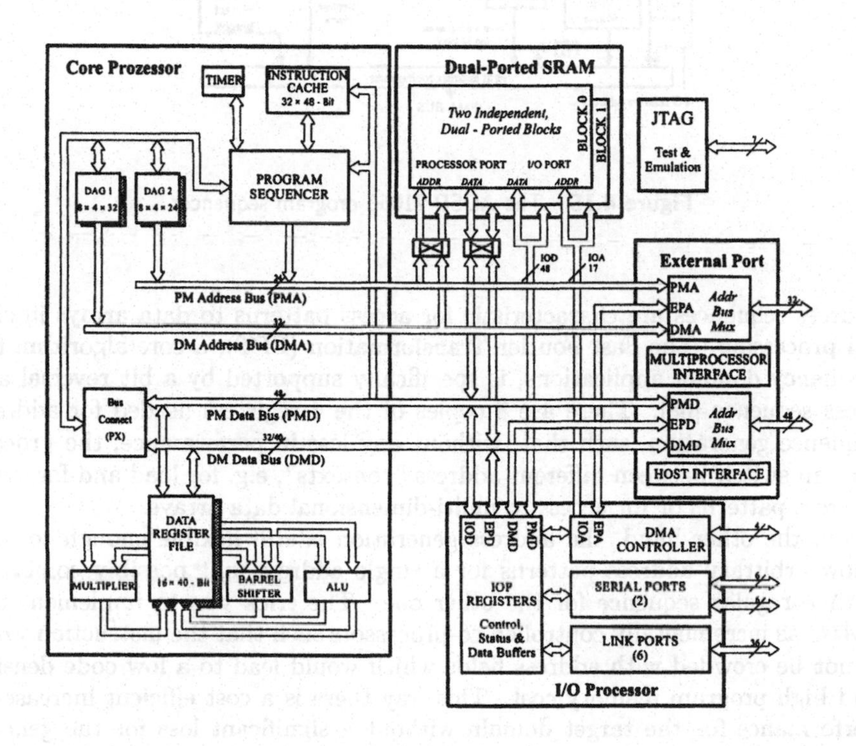


**Figure 4.14** The ADSP 21060 SHARC.

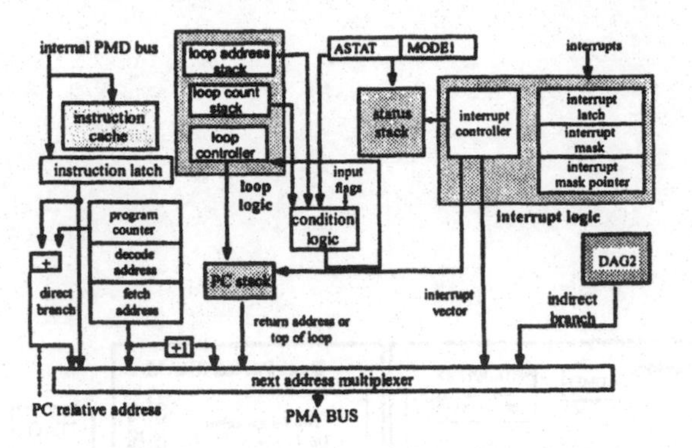


**Figure 4.15** The ADSP 21060 program sequencer.

address sequences are characteristic for access patterns to data arrays in signal processing. The Fast Fourier Transformation (FFT), a core algorithm for frequency domain applications, is specifically supported by a bit reversal address sequence [68]. There are 8 copies of the 4 registers needed for address sequence generation, such that, without any loss in performance, the processor can switch between different address "contexts", e.g. for load and for store address patterns or for access to multi-dimensional data arrays.

On the other hand, the address generation is still flexible enough to also allow arbitrary address patterns for a single address unit possibly combined with a regular sequence for the other one. The trick was to implement the *DAGs* as incrementally controlled co-processors such that the instruction word is not be crowded with address fields which would lead to a low code density and high program memory cost. This way there is a cost efficient increase in performance for the target domain without a significant loss for the general case. Two address buses connect the *DAGs* to the internal memory blocks and to an external host port. To summarize, the *DAGs* are co-processors which exploit special static address sequences with specialized function units, data path, memories, and instruction set. The *DAGs* are used to specialize the *core processor* control and are connected by specialized buses.

The **program sequencer** demonstrates an even more advanced specialization. Figure 4.15 gives a block diagram. Using a *loop stack*, it supports the execution of nested loops with a fixed number of iterations with zero control overhead for loop iteration, except for loop initialization. There are special hardware registers and stacks for fast subroutine calls and context switches, but with a very limited depth (*PC stack*: 30 entries).

Given the large number of components, the 48-bit instruction word must be considered small, since all instructions have a unit length of a single word which simplifies program sequencing. The control unit has an extremely small *program cache* of 32 instruction words only with a block size of 1. This cache shall set the *PMD bus* free for 2 parallel data transfers. While this limited cache size would be unacceptable for a general-purpose processor, it seems sufficient for speeding up the small and often iterated program segments in inner loops. Clearly, speeding up small program parts in loops, as found in typical DSP applications, was the major concern of the sequencer design. Nevertheless, there is enough flexibility not to slow down other program parts, where the added functionality is just not effective. The sequencer is an example of highly *specialized control*.

The **I/O processor (IOP)** is a DMA processor which controls 2 serial I/O ports and 6 *link ports* for 4 bit-serial communication with other DSPs. The bidirectional *link ports* are buffered and allow a peak transmission rate of 40 Mbytes/s each. While the channels have their own register sets and are individually controlled by an internal time base or external handshake signals, there is only a single *DMA controller* which controls and transfers the data for all the channels based on a prioritized round robin scheduler. There is an obvious similarity between the *TPU* and this *IOP*, even though their applications are completely different. The similarity is the underlying global control mechanism. Both are partially dependent co-processors with a hardware scheduler. Like the TPU, the IOP is programmed in "transfer control blocks" consisting of the function and a parameter block which can be chained as shown in Figure 4.3. Like in the TPU, serialization is a bottleneck, in this case due to the single memory access port.

The system memory architecture is dominated by two large dual-ported single-cycle RAMs which are organized in two memory banks. One of the ports of each memory bank is dedicated to the *I/O processor*, the other one is shared between the two address and data buses for program and data. A crossbar switch with internal queues is installed to give access from bus to memory block. The queues are necessary to handle access conflicts between instruction cache and data load/store traffic. Such a *specialized memory architecture* of several banks is standard even in smaller DSPs, such as the Motorola 56000 [85], and efficient techniques are known in the signal processing community to highly utilize concurrent access.

Not only the system memory but also the address space is adapted. Each ADSP21060 has an address space which is interleaved with that of 6 other processors which communicate over a single external bus through the *multiprocessor interface*. The address mapping is programmable and done at the external ports which are FIFO buffered. This feature supports buffered shared memory communication between up to 7 processors with minimum external circuitry. It allows densely packed DSP multiprocessor systems which has spurred a number of commercial multiprocessor board designs. Compared to more general-purpose multiprocessors, the memory size is extremely small and all address spaces are overlapping, unprotected and without any caching. So, the resulting multiprocessors are significantly specialized.

On the other hand, the *link ports* support point-to-point buffered communication. Using the external ports for local memories, or possibly a host processor, this permits larger multiprocessor configurations with larger memories at higher costs. Here, it can even be used as a communication co-processor itself. Communication through link ports has been successfully introduced in the Transputer [86] and was already adapted to the TMS320C40 [87].

Given all this rich variety of communication features which are extremely valuable for DSP multiprocessors, in fact, analysts consider it the dominant VME bus architecture [88], it is still no replacement for a general-purpose processor. Apart from the limitation in the address spaces and memory sizes, a major reason are indirect limiting effects of specialization. The approach in the ADSP21060 was to support application typical functions without introducing major performance losses to less typical parts of an algorithm to reach a high degree of flexibility. This approach was paid by numerous specialized memories and registers which are now spread all over the components. Typical for a DSP, which is primarily intended to implement flow graphs, buffers can be found everywhere. Buffers would certainly be much less attractive for reactive systems with short latency times. Much worse, however, a context switching mechanism which would force all these registers to be saved in a common place would be slow. So the approach the designers took in the ADSP21060 was to save the values in place by doubling the register sets, i.e. in the CU, in the DAGs, and in the sequencer (stack), while IOP registers are not doubled probably assuming that communication scheduling is not preempted. So, a single fast context switch is supported, but then flexibility comes to a hard end. Similarly, variable saving for subroutine call is highly limited by the hardware stack size. Such a limitation would be unacceptable in general-purpose processing.

In summary, the ADSP21060 illustrates most of the component and system specialization techniques applied to DSPs. It demonstrates the necessity to include system level consideration to fully understand and control the effects of specialization on performance and flexibility. Some similarities between architectures (TPU) only appear at the system level. The example also shows that flexibility has very different aspects, since some of the features which were introduced to increase or at least not limit flexibility have decreased it from a different perspective (registers and context switch).

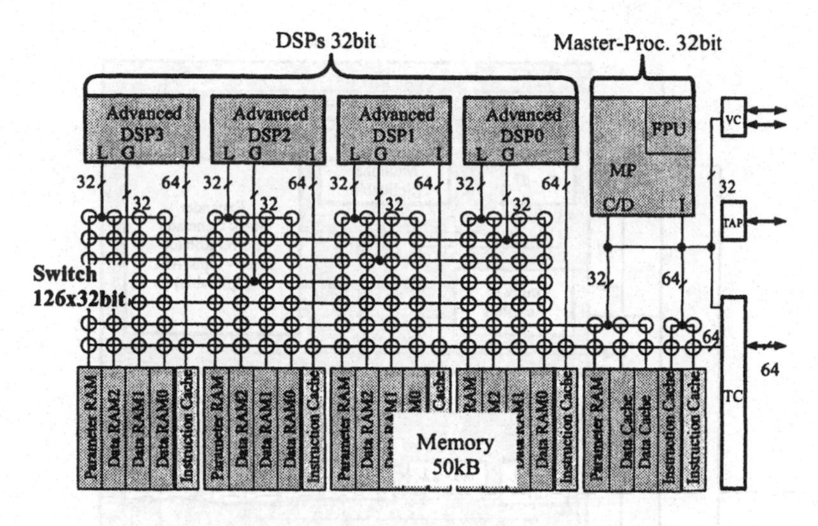


**Figure 4.16** The TMS320C80 MVP.

### *4.6.2 TMS320C80 MVP*

The second DSP example is a single chip multiprocessor, the **TMS320C80**, shown in Figure 4.16. The target domain is narrowed down to video applications. This multiprocessor IC consists of four 32 bit integer DSPs (*Advanced DSP*) with 4 single-cycle RAM blocks of 2 Kbytes each and a 2 Kbytes instruction cache. The processors are connected by a programmable crossbar switch. A single master processor (*MP*) for general-purpose computing and floating point operations is coupled to this processor array. The transfer controller (*TC*) is a DMA processor which is programmed through the *MP* and is responsible for the data and program I/O over a 64-bit external bus. For highly integrated solutions, a video controller (VC) is implemented on chip. The clock frequency is 50MHz.

Data path and function units of the 4 *Advanced DSPs* are more specialized than in the previous example. Floating point operations are not supported, but there are special image function units and operations, such as for pixel expand and masking which were added to the standard DSP function units, i.e. barrel shifter, ALU and multiplier/accumulator. The ALU has 3 inputs to support sum operations on arrays. Several multiplexers in the data path allow chaining of operations. Moreover, the 32 bit words can be split in smaller segments of 8 bits which can execute the same operation on all four 8-bit subwords. This

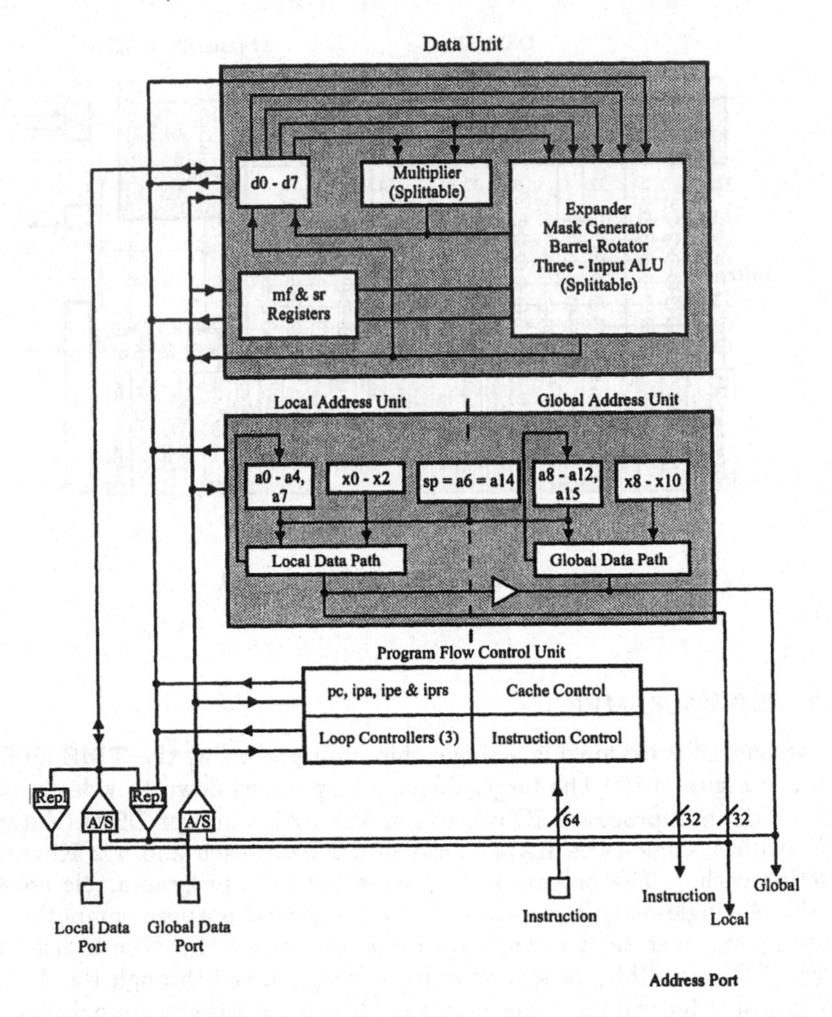


**Figure 4.17** An Advanced DSP from the TMS320C80.

kind of instruction and data path specialization can meanwhile be found in many general-purpose processors, as well, such as the MMX for the Pentium architecture or the VIS of the ULTRA SPARC [89].

Each *Advanced DSP* (Figure 4.17) has two address ALUs for index or incremental addressing corresponding to the 2 data buses of each *Advanced DSP*.

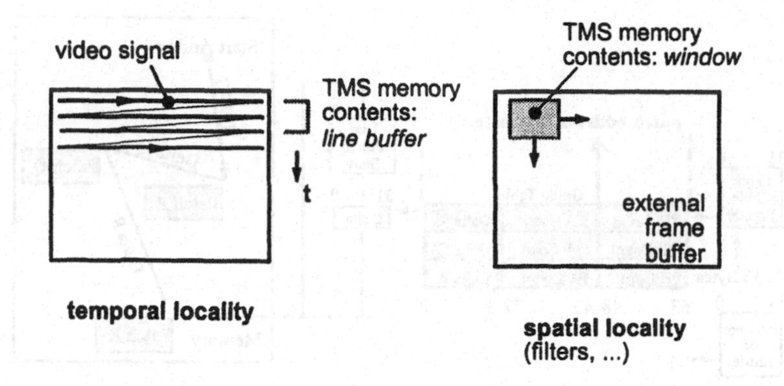


**Figure 4.18** Locality in video algorithms.

Again, they are incrementally controlled co-processors. The *Advanced DSP* sequencer supports zero overhead loops with a loop stack and loop operations. The sequencer is even more limited than in the previous example allowing only three levels of nested loops and a single subroutine call. Like in the other DSP example, there is a second register set for fast context switching.

Like in the ADSP21060, all instructions are given in a single 64-bit instruction word, but in a fixed format of three independent fields, one of them for the Advanced DSP operation, one for a local data transfer, and one for a global data transfer from the Advanced DSP to any of the memory blocks. Therefore, TI regards the Advanced DSP as a VLIW processor. Each Advanced DSP has a small cache of 256 instruction words. Unlike the ADSP21060, however, where the cache is used to reduce the internal bus load, there is no on-chip main memory in the TMS320C80, and all instruction loads must be routed through the transfer controller, such that cache misses are expensive and block the I/O data transfers.

The *memory architecture* with a crossbar interconnect seems very flexible. Obviously, there was high emphasis on minimum memory access time. On the other hand, the memory architecture is extremely specialized. Depending on the resolution, video frame sizes are on the order of several hundred Kbytes, such that not even a single frame could be stored in all RAM memories together. Since all RAM blocks can only be accessed through the *TC* bottleneck, this would be an unacceptable limitation in general-purpose signal processing. So there must be an algorithmic justification for this memory size. In fact, video algorithms typically use some kind of spatial or temporal localitywhich is exploited here. As shown in Figure 4.18, the video signal scans the image,

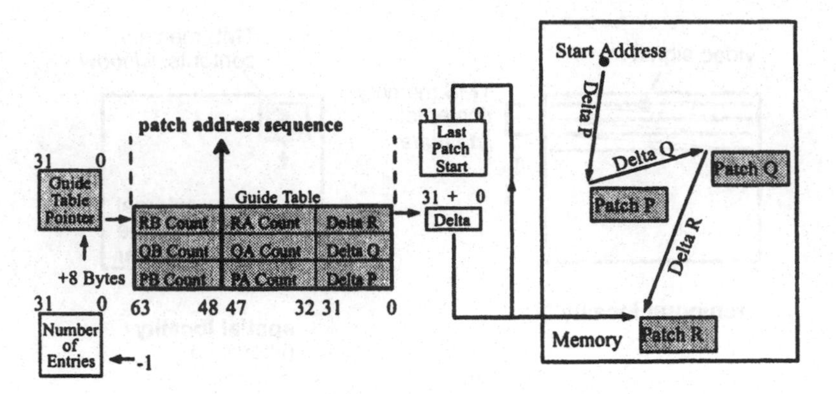


**Figure 4.19** Patch addressing in the TMS320C80.

such that algorithms which operate on few neighboring lines can use a line buffer keeping a small number of recently scanned lines. Such a buffer could fit into one RAM block. If the algorithm operates on windows, such as in standard image compression applications ([90]), then only the current window could be loaded into internal RAM blocks. This also gives an impression of the prospective role of these RAM blocks. They can be used as buffers between the operations of a signal flow graph (SFG) [84] or a similar representation, such as a synchronous data flow graph (SDF) [3], while the operations or functions are mapped to Advanced DSPs. The crossbar interconnect allows to implement arbitrary flow graph topologies as long as the buffers are large enough. In this context, the TC will have to read and write the windows or lines to or from an external buffer (or the internal video interface) and, at the same time handle potential instruction cache misses. This is certainly a potential bottleneck even if the internal data flow is larger than the external one. So, again, we observe flexibility in the one place (internally) and constraining effects of specialization on the other, this time in system interconnect. Here, only interconnect and memory specialization together make sense. So when specializing, the designers need a clear understanding of the application domain, which in this case leans itself to a data decomposition even below the level of an algorithmic object which would be a video frame.

To reduce the constraining effects of the single I/O bus, the *TC* must be a powerful and versatile I/O processor, and being familiar with the ADSP21060, the reader would meanwhile reasonably think of a partially dependent coprocessor. In fact, the *TC* has a similar DMA functionality as the IOP but is programmed in a very compact "Guide Table" where the windows to be transferred are defined as "patches" (Figure 4.19). While the ADSP21060 has a resource conflict when several of its channels want to access the internal memory, here, the TC could possibly split the data bus and handle several memory transfers at a time, however, at the cost of a complex TC and, more important, of additional address buses. Maximum throughput I/O or multiprocessing has obviously not been the primary goal of the designers but high integration and moderate packaging cost. This reflects the price sensitivity of the video market. Since the TC must manage cache misses and data transfers, it needs an internal context switching mechanism, which must support external memory delays. So, extra buffers are inserted in the TC.

The TMS320C80 is a good example for how to apply the classification scheme of system specialization techniques to understand adaptation and flexibility of an architecture.

## 4.7 MIXED SYSTEMS AND LESS SPECIALIZED SYSTEMS

Many systems, such as for mobile communications where signal processing must be executed in parallel to an event driven complex protocol [91], and contain interconnected data-dominated and control dominated functions. There are two approaches to such systems.

One approach is to design **heterogeneous systems** consisting of *independently controlled, communicating specialized components*, such as DSPs, microcontrollers and peripheral processors. This approach has been known in board level design for a long time and is migrating to systems-on-a-chip [92]. It requires that such a decomposition in different functions is possible and efficient.

The second approach is to limit specialization. Dependent on the design objectives, such **less specialized systems** can still be tailored to an application domain. We can distinguish:

- **computation applications without specific specialization potential** Application examples are printer or scanner controllers, game controllers, etc. Such systems often use modified versions of standard general-purpose architectures which are selected according to the computation requirements, such as the **M68000**, the **MIPS3000**, the **PowerPC** architecture, e.g. , the **MPC8xx** (RCPU) [93], or the **Alpha** 21066 [94].

- **applications with specific non-functional constraints** Most popular are low power applications (portable systems). Examples are general-purpose architectures which are tailored to minimize power consumption and cost (cost sensitive applications) for a required level of performance, such as the **ARM** family of processors [95], or the Motorola **ColdFire** family with an instruction set which is upward compatible to the M68xxx [96].

- **mixed systems** A widely used example of a processor family for mixed application is the i960, which shall be explained in more detail.

The **Intel i960** is a family of different processors with a common instruction set processor model [97] but different processor structures [98] much like [95, 96]. The 32 bit instruction set architecture combines features for computation performance in data dominated or general-purpose functions, features for reactive parts, and support of real-time operating systems. The i960 designers applied a subset of the specialization techniques, i.e. specialized instruction set, specialized memory, and, to a small degree, control specialization.

Data-dominated and general-purpose parts are supported by a load/store architecture and RISC-type 3 address instructions. It has two types of register sets, a global register set and a local register set with windowing (cp. Siemens x166). The i960 even permits static branch prediction under user or compiler control [65]. Some of the processors contain an floating point unit.

Reactive functions and I/O are supported by single-bit operands and bit operations (cp. 8051). The architecture has an interrupt system with 32 priority levels and 248 interrupt vectors. This rich set of interrupt vectors is also useful for real-time operating system scheduling, especially with static priority, such as rate monotonic scheduling(RMS) [12].

The operating system support is specialized to embedded applications. There are user and supervisor modes but the flat 32 bit address space has no data access protection.

Code density was a primary concern in the architecture definition since program memory can be a significant cost factor for larger mixed applications. All instructions are single word (32bit). Few frequent instruction sequences are encoded in a single instruction, which would be a characteristic of a microcoded CISC processor. Examples of such instruction sequences are subroutine call with an automatic caller saving mechanism and register spilling control. This makes the use of subroutines very attractive which, in turn, reduces memory size.

The overview on less specialized architectures shall conclude the tour through selected and typical target architectures. The last chapter will derive some important co-design problems which must be approached if computer aided co-design shall cover a substantial part of the solution space of manual hardware/software co-design.

## 4.8 SELECTED CO-DESIGN PROBLEMS

The classification of system specialization techniques has given structure to the confusing variety of seemingly unrelated embedded architectures. Understanding embedded target architectures is prerequisite for useful and accurate modeling and estimation.

We should also find ways to apply this knowledge to system synthesis. On the one hand, the system specialization techniques which were derived from manual designs could be applied as optimization strategies of a synthesis and code generation process. Current research approaches include **instruction set definition** [99, 100], **instruction encoding** for code compression [101] and **memory optimization** [102, 72, 103]. The other techniques are less examined, in particular the system level techniques are largely unexplored. Just to highlight one of the problems, **global control and data flow optimization** would be a worthwhile topic which goes beyond the work in data-dominated systems which has been mentioned above. Related to this problem are **communication channel selection** and **communication optimization. Component interface synthesis** is already an active research topic [45, 104, 81]. Specialization for **power minimization** would clearly be another topic of interest. All problems should be approached under **flexibility requirements.**

On the other hand, the classification of specialization techniques could be applied to support **component selection** and **component reuse**. In this case, we could characterize a component with respect to the specialization techniques and then compare if this specialization fits the application. Such an abstract approach could also include abstract flexibility requirements while synthesis would require a more formal definition of flexibility.

## 4.9 CONCLUSIONS

There is a large variety of target architectures for hardware/software co-design. This variety is the result of manual architecture optimization to a single application or to an application domain under various constraints and cost functions. The optimization leads to specialized systems. Target system optimization can therefore be regarded as a specialization process. We have identified a set of system specialization techniques for components and systems which can serve to describe and to classify the resulting target systems. A number of target system examples for very different data-dominated, reactive and mixed application domains with different models of computation have been analyzed based on this classification. We recognized various degrees of specialization and discussed the conflict of specialization and flexibility. Some of the designs exploit few specific algorithm properties with a complex combination of specialization techniques. System specialization has a large impact in practical system design and a number of co-design problems ranging from target architecture modeling to systems synthesis with reuse has been outlined.

# 5 COMPILATION TECHNIQUES AND TOOLS FOR EMBEDDED PROCESSOR ARCHITECTURES

**Clifford Liem and Pierre Paulin**

Laboratoire TIMA
Institut National Polytechnique de Grenoble
Grenoble, France
*and*
SGS-Thomson Microelectronics
Crolles, France

## 5.1 INTRODUCTION

Compiler technology and firmware development tools are becoming a key differentiator in the design of embedded processor-based systems. This chapter presents a look at trends in embedded processor architectures. High-performance applications such as multimedia, wireless communications, and telecommunications require new design methodologies. The key trend in embedded CPUs is the **Application Specific Instruction-Set Architecture**



*J. Staunstrup and W. Wolf (eds.), Hardware/Software Co-Design: Principles and Practice*, 149-191.

or **ASIP**—a processor designed for a particular application or family of applications.

The next section considers the requirements on embedded processors and the motivation for ASIPs. Section 5.3 reviews the characteristics of embedded architectures important for applications such as multimedia and telecommunications. Section 5.4 describes the requirements of software support tools for embedded architectures. Section 5.5 describes compiler algorithms which handle the particular characteristics of embedded architectures. Section 5.6 describes some of the system integration and other practical problems attendant to the introduction of software tools for embedded architectures.

## 5.2 CONTINUED INTEGRATION LEADS TO EMBEDDED PROCESSORS

Advances in VLSI technology are pushing many systems onto single-chip implementations known as **systems-on-silicon**. Nevertheless, the applications aimed for these complex systems have become a moving target. Standards organizations such as ISO (International Standards Organization) and ITU (International Telecommunications Union) set forth proposals which are continually adapting to a diverse set of needs. This constant evolution in standards has become a typical design constraint and can no longer be handled by standard ASIC design practices, since fully hardwired designs cannot easily adapt to specifications which are modified late in the design stage. As a result, programmable instruction-set processors are increasingly used in embedded computing systems. A programmable solution provides the capability to track the evolving standards using software for late design changes. Furthermore, a custom instruction-set core does not compromise high speed, low cost, and low power which could be the case with a general-purpose processor. Additionally, embedded processors make it easier for design teams to carry out concurrent engineering practices. The instruction set of the processor serves as a formal contract between the two teams to carry a product to market in an efficient time cycle. Furthermore, as higher levels of integration make current design practices more complex, reuse becomes a key methodology. Embedded cores are convenient macro blocks which can be reused in system chip designs.

While the design with embedded processor cores has many advantages, the design flow which supports their use is much different than the standard hardware design flow as well as the design flow for general purpose processors. A key technology in the design for embedded processors is retargetable compilation; however, the techniques in this area are just beginning to appear.

This chapter examines the design tool needs for deeply embedded processors such as DSPs (Digital Signal Processors) and MCUs (microcontrol units) used in the the application domains of wireless, telecommunications, and multimedia, where many or all of the following criteria are present:

- The system runs embedded firmware and requires real-time response.
- Low cost and low power are of critical importance.
- Products are sold in high volumes.
- The processor can be used on-chip as an embedded core.

The discussion will not include the requirements and tools for general-purpose microprocessors used for workstations, personal computers, and highly parallel applications. These type of machines have a very different set of application and market constraints resulting in different design tool needs.

## 5.3 MODERN EMBEDDED ARCHITECTURES

This section presents a description of trends in architectures used in today's products in multimedia, wireless communications, and telecommunications. Increasing product complexity is driving architectures toward increasing reliance on CPUs; furthermore, customized cores are increasingly used to meet cost/performance goals.

### *5.3.1 Architectures in Multimedia, Wireless, and Telecommunications*

The market for workstation/PC microprocessors is driven by compatibility. In contrast, embedded architectures show a very different picture—custom programmable architectures tuned to application types (ASIPs) are proliferating in many application domains.

In multimedia applications such as MPEG-1 and 2 encoding and decoding, Dolby AC-3, and videotelephones, many companies are turning toward custom programmable solutions [105]. Some specific examples are: the Philips TM-1, a custom VLIW processor for MPEG-1 and 2 coding and decoding [106], the Chromatic Mpact media processor which performs MPEG-1 and 2 video decoding, the NTT MPEG-2 video encoder chip set which uses a set of custom RISC processors [107], and the TCEC VLIW processor for MPEG-2 Audio, Dolby AC-3, and Dolby Pro-logic decoding [108]. These are just a few of the growing number of ASIP designs being used in multimedia.

Custom programmable architectures are popular in multimedia systems for three main reasons. First, a tailored architecture is typically more cost-effective than a general-purpose architecture. This is particularly true for a second, cost-reduced design which can often be the major source of revenue. Second, designers can make considerable performance gains with special purpose memories and data-paths. Third, instruction-set compatibility is not an issue for a chip which executes a small set of specialized routines, so specialized instruction sets which save memory and/or increase performance are attractive.

A common argument against the use of ASIPs is that many recent applications which required an ASIP solution could be replaced by the latest general purpose processor. Unfortunately, the applications themselves do not stand still, and by the time MPEG-1 audio decoding can be performed by a standard processor, the market demand is for low-cost DSPs which perform MPEG-2, Dolby AC-3, and MPEG-4. This may explain why the native signal processing (NSP) initiative for signal processing on x86 architectures has had few big commercial applications so far [109]. The increasing demands for multimedia continually require a higher end X86 processor, and the same function can be offered more cost-effectively by a dedicated multimedia processor. Manufacturers of X86-class processors have responded with specific multimedia extensions (MMX) for its new processors [109]. Nevertheless, there is still an estimated factor of ten performance speedup needed beyond MMX to handle high-quality game programs or professional-level 3D graphics [110].

Taking a look at the designs for GSM handsets and base stations in major telecommunication system houses shows a number of custom DSP solutions appearing [105]. Some examples of these include GSM designs for mobile handsets done at CNET, the R&D group of France Telecom [111] and Alcatel [112], where each has a custom DSP solution claiming a reduction of power consumption by up to one half of a commercial DSP solution.

Italtel uses 2 in-house VLIW DSPs [113] to replace a GSM base station equalization system that would have required 8 commercial DSPs. The 109 bit instruction-word of the VLIW machine provides many parallel execution units, providing throughput that cannot be matched by off-the-shelf components.

In telecommunication applications, Northern Telecom (Nortel) has designed a custom DSP for a key system unit (i.e. a local telephone switch) [114, 115]. Using an architecture which has been simplified according to the needs of the application, this ASIP outperforms standard DSPs. This is especially true for the areas of power and speed, achieved through the reduction of busses and multiplexers. Furthermore, a 40-bit instruction word allows the designers to incorporate the required parallelism for the system performance.

These examples demonstrate reasons for the popularity of a custom ASIP solution in the areas of wireless and telecommunications, which can be summarized as follows:

- **performance** Stringent constraints on speed, especially for wireless applications mean that a complex network of many commercial DSPs is needed for the required throughput. A custom solution reduces the number of components and the complexity of the communication between components.

- **power** For handheld applications, low power is essential. Architectures containing extra features which are not used imply unnecessary multiplexing of hardware and consumption of power.

- **cost** Many wireless and telecommunication applications cannot afford the luxury of a many chip solution. Especially in high volume products, it is more cost effective to have a custom chip set which is few in number.

### 5.3.2 *Examples of Emerging Architectures*

Using an instruction-set architecture in an embedded application opens doors to a wide range of design styles. Bolsens et. al. [116] categorize embedded programmable blocks as Highly Multiplexed Data Path (HMDP) processors, of which three types are emerging:

- **Commercial, programmable DSPs** These architectures are typically characterized by a Harvard architecture [117] with separate program and data memories, a fixed point-core, and fixed I/O peripherals. The instruction-set is highly encoded and tuned to multiply-accumulation operations used for convolution algorithms common in DSP.

- **VLIW processors** As described in Chapter 4, this type of architecture is characterized by a network of horizontally programmable execution units.

- **ASIPs** As described above, ASIPs are processors specialized to a specific domain. An ASIP can be optimized for the performance, speed, and power characteristics of the application and can be reused for variations of the application by reprogramming.

While these may be regarded as base categories, practical examples may display characteristics of more than one of these types. We illustrate the principles of embedded processors using some concrete industrial examples.

In multimedia, memory manipulation and operation throughput are important criteria. The parallelism of a VLIW architecture can provide the needed speed, while the orthogonality of the instruction-set also allows the compilability.

An example architecture is shown in Figure 5.1, the MMDSP, a processor core designed at Thomson Consumer Electronic Components (TCEC) that is used to perform MPEG-2, Dolby AC-3 and Dolby Pro Logic audio decoding [108]. This is an evolution of the architecture of the MPEG-1 audio decoder used in the satellite to set-top box DirecTV application [118]. Other target applications for this architecture include DVD (Digital Versatile Disk), multimedia PC, High Definition Television (HDTV), and high-end audio equipment. This is an example of a high-volume multimedia application for which low cost is an essential feature.

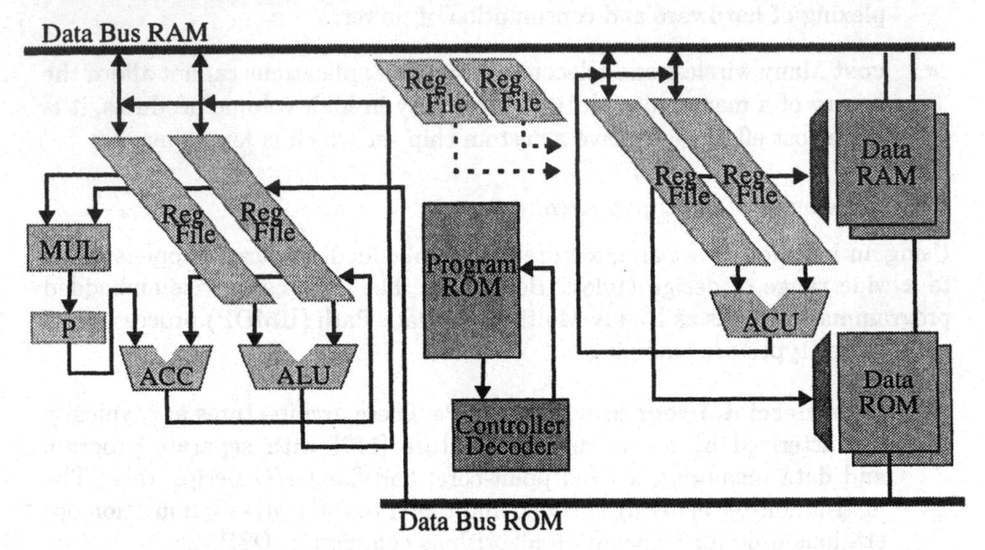


**Figure 5.1** The TCEC MMDSP for MPEG audio applications.

The architecture shown in Figure 5.1 is the result of a thorough analysis of the time-critical functions of the MPEG2 and Dolby AC-3 standards. At first glance, it is similar to many commercial DSPs. It is based on a load/store, Harvard architecture. Communication is centralized through a bus between the major functional units of the ALU (Arithmetic and Logical Unit), ACU (Address Calculation Unit), and memories. The controller is a standard pipelined decoder with the common branching capabilities (jump direct/indirect, call/return), but also including interrupt capability (goto/return-from interrupt) and hardware do-loop capability. Three sets of registers are used to provide three nesting levels of hardware loops.

Several features of the architecture are optimized for this application. The post-modify ACU includes custom register connections and a rich set of increment/decrement capabilities which allow it to efficiently access the special memory structures. The instruction encoding is designed so that a carefully selected subset of the ACU operations can be executed in parallel.

The ACU has been designed to work in concert with the memories. The memory structure has been developed around the data types needed for the application. A first partition separates memory into ROM which is used mostly for constant filter coefficients, and RAM to hold intermediate data. For each of these memories, several data types are available, some are high precision for DSP routines, others are lower precision mainly for control tasks. This choice of data types is key to the performance of the unit.

The MAC (multiply-accumulate) unit was designed around the time-critical inner-loop functions of the application. The unit has special register connections which allow it to work efficiently with memory-bus transfers. In addition, certain registers may be coupled to perform double precision arithmetic. Finally, the very large instruction word format (61 bits) allows for far more parallelism than most commercial DSPs. This is crucial in obtaining the required performance for the MPEG-2 and Dolby AC-3 audio standards.

In comparison with the previous MPEG-1 decoder [118], this architecture contains significantly more functionality. Furthermore, it supports many more application algorithms. Consequently, the former solution of programming in an in-house macro-assembler, known as TTL-C [118], is no longer productive. The basic VLIW principles of the machine mean that the development of a C compiler was possible [108]. However, because the architecture was custom, the support software (including source-level debugger, and bit-true library) required substantial effort.

The SGS-Thomson integrated video telephone [119, 120] presents an interesting approach to macro-block reuse of an ASIP core. This system-on-a-chip contains several block operators which communicate through a set of busses, shown in Figure 5.2. Each of these operators can perform a number of functions dedicated to a certain task. This design increased the number of embedded processor cores to six from the two used in the previous generation video codec [121]. Some of these cores directly replace dedicated hardwired realizations of similar functions: (e.g. the BSP (Bit-Stream Processor), the HME (Hierarchical Motion Estimator)). Other cores provide new functionality. The design constraint which forced this transition is the need for flexibility to adapt to the H.263 standard evolution, and to accommodate any late specification changes. The new architecture also performs a larger number of functions on the internal core rather than the host processor.

The key to this design is the significant reuse of a template control structure as well as the bus communication protocols. Four of the ASIP blocks are based on the same controller architecture: the MSQ (MicroSequencer), the BSP (Bit Stream Processor), the VIP (VLIW Image Processor), and the HME (Hierarchical Motion Estimator). In addition, these operators communicate to one another through the same well-defined bus interface protocol. Each of these blocks has a template architecture of which only the data-path, data bit-widths, and memories differ. This level of reuse simplifies the hardware design, but also

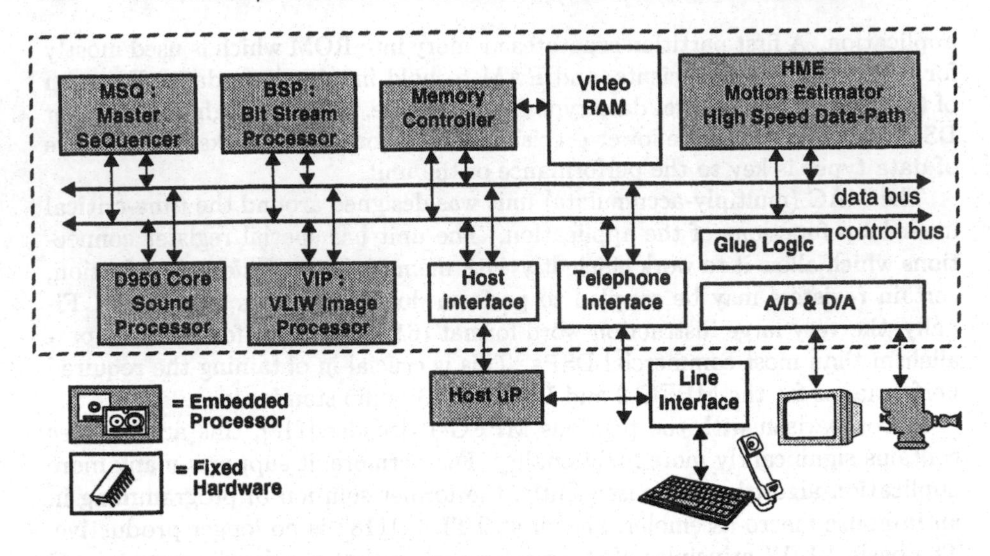


Figure 5.2 Reuse of programmable blocks in the SGS-Thomson Integrated Videophone.

has a great impact on the development of the compilation environment. The compiler for each ASIP could be based on an original and modified depending on the needs of each new block.

One of the most complete portfolios in DSP offerings is that of Motorola. The Motorola 56K series of DSPs is categorized in five main families: the DSP56000 for digital audio applications, the DSP56100 for wireless and wireline communications, the DSP56300 for wireless infrastructure and high MIPs applications including Dolby AC-3 encoders, the DSP56600 for wireless subscriber markets, and the DSP56800 for low cost consumer applications [122]. These architectures resemble one another, but have different characteristics based on the target application area and the cost, performance, and power requirements of those domains.

It is clear that for the embedded processor market, it is not enough to have a single product with a fixed architecture. A solution is competitive because it is specialized for the application domain and the architecture is refined for the type of algorithms to be executed. Furthermore, the instruction-sets of the architectures have been encoded using methods which allow the costly program memory to remain at a narrow width. However, it is also done in a manner which maintains the performance for the application domain of interest. This is a difficult balance to achieve, but is key for the success of a product.

Let us consider one part of the Motorola DSP architecture to illustrate some of the points. The Address Calculation Unit (ACU) of the Motorola 56000 is based on a post-modification operation, and is shown in Figure 5.3. It contains two identical halves, each with an arithmetic unit which performs operations on separate sets of registers. The two halves exist mainly for the two memories X and Y addressed by the address busses XAB and YAB. Therefore, both halves of the addressing unit may be active in parallel with the central data calculation unit (DCU) and accesses to each of the memories.

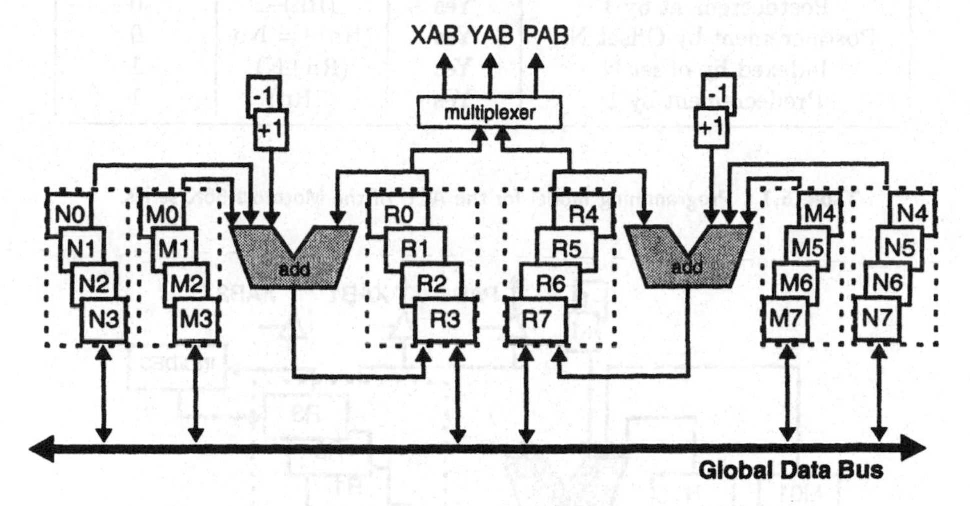


**Figure 5.3** Address Calculation Unit of the Motorola 56000 DSP.

Registers are treated as triplets (i.e. R0:N0:M0, R1:N1:M1, etc.). An address register, Rn, may be post-incremented only with its respective index register, Nn. The available operations are summarized in the programming model of Table 5.1. Post-increment and decrement operations are available for the constant 1 or a value within the Nn register. The Mn register determines the type of address arithmetic: linear, modulo, or reverse-carry.

Now consider the comparison to the Motorola 56800, a low-cost 16-bit processor that Motorola has recently introduced for consumer applications where price is critical. It is a marriage between a micro-controller and a DSP aiming for applications like digital answering machines, feature phones, modems, AC motor control, and disk drives.

Isolating the ACU of the 56800, one can see that it is based on the functionality of the 56000 ACU, but with variations. Shown in Figure 5.4, the ACU has five address registers, one index register, and one modification register.

The available operations of the unit are shown in Table 5.2. Since there is one memory, only one half of the ACU of the 56000 is available. Although the unit has more addressing modes, there is an instruction-cycle penalty for the

| Description of ACU Operation | Uses Mn Modifier | C-like operation | Additional Instruction Cycles |
|---|---|---|---|
| No Update | No | (Rn) | 0 |
| Postincrement by 1 | Yes | (Rn)++ | 0 |
| Postdecrement by 1 | Yes | (Rn)– | 0 |
| Postincrement by Offset Nn | Yes | (Rn)+= Nn | 0 |
| Indexed by offset N | Yes | (Rn+N) | 1 |
| Predecrement by 1 | Yes | –(Rn) | 1 |

**Table 5.1** Programming model for the ACU of the Motorola 56K series.

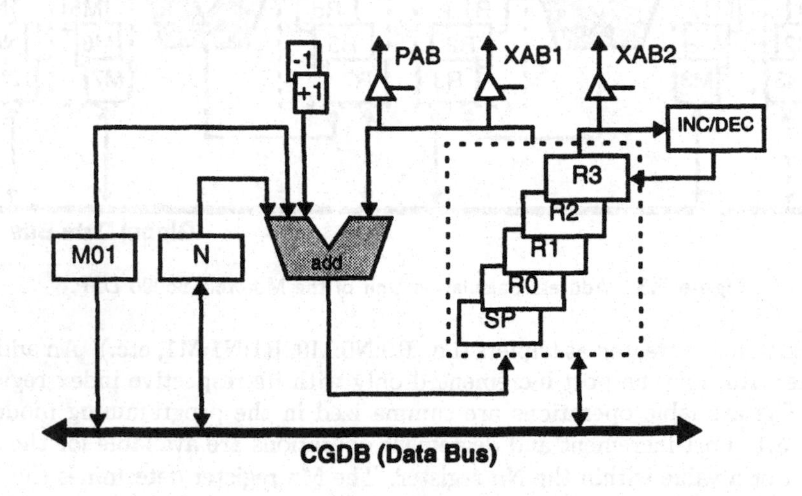


**Figure 5.4** Address Calculation Unit of the Motorola 56800 processor.

indexed modes. The post-modify modes remain the more powerful. Note the restrictions in register uses: only R0 and R1 can perform modulo addressing (bit-reverse is not available); only R2 and SP can perform the short indexed mode. In addition, register R3 has a special property as shown in Figure 5.4. An ACU operation may be performed on R3 in parallel to ACU operations on another address register in addition to up to 2 reads from the double-port data-memory.

The major differences between the 56000 and the 56800 can be summarized as follows:

| Description of ACU operation | Uses M01 Modifier | C-like operation | Additional Instruction Cycles |
|---|---|---|---|
| No Update | No | (Rn) | 0 |
| Postincrement by 1 | R0, R1 optionally | (Rn)++ | 0 |
| Postdecrement by 1 | R0, R1 optionally | (Rn)– | 0 |
| Postincrement by offset N | R0, R1 optionally | (Rn)+= N | 0 |
| Indexed by offset N | R0, R1 optionally | (Rn+N) | 1 |
| Indexed by short (6-bit) | No | (R2+xx) (SP+xx) | 1 |
| Indexed by long (16-bit) | R0, R1 optionally | (Rn+xxxx) | 2 + extra word |

**Table 5.2** Programming model for the ACU of the Motorola 56800 processor.

- The 56800 has fewer address and indexing registers, thus requiring fewer encoding combinations for the instruction-set. This results in an efficient program word width and lower cost.
- Short and long constants allow the more efficient use of program memory, especially with a compiler.
- Several resources are inserted into the 56800 for use by the compiler. The register, SP, is designed to be used as a stack pointer. Furthermore, additional pre-indexed addressing modes are made available even though they cost additional instruction cycles. Whereas a compiler can more easily utilize these addressing modes, they have a performance penalty but still allow good code density.

New compiler techniques are just beginning to appear to make use of post-indexing addressing modes [123] instead of the expensive pre-indexing addressing modes. Cost-sensitive application areas are creating the need for effective compilers. At the same time, architectures designs are taking compilers into account. Compilation technology is lagging behind the need in embedded architectures.

Today, a wide variety of instruction-set architectures are being used for embedded applications. A clear trend in the processor architectures is the specialization of hardware for application domains. Flexibility requires programmable architectures, but application-tuned architecture are required by competition on cost and performance. Architectural specialization is a key competitive advantage in embedded markets. ASIPs provide a good balance between performance and cost for their target domains.

The concept of architecture customizing has been taken even further in some companies such as Philips who have designed the flexible EPICS DSP core [124] for a range of products including digital compact cassette players (DCC), compact disc players, GSM mobile car telephones, and DECT cordless telephones. Flexibility in the EPICs architecture includes the customization of word-lengths, peripherals, memory types, memory dimensions, and register sets.

## 5.4 EMBEDDED SOFTWARE DEVELOPMENT NEEDS

The design tools needs for embedded processor systems is quite different from those in a standard ASIC design flow. They also differ from the tools used solely for general-purpose processors. This stems not only from the wide variety of architecture styles, but also the desire to customize architectures in an application domain. In this section, we discuss current commercial support for embedded processors followed by a picture of the design tool needs we envision for embedded processors and ASIPs.

### *5.4.1 Commercial Support of Embedded Processors*

In recent years the EDA (Electronic Design Automation) industry has begun to offer tools for commercial processor cores. This includes a number of HW/SW co-simulation offerings [125] which allows the co-verification of embedded processor software running on cores with associated hardware. However, few commercial higher-level design tools for embedded processors are available at this writing.

On the other hand, the semiconductor companies who offer commercial DSPs and MCUs are seeing the need to work with closer with compiler companies [126], or to provide internal compiler expertise through development [127] or acquisition [128].

Nevertheless, the current situation regarding the quality of the code produced by commercial C compilers is not promising. The DSPStone benchmarking activities [129, 130] of the commercial compilers for many fixed point DSPs (Motorola 56001, TI TMS320C5x, ADI ADSP2101, NEC uPD77016 and Lucent DSP1610) demonstrate a significant failure of these compilers to produce acceptable code quality. This is true for both DSP and control-oriented applications, although control code fared slightly better. For these compilers, the DSPStone benchmarking showed that machine code runs anywhere from 2 to 12 times slower than a hand-crafted program.

Although low-cost, fixed-point, register-poor DSPs are notoriously difficult for which to develop efficient compilers, even the compilers for the higher-end, floating-point DSPs need work. Some companies claim a two to one execution speed degradation for existing floating-point DSP compilers [131]. When compared to the requirements of real-time systems, these are quality levels which are unacceptable. A designer typically needs a performance measure within 25% of what can be done by hand, otherwise programs will continue to be written in assembly code. In addition to compounding maintenance problems, assembly code also locks designs to old architectures.

### 5.4.2 *Design Tool Requirements*

From the growing popularity of custom instruction-set architectures coupled with the unacceptable state of todays commercial compilers for embedded processors, we can easily see that quality compilers are an important need. This has been further supported by a set of surveys of DSP designers [132] and embedded processor users [105] at Northern Telecom, which shows compilers as the number one need followed by instruction-set simulation, multilevel co-simulation, and source-level debugging.

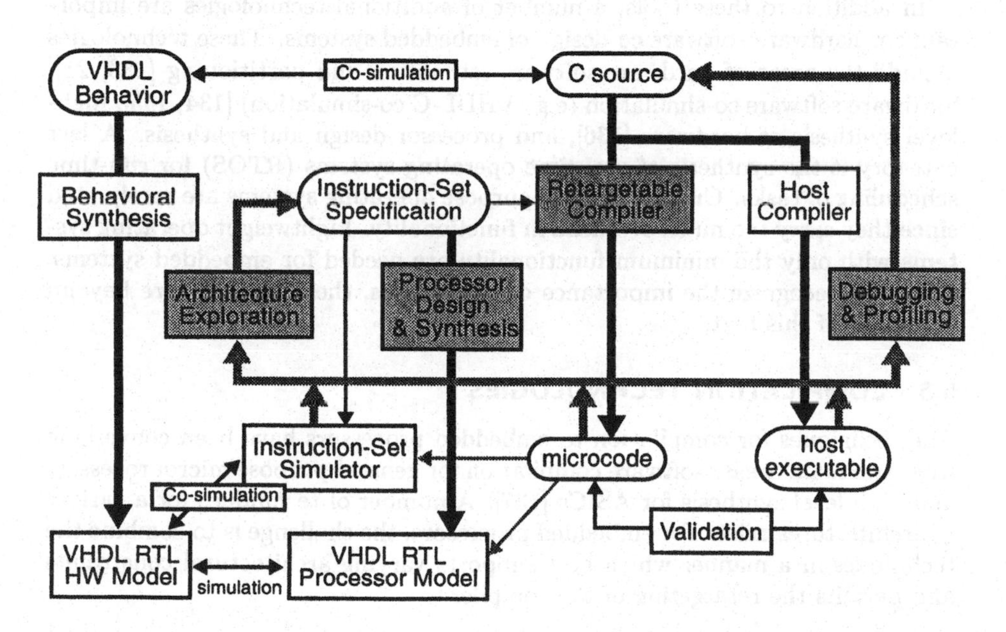


**Figure 5.5** Design tools for embedded processors.

Figure 5.5 presents a principal set of design tools envisioned for embedded processor systems. The core technology is retargetable compilation, which is driven by an instruction-set specification. Modifications to this specification allow the compiler to be reconfigured for changes to the processor. Furthermore, an instruction-set simulator, or even a hardware description of the processor itself could be generated from the specification. The instruction-set specification is the principal implementation means to support a variety of architectures as well as processor evolution and reuse.

The use of a host compiler (e.g. workstation or personal computer) serves multiple purposes. The first is early functional verification of the source algorithm even before a processor design is available. In addition, standard debugging tools are also available. The second purpose is validation of the targeted compiler, by simulation and comparison of a test suite on both the host and an instruction-set simulator.

By consequence, the presence of both a retargetable compiler and host compiler also allows further possibilities. They can be used for debugging in various forms and for architecture and algorithm exploration. Profilers can be used for both algorithm refinement as well as the tuning of an architecture. Furthermore, execution-based optimization strategies for the retargetable compiler are also possible.

In addition to these tools, a number of additional technologies are important for hardware-software co-design of embedded systems. These technologies include the areas of hardware-software estimation and partitioning [133, 21], hardware-software co-simulation (e.g. VHDL-C co-simulation) [134, 135], high-level synthesis of hardware [136], and processor design and synthesis. A last category is the synthesis of real-time operating systems (RTOS) for run-time scheduling of tasks. Current general-purpose operating systems are rarely used since they carry too much overhead in functionality. Lightweight operating systems with only the minimum functionality are needed for embedded systems. While we recognize the importance of these areas, these subjects are beyond the scope of this text.

## 5.5 COMPILATION TECHNOLOGIES

The techniques for compilation to embedded processors have been converging from two main areas: software compilation for general-purpose microprocessors and high-level synthesis for ASICs [137]. A number of techniques for a variety of architectures exist. For embedded processors, the challenge is to combine the techniques in a manner which best supports varying architectural constraints and permits the retargeting of the compiler.

### *5.5.1 Are Traditional Compilation Techniques Enough?*

Aho, Sehti, and Ullman [138] define compilation as the translation of a program in a source language (e.g. C) to the equivalent program in a target language (e.g. assembly code and absolute machine code). This translation is typically decomposed into a series of phases, as shown in Figure 5.6.

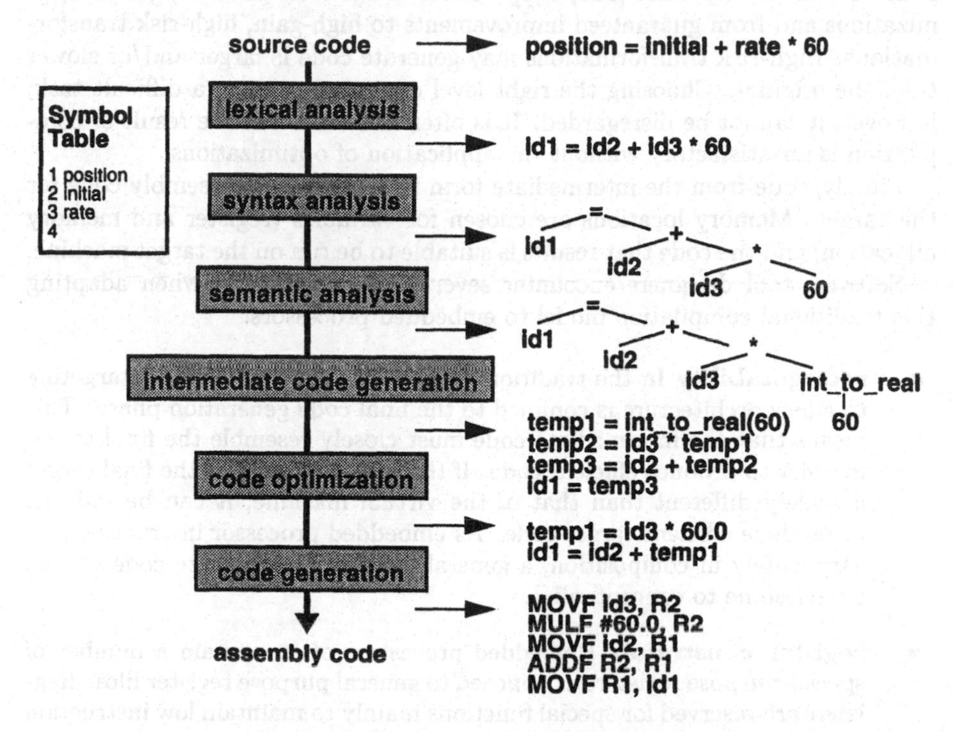


**Figure 5.6** Traditional steps in compilation.

The first two phases are lexical analysis, which tokenizes the program source and syntax analysis, which parses the program into grammatical units. The result of this is an intermediate representation of the source code. A typical example of this representation is a forest of syntax trees (Figure 5.6). For each tree, a node represents either an operation (e.g. =, +) which is to be executed upon its children nodes, or the identifier of a symbol. The third phase is semantic analysis of the intended meaning of the language. It statically determines that the semantic conventions of the source language are not violated. Examples of semantic checks are: type checking, flow-of-control checking, and symbol name checking.

Following these phases, many compilers produce an intermediate code, which can be thought of as code for an abstract or virtual machine. A common form is known as three-address code (or tuples), which simply means that each instruction has at most three operands: 2 sources and 1 destination. The naive intermediate code can be improved using code optimizations of which a large number and variety exist [138, 139]. These range from local to global optimizations and from guaranteed improvements to high-gain, high-risk transformations. High-risk transformations may generate code is larger and/or slower than the original. Choosing the right level of optimization is a difficult task; however, it cannot be disregarded. It is often the case that the result of compilation is unsatisfactory without the application of optimizations.

Finally, code from the intermediate form is translated to assembly code for the target. Memory locations are chosen for variables (register and memory allocation) and the code that results is suitable to be run on the target machine.

Software tool designers encounter several major problems when adapting this traditional compilation model to embedded processors:

- **retargetability** In the traditional approach to compilation, retargeting to a new architecture is confined to the final code generation phase. This means that the intermediate code must closely resemble the final target in order to produce efficient code. If the instruction-set of the final target is widely different than that of the virtual machine, it can be difficult to produce efficient target code. As embedded processor instruction-sets vary widely in composition, a general form of intermediate code can be troublesome to conceptualize.

- **register constraints** Embedded processors often contain a number of special-purpose registers as opposed to general purpose register files. Registers are reserved for special functions mainly to maintain low instruction widths. The instruction width reflects directly into costly on-chip program space. The constraints of register assignment can affect all the phases of compilation.

- **arithmetic specialization** Three-address code artificially decomposes data-flow operations into small pieces. Arithmetic operations which require more than three operands are not naturally handled with three-address code. Such operations occur frequently on DSP architectures [108].

- **instruction-level parallelism** Architectures with parallel executing engines require different compilation techniques. For example, a DSP often has both data calculation units (DCU) and address calculation units

(ACU). A compiler should take into account the possibility to perform operations on different functional units, as well as choose the most compact solution.

- **optimizations** Real-time embedded firmware cannot afford to have performance penalties as a result of poor compilation. Efficient compilation is only arrived upon by many optimization algorithms. Optimizations which are restricted to intermediate code (e.g. three-address code) work most efficiently on a local area. Global optimizations which use data-structures (e.g. arrays, structures), data-flow, and control-flow information would be more naturally suited to a higher level of abstraction, closer to the source program structure.

An example of a widely used traditional compiler is gcc, distributed by the Free Software Foundation [140]. With the free distribution of its C source code, it has been ported to countless machines and has been retargeted to even more. For examples of embedded systems, the gcc compiler is offered as a commercial retarget to several DSPs including the ADI 2101, the Lucent 1610, the Motorola 56001, the SGS-Thomson D950, and the DSP Group Pine and Oak cores. It has become the de-facto approach to develop compilers quickly from freely available sources. One of the well-known strengths of gcc is its complete set of architecture independent optimizations: common subexpression removal, dead code elimination, constant folding, constant propagation, basic code motion, and others. However, for embedded processors, it is extremely important that optimizations be applied according to the characteristics of the target architectures. Unfortunately, gcc has little provisions for which optimizations may be applied according to the target machine.

In the area of real-time DSP systems, the performance of many of the gcc-based compilers fall short of producing acceptable code quality. This has been evaluated with the DSPStone benchmarking activities, mentioned earlier [129, 130]. The underlying reason for the poor performance is made clear in the document distributed with gcc [140]. It states that the main goal of gcc was to make a good, fast compiler for machines in the class that gcc aims to run on: 32-bit machines that address 8-bit bytes and have several general registers. Elegance, theoretical power and simplicity are only secondary. Embedded processors usually have few registers, heterogeneous register structures, unusual wordlengths, and other architectural specializations. Compiler developers using gcc for such embedded processors are faced with two choices: lower code quality or a significant investment in custom optimization and mappings to the architecture. In the latter case, ad-hoc parameters in the machine description and machine-specific routines are needed. Naturally, this greatly reduces the compiler retargetability.

### *5.5.2 Retargetability, Specification Languages, and Models*

Retargetability has been a topic of interest for compilers for many years [141]. Nonetheless, a formal retargetability model has never been fully adopted because general-purpose CPUs have long lifetimes that do not justify the effort required to make their compilers retargetable. For embedded processors, the renewed interest in retargetable compilers is two-fold:

1. Retargetability allows the rapid set-up of a compiler to a specific processor. This can be an enormous boost for algorithm developers wishing to evaluate the efficiency of application code on different existing architectures.

2. Retargetability permits architecture exploration. The processor designer is able to tune his/her architecture to run efficiently for a set of source applications in a particular domain.

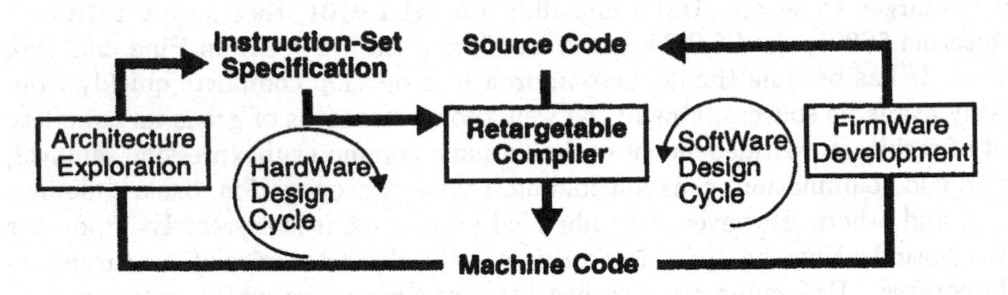


**Figure 5.7** The retargetable compiler concept.

Figure 5.7 shows two design cycles: the software and the hardware design cycle. The right side of the figure is the familiar firmware development cycle, where the programmer uses the compiler to develop software. The left side of the figure shows the retargetable compiler being used as a design tool to explore the processor architecture. The ideal user-retargetable situation is where the instruction-set specification describes completely the processor mechanics in a manner which is simple enough, so that the compiler user is able to make changes himself. Therefore, he/she is also able to explore different architectures simply by changing the specification.

The most promising avenue for supporting the retargetability of compilers for embedded processors is the work on instruction-set specification languages and models which allow a user to describe the functionality of a processor in a formal fashion. The transformations of a compiler may be retuned according to the architecture by means of an architecture model. While an architecture model need not be generated by a specification language, it is the most natural interface to the user.

The MSSV/Q compilers [142] represent early work on mapping high-level algorithms to structural representations of processors. The structure of the processor is described in a hardware description language called Mimola. This netlist displays an explicit activation of functional components by bits in the instruction word. The algorithm language is the Pascal-like subset of Mimola. After the application of a set of target-dependent, user-definable program transformation rules, the algorithm is matched to the target structure. The compiler uses a recursive descent algorithm matching operations to functional units, and constants/variables to memory locations. During this execution, paths are matched using reachability analysis of the target structure. For optimization reasons, several instruction versions are generated and bundling is performed to reduce the number of final instructions.

The more recent generation of compilers from University of Dortmund is the Record compiler [143]. It again uses Mimola as the processor description language, but uses another approach for compilation. A pre-compilation phase called instruction-set extraction automatically generates a compiler code selector from a hardware description model of the processor. The advantage of this approach is the link to the use of efficient tree pattern-matching methods for the main instruction-set selection phase of compilation. In addition to this are a number of additional compiler transformations which improve instruction-level parallelism, including address generation and compaction.

The strength of a Mimola-based compiler is the direct description of the processor architecture. The compilers work directly with the physical structure of the hardware, which leads to a fair level of retargetability. At the same time, this is a weakness of the compilation approach. The processor description requires intimate detail of the decoding strategy of the entire architecture. In the case of commercial processors, for example, detailed information of the hardware would not be available, only a programmers manual of the instruction-set.

The CBC compiler [144] is a project that inspired the development of a processor specification language known as nML [145]. The language is intended to describe the behavioral mechanics of a processor rather than the structural details, by means of the execution mechanics of that instruction-set. The key elements of the language are the description of operations, storage elements, binary and assembly syntax, and an execution model. This combined with some features such as the derivation of attributes allows the full description of an instruction-set processor without the detailed structural information of a netlist. The level of information is comparable to a programmer's manual for the processor. The nML language is based on a synchronous register-transfer model, allowing also the description of detailed timing including structural pipelining. For embedded processors, the user is able to capture all the functionality, execution, and encoding of the machine. The strong point of nML is that the language is not tied to the implementation of a compiler or simulator, which is the case for many machine descriptions [140].

One approach to retargetability is the use of a processor model, such as the Instruction Set Graph (ISG) used in the Chess compiler [146]. The representation is an example of a model that associates behavioral information of the processor with structural information. Making use of nML as the description language, the ISG model is generated automatically and encapsulates the functionality of the processor together with the instruction-level semantics. The principal elements of the ISG are shown in Figure 5.8. The graph contains two types of storage elements: static resources such as addressable memory or registers with explicit bit-widths, and transistors which pass values with no corresponding delay. Storage elements are interconnected by micro-operations which correspond to specific operations which may be executed on a functional unit of the processor by an instruction code, or to connectivity to other storage elements. Each micro-operation contains a list of legal instruction-bit settings, which in principle activate a connection between storage elements. Therefore, a legal micro-instruction is constructed by forming a path through the ISG, keeping structural hazards in mind.

This approach allows a convenient encapsulation of the operations of the processor while keeping an active record of the encoding restrictions defined by the instruction-set. In the Chess compiler, the ISG serves a base model for all the phases of compilation (pattern matching, scheduling, and register assignment) to form the mapping from source algorithm to microcode implementation.

The CodeSyn compiler developed at Bell-Northern Research and Northern Telecom also uses a mixed structural and behavioral level model to describe the target instruction-set processor. The instruction-set specification consists of a mixed behavioral and structural-level model composed of three main parts: a pattern set of microinstructions; a structural connectivity graph; and a classification of the resources in the structural graph. A pattern is a behavioral level representation of an instruction, comparable to the description of an instruction in an assembly programmers manual. Microinstructions are described as small pieces of control-data flow graphs, and can be categorized in three ways: pure data-flow (containing arithmetic, logical, relational operations, and address calculation); pure control flow (containing hardware loops, unconditional jumps and branches); and mixed data/control flow: (containing conditional jumps and branches). Assembly and binary instruction formats are associated with each pattern.

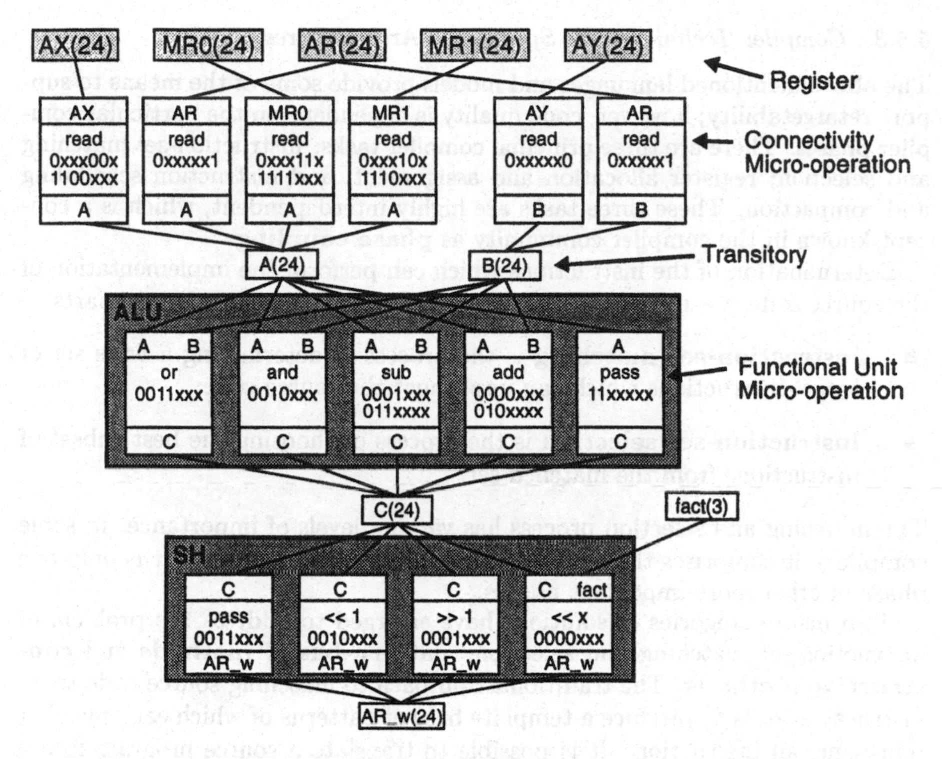


**Figure 5.8** Principles of the Instruction Set Graph (ISG) model of the Chess compiler.

A structural graph is used by the compiler to determine the possible data movement through the processor. The register classification typically categorizes the architectural registers on two levels: a broad classification of general function and a small classification of specific function. The three parts of the instruction-set specification are inter-related. The structural graph is built through a user specification of the relationship between register classes and functional units. Register class annotations are associated with the input and output terminals of each pattern indicating data-flow between classes. As well as allowing proper pattern matching, the register class annotation guides the register assignment algorithms to bind reads and writes to physical registers in the architecture. One last relationship is the correspondence between operations in the pattern set with the functional unit which performs the operation in the structural graph.

### *5.5.3 Compiler Techniques for Specialized Architectures*

The above-mentioned languages and models provide some of the means to support retargetability; however, code quality is dependent on the particular compiler phases. There are three principal compiler tasks: instruction-set matching and selection; register allocation and assignment; and instruction scheduling and compaction. These three tasks are highly interdependent, which is a concept known in the compiler community as **phase coupling**.

Determination of the instructions which can perform the implementation of the source code is a main task of compilation, which we define in two parts:

- **instruction-set matching** is the process of determining a wide set of target instructions which can implement the source code;

- **instruction-set selection** is the process of choosing the best subset of instructions from the matched set.

The matching and selection process has varying levels of importance: in some compilers, it comprises the entire compilation process; in others, it is only one phase of other more important phases.

Two main categories of solutions have emerged to address the problem of instruction-set matching and selection: **pattern-based methods** and **constructive methods**. The traditional approach to matching source code to an instruction-set is to produce a template base of patterns of which each member represents an instruction. It is possible to translate a source program into a forest of syntax trees, which are then matched to the pattern set of syntax trees. A subset of all the matched patterns are selected to form the implementation in microcode. Dynamic programming [138] can be used to select a cover of patterns for the subject tree. It is a procedure with linear complexity that selects an optimal set of patterns when restricting the problem to trees and a homogeneous register set. However, many embedded processors have heterogeneous register sets and instructions best described by graph-based patterns, meaning the advantage of dynamic programming is diminished for embedded processors.

Tree-based pattern selection extensions which allow the handling of heterogeneous register sets have been formulated in the work of Wess [147]. In this approach, register constraints are encapsulated by a trellis diagram, and following, the code selection process is considered as a path minimization problem. Other pattern-based methods include the the SPAM (Synopsys Princeton Aachen MIT) project [148] whose members have been able to apply the principles to the Texas Instruments TMS320C25 DSP. Using the Olive code generator generator, a grammar for the TI C25 allows the user to be hidden from the details of the parser whose pattern set treats the restrictions of registers for the architecture. The CodeSyn compiler also uses pattern matching for instruction-set selection, based on control-data flow graphs (CDFGs). Pattern matching is formulated as the task of determining isomorphic relationships between instruction-level CDFGs and the source CDFG. An organization scheme allows the number of attempted matches to approach linearity with the number of nodes of the source graph [115]. The approach is an efficient method to determine all the possible pattern implementations at all the points in the subject graph to allow an exploration of possible pattern selections. It also takes into account the presence of overlapping register classes. An important feature of the approach is that it provides pattern matches in a fashion that allows the matches to be propagated to other phases of the compiler to find the best coverings.

Another pattern-based approach is virtual code selection, used in the FlexCC compiler of SGS-Thomson Microelectronics [108]. Building on the traditional view of a compiler (Figure 5.6), this is a developer's environment. The developer defines a virtual machine which resembles in functionality the instruction-set of the real machine, but is sequential in operation. Those processors with parallel execution streams would be simplified to one stream. The virtual machine description contains: a description of resources including register sets and addressing modes; and a set of code selection rules.

For the code selection rules, the developer defines the mapping between the C code onto the virtual machine instruction set. The compiler developer has at his/her disposal a programming language which contains a set of high-level primitives corresponding to information which is generated as syntax trees of the source program. For each operation which may occur in a syntax tree, the developer provides a rule for the emission of code for the virtual machine. This rule will be triggered upon matches to the source code and executed at compile time. This allows the developer to provide standard rules for the most common cases and more sophisticated rules for areas of architecture specialization.

Some approaches have been introduced which base pattern matches on structural connections of the processor. Using the Mimola model as described earlier, MSSV/Q determine valid patterns by verifying against the structure. Similarly, the Chess compiler uses a bundling approach [146] which couples nodes of a control-data flow graph (CDFG) based on the instruction-set graph model. This also allows the selection of bundled patterns which are heavily restricted by the encoding of the instruction-set. The validity of patterns is determined directly by the architecture model. Two advantages of this approach are that the pattern set need not be computed at pre-compile time, and the bundling of patterns can possibly pass control-flow boundaries in the source code.

Mapping to architecture storage units is a principal means of compiling efficient code. **Register allocation** is the determination of a set of registers which may hold the value of a variable. **Register assignment** is the determination of a physical register which is specified to hold the value of a variable. Register allocation and assignment for embedded processors is further complicated by special-purpose registers, heterogeneous register files, and overlapping register functions. However, as a large base of work already exists, much of the today's research builds upon techniques of the past. A survey of register allocation methods until 1984 can be found in [149], and classic approaches are discussed in [138].

Pioneering work on register allocation by Chaitin [150] introduced the notion of **coloring** to determine the number of registers needed for a program's variables. The process of coloring proceeds in two steps. The first step is to build an interference graph whereby the nodes of the graph represent the variables and a set of edges connecting the nodes. An edge represents an overlap of lifetimes between the two variables, meaning that the two variables cannot use the same register. The second step is to assign a color to each node of the interference graph, such that no two connecting nodes have the same color. The number of colors used in the graph is the number of registers needed in the program. In a real program, the coloring formulation is entangled with control-flow constructs such as if-then-else conditionals, loops, case statements, function calls and local/global scoping. The formulation using coloring for a real application must be extended to handle these cases. In addition, the heterogeneous registers and overlapping functions significantly change the nature of the formulation.

Extensions to Chaitin's formulation which use exact lifetime information has been studied by Hendren et. al. [151]. Their formulation also includes careful treatment of cyclic intervals, which colors variables whose lifetimes extend over several iterations of a loop. Unfortunately, it does not handle heterogeneous register files.

One approach to handling heterogeneous register sets is **register classes** [140, 152]. In general, a register may belong to one or more register classes of overlapping functionality. By these means, the compiler is able to calculate those registers which are most needed for a specific function to guide register allocation and assignment.

In both the CBC compiler [144] and the GNU gcc compiler [140], the concept of a symbolic or pseudo register is used so that the compilation may proceed in two steps. During instruction-set selection, a symbolic register is assigned for each program storage element. Next, symbolic registers are organized by means of the register classes in a register allocation phase. Following, detailed register assignment of each symbolic register to a real (i.e. physical, hard) register is performed.

The CodeSyn compiler builds upon the concept of register classes for special-purpose registers [153]. Benefiting from the annotation of register classes on the input and output terminals of patterns, the allocation of registers is done by calculating overlaps in the classes following the data-flow. Candidate register sets are calculated between operations from the intersection of registers in each register class. The assignment procedure then begins with the mostly highly constrained allocated registers to the least constrained.

An example is shown in Figure 5.9, where a control-data flow graph is annotated by register classes corresponding to the architecture. Two heavily constrained classes, MemoryStore and DataToMemAddr, have only one register in their respective candidate lists. Propagation of this constraint to the surrounding data-flow means that a resolution can take place only by the insertion of a data-move operation. The most constrained classes are taken into account first, then the others classes. In this manner, the overlapping register roles are handled.

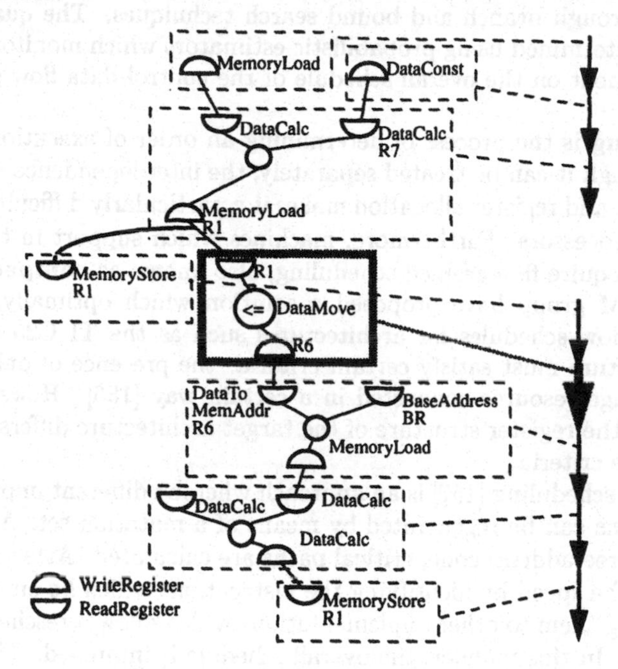


**Figure 5.9** Register assignment using classes in the CodeSyn compiler.

To deal with the distributed nature of registers in DSPs and other embedded architectures, some have reformulated register assignment as a problem closely tied to the architecture structure. This is known as data-routing, where the goal is to determine the best flow of data through the architecture minimizing the time. While data routing techniques are generally more time consuming than other register assignment approaches, they usually provide solutions for architectures with very heavily constrained register resources and distributed register connections. Some of the previously mentioned register assignment techniques can fail for these architectures.

Rimey and Hilfinger [154] introduced an approach known as lazy data routing for generating compact code for architectures with unusual pipeline topologies. The idea is to schedule instructions as compact as possible and to decide on a data-route only after an operation is scheduled. A spill path to memory is always guarded to guarantee that no deadlocks will occur. A similar approach was used by Hartmann [155], where a complicated deadlock avoidance routine was incorporated for architectures with very few registers.

The Chess compiler uses a branch and bound approach to data-routing [146]. The approach determines a solution from a set of alternative candidates which are found through branch and bound search techniques. The quality of each solution is determined using probabilistic estimators which monitor the impact of an assignment on the overall schedule of the control-data flow graph of the source code.

**Scheduling** is the process of determining an order of execution of instructions. Although it can be treated separately, the interdependence with instruction selection and register allocation makes it a particularly difficult problem for embedded processors. Furthermore, machines which support instruction-level parallelism require fine-grained scheduling, also known as **compaction**.

The SPAM group have proposed a solution which optimally solves tree-based data-flow schedules for architectures such as the TI C25 architecture. The architecture must satisfy certain criteria: the presence of only single and infinite storage resource connected in a certain way [156]. However, they do not apply if the register structure of the target architecture differs widely from these storage criteria.

Mutation scheduling [157] is an approach whereby different implementations of instructions can be regenerated by means of a mutation set. After the generation of three-address code, critical paths are calculated. Attempts are made to improve the speed by identifying the instructions which lie on critical paths and mutating them to other implementations which allow a rescheduling of the instructions. In this manner, the overall schedule is improved. The advantage of this approach is that it works directly on critical paths and improves timing on a level of the code which is very close to the machine structure.

Wilson et. al. [158] use integer linear programs (ILPs) to solve compilation problems for embedded processors. They propose a compilation model which integrates pattern-matching, scheduling, register assignment and spilling to memory. The ILP solver dynamically makes trade-offs between these four alternatives based on an objective function and a set of constraints. The objective function is usually a time goal which is iteratively shortened until further improvement is minimal. The set of constraints includes architecture characteristics like the number of accumulators, other registers, and functional units. A clear advantage is the concurrent treatment of the main phases of compilation to arrive at the best solution.

Compaction is a form of scheduling designed to increase parallelization in microinstruction words. Lioy and Mezzalama [159] have approached the compaction problem by defining pseudo microinstructions and sequences of micro-operations with source and destination properties. These sequences can then be packed into and upward past pseudo microinstructions to form real microinstructions. This packing takes into account the resource conflicts of the machine, such as register dependencies and the use of functional units.

The compaction phase of the rule-driven approach [117] is based upon on similar methods using a practical reprogramming approach which allows retargeting to new architectures. The compactor attempts to pack micro-operations as tightly as possible within the given constraints which come in two forms: the bit fields in the microinstruction-word and a set of resources defined by the programmer. These resources usually include architectural storage units such as the registers and memories, as well as shared transitional units such as busses. In the definition of micro-operations, the programmer must indicate which resources are written-to, read-from, and/or occupied. Consequently, the compaction algorithm is able to obey all the data-flow dependencies as well as the resource restrictions of the machine.

An example is shown in Figure 5.10, where three types of microoperations (*mops*) are declared, each defined to read, write, and occupy certain resources. As the compactor proceeds through the list of microoperations 1 to 5, it pushes each mop into its respective field. If there is no resource conflict, a mop can be placed in parallel with (e.g. Figure 5.10: mop 1 and 2) or even past a previously placed mop. Resources include registers and memories which keep data-dependencies in order, but also include architecture constraints like the occupation of functional units or busses (e.g. Figure 5.10: mop 2 and 3). Furthermore, non-existent resources may be defined to specify unusual architecture constraints. While giving a flexible view to the programmer on how the compaction procedure is executed, the onus is on the developer to validate that all the resources have been correctly declared and used in each micro-operation.

However, granted that the approach is flexible as a compaction procedure especially for VLIW machines, it does not provide a straightforward solution to highly encoded microinstruction words since the bit fields represent the first constraint of the compaction algorithm. Parallelism in a machine architecture

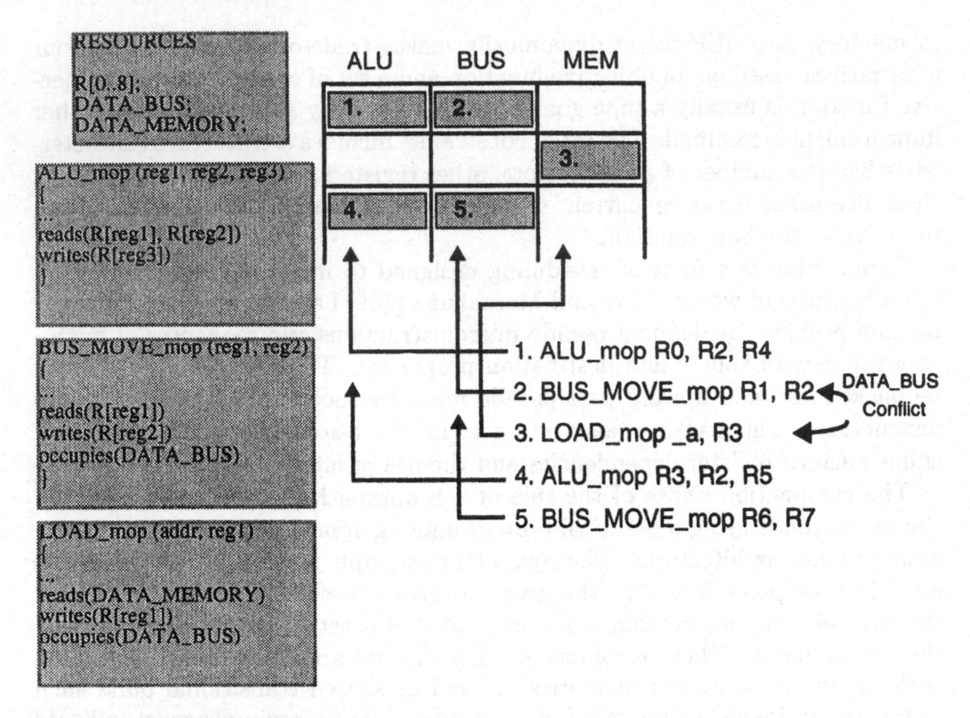


**Figure 5.10** Reprogrammable micro-operation compaction.

does not always imply that the instruction-word supports that parallelism. Again, this stresses the point of phase coupling with the other tasks of compilation like instruction-selection and register allocation. To address highly encoded instruction sets, a compaction procedure which takes into account the *meshing* of instruction formats would be needed.

### *5.5.4 Optimizations for embedded processors*

The subject of compiler optimizations for embedded processors rests largely an open problem. While a large amount of optimization theory exists for general computing architectures [138, 160, 139], the topic is not well understood for embedded real-time architectures. This is primarily because of the infancy of the standard mapping techniques for embedded architectures; and secondly, because of the amplitude of constraints that embedded processors impose on standard optimization techniques. This section covers only a subset of optimizations which are relevant to the characteristics of today's embedded architecture. A larger set is discussed elsewhere [117].

A large number and variety of standard optimizations exist, for example: constant propagation, constant folding, common subexpression elimination, and strength reduction. In the past, many of these have been regarded as architecture-independent transformations; however, for embedded processors, this is no longer true. Even for very simple architectures, these optimizations can produce less efficient code [117]. For example, common subexpression elimination removes expressions which are common among different parts of code. When there are only a few registers, the intermediate variables required by this transformation increase the register pressure, which may adversely effect the efficiency of the final code. An effective approach must take all the local storage requirements into account. Optimizations on constants can also cause problems. For embedded processors, the instruction-word width is one of the most important resources. Immediate constants are traditionally coded within the instruction-word and can therefore be a large source of program usage. Optimizations such as constant propagation seek to push and duplicate the occurrence of constants to later points in the program. For an architecture with a narrow instruction word, this optimization may reduce the efficiency of the code.

Despite the fact that a multitude of standard optimizations prevail, their effects on microcode for real-time architectures can sometimes be contrary to intuition. The important lesson is that perhaps new strategies should be developed which allow the application of these standard optimizations depending on the family of architectures being targeted. An effective compiler could apply a set of optimizations based on characteristics of the architecture. Furthermore, provisions could be implemented so that the programmer is allowed to control where and when optimizations are applied.

In real-time applications, loops are the critical regions of code, since most of the time is spent therein. General optimization theory for loops can be found in [160, 161]. Streamlining the retrieval of data from and the storage of data to memory elements can produce substantial gains. Transformations from higher level language constructs like array and structure references to efficient machine-specific address generation are an important technology [123]. These type of transformations are most effective when they take the processor characteristics into account with an architectural model.

Loop restructuring is the term for transformations which change the structure of loops without affecting the computation of a loop. Loop unrolling [160] reforms loops by replicating the loop bodies for an unrolling factor, u, and iterating over the new step u instead of the original step 1. Unrolling can reduce the looping overhead and increase instruction-level parallelism. Moreover, for loops with few iterations, it can completely eliminate the loop structure.

Loop pipelining (or software pipelining) is a related restructuring procedure which improves the instruction-level parallelism of code within loops: first, operations are forced to place values in registers entering and exiting a loop; the loop is then rearranged to allow parallelization. Many commercial architectures such as DSPs benefit from software which has been loop pipelined. This optimization analysis involves all the key compiler phases: instruction selection, scheduling, and register allocation. Furthermore, compaction of the instructions to meet the architecture constraints must be considered on the microcode level.

Previous work on the software pipelining subject includes the scheduling approach by Lam for VLIW machines [162]. The procedure includes first unrolling the loop body, then rescheduling the remaining instructions. Software folding is another concept [163] used for pipelining DSP hardware architectures and is equally applicable to software.

Another type of loop optimization is loop-invariant code motion [160, 139]. This analysis determines whether a computation within a loop can be executed outside of a loop. Code hoisting is the general term for moving code to an earlier execution point. Loop unswitching moves conditional tests outside of a loop by repeating the loop structure for each condition. Other loop reordering methods can be used to improve the characteristics of loops so that other optimizations like loop-invariant code motion can have a better effect: loop interchange, loop skewing, loop reversal, loop distribution, etc. Each of these are behavior-preserving transformations which allow other manipulations to be done.

Memory optimizations are important because program memory can be a particularly expensive part of an embedded architecture, especially for single chip solutions. Efforts such as the narrowing of instruction words through encoding implies continual difficulties on compiler methods. An illustration of this is the limitation of absolute program memory addresses. Because of short instruction words, an instruction-set which uses exclusively absolute memory addresses is limited in program size. Embedded processor designers have overcome this limitation in a number of ways. One approach is to provide near and far program calls and branches. A program memory can be organized in a set of *pages.*

While a paged program memory is a good solution for the hardware, it poses a number of challenges to the compiler developer. For code with good performance, subroutines need to be allocated in memory in a fashion which reduces changing pages. Furthermore, long subroutines must be broken into smaller pieces so that each block fits into a page. Solutions to the problem are straight forward; however, an optimal solution is non-trivial. One solution is to store and manage subroutine page addresses in a branch table. Each time one of these subroutines is called, the return page address is kept on the run-time stack. The equivalent to program pages can also occur in data memory, when a large amount of data memory is needed. For example, data *windows* can be used to organize the memory. Similar types of considerations must be taken in the compiler to minimize the time needed for the global retrieval and storage of data.

Another hardware solution to improve memory retrieval caching. A cache is a temporary buffer which acts as an intermediary between program or data memory, improving the local nature of the data. These are common to general-purpose computing architectures and appear on more sophisticated embedded processors. Approaches to improve the cache hit/miss ratio are beginning to appear for both program caches [164] and data caches [165] in embedded processors.

Allocation to multiple memories is a topic which arises for some DSP architectures, such as the Motorola 56000 series and the SGS-Thomson D950. For these architectures, the parallelism is improved by allowing independent retrieval and storage operations on each memory. The implication for compilers is the need for memory allocation strategies which make best use of these resources based on the data-flow in the source program. This problem has been addressed in approaches which improve upon previously generated or hand-written assembly code balancing the data in two memories [166]. However, techniques which are incorporated into the analysis phase of compilers are also needed. Still, some practical issues such as the support for a linking phase need to be considered before automatic memory allocation can be supported.

## 5.6 PRACTICAL CONSIDERATIONS IN A COMPILER DEVELOPMENT ENVIRONMENT

This section discusses pragmatic issues in setting up a compiler environment for embedded systems. While the techniques presented earlier form the basis of a compiler system, many other factors must be taken into account for a usable development environment:

- **language support** What ingredients of a programming language should be provided to the user?

- **embedded architecture constraints** What facilities should be provided to the user to control the specialized architecture?

- **coding style** What abstraction of coding style should be supported? What are the trade-offs?

- **validation** What level of confidence will be provided with a retargeted compiler?
- **source-level debugging** How does debugging on the host fit in with debugging on the target?
- **architecture and algorithm exploration** How well does my architecture and instruction-set fit the application?

Language support is one of the most visible choices to the designer working with a CPU, since the programming language provides the interface to the machine. In the embedded industry today, the language of choice is C. While there exist languages more suitable for certain domains (e.g. Silage/DFL [167] in the DSP domain), C remains the most widely used high-level language in embedded processors mainly because of the wide availability of compilers and tools (linkers, librarians, debuggers, etc.) on workstations and PCs. Furthermore, many standards organization such as ISO (International Standards Organization) and the ITU (International Telecommunications Union) provide executable models in C. Some examples of these are: GSM (European cellular standard), Dolby (audio processing), MPEG (video and audio processing), H.261/H.263 (videotelephony), and JPEG (still picture processing).

C's limitations as an embedded programming language include:

- **limited word-length support** The fixed point support in C is limited to 8 bit (char), 16 bit (short int), and 32 bits (long int). While this is sufficient in applications such as speech processing, it is insufficient in many other embedded applications, for example in audio processing where typically 24 bit types are needed and image processing where much large data types are needed.
- **a limited set of storage classes** In many DSP systems, in addition to multiple register files, there are two or more data memories as well as a program memory. Moreover, specific addresses need to be distinguished for items such as memory-mapped input-output (MMIO). ANSI C provides only the auto, static, extern, and register storage classes [168], which are insufficient in providing the user control over where to place data.
- **a fixed set of operators** Embedded systems may have hardware operators which do not correspond to the classical operations found in C.
- **separate compilation and linking** C allows modules to be compiled separately; the modules are then linked together in a separate phase. In the presence of limited register resources, this imposes an obstacle to

efficient interprocedural optimization, such as the passing of arguments in registers.

The practical solution to these problems is to work within the constraints to provide the right compiler support for the architecture at hand. In many cases, the above-mentioned constraints are not limiting, while in others the difficulties can be managed. The way to work within the limitations is to make good choices about the levels of support. For example, a subset of C could be chosen to allow a certain optimization. In other cases, a minimal extension to the C language gives the features desired. If handled carefully, this does not destroy the compilability of the code with other C compilers.

Other languages include C++ which has seen some use in larger embedded processors, the programming benefits being the object-oriented capabilities. However, sometimes the overhead of carrying extra information such as a virtual class table is not beneficial. Java is appearing on the horizon as an alternative to C++, with the attractive feature of being much simpler. However, some issues need further progress, such as the viability of an operating system [169].

Data types must be chosen carefully in an embedded architecture: not providing the right data types incurs run-time costs while providing unnecessary data types increases system cost. The design of a retargetable compiler is made simpler if the number of supported data types are kept to a minimum, since they must finally be mapped onto the data types supported by the architecture. While it is possible to support larger data types by providing libraries of larger data types built upon smaller data types, this elevates all the tasks in developing, maintaining, and validating the firmware development environment. Furthermore, the embedded processor programmer is most concerned with performance. Working at a level where the data types match the register and memory widths of the architecture is the most natural level at which the designer can guarantee real-time performance. While working at a higher level (e.g. larger data types) may simplify the programming, it is a secondary concern of the designer after the guarantee of meeting real-time performance constraints.

While there is a limited number of fixed data type support in C, it is not often the case where the architecture supports an extensive number of fixed data types. It is therefore possible to re-map certain C data types to the types needed for the application. For example, in audio applications, a word-length of 24 bits is commonly used for sampled data. Whereas the 24-bit data type does not exist in C, the 32-bit data type may be used instead. This is assuming that the 32-bit type is not needed as well. The data type may be reinterpreted by the retargetable compiler. Unfortunately, breaking the synchronization between the host and target compilers introduces problems. We can solve the problem in either of two ways:

1. The system can provide built-in functions a bit-true library. Built-in functions provide a common interface for compilation in both paths, allowing the correspondence between host compilation and target compilation to remain intact.

2. The host compiler can be extended to use C++ data types and operator overloading. C++ operators can be overloaded to provide the bit-true operations depending on the data type. While, the retargetable compiler still treats the source as C, the mapping of the data types is interpreted according to context. The equivalent compilation on the host is interpreted using C++.

An increasingly common method to improve memory support in C is to extend the storage classes in ANSI C to those needed for the architecture. For example, if two RAM data memories exist on the architecture, a storage class specifier _MEMORY1_ could indicate the first memory and _MEMORY2_ could indicate the second memory. This allows the designer to choose the location of his data variables. This is a pragmatic solution in contrast to providing a memory allocation algorithm in the compiler. In the case of separate memories of RAM and ROM, it is also possible to use the type qualifier const to identify ROM values, instead of specifying the storage class. The compiler simply needs to intercept these variables for placement in ROM. An example is shown in Figure 5.11.

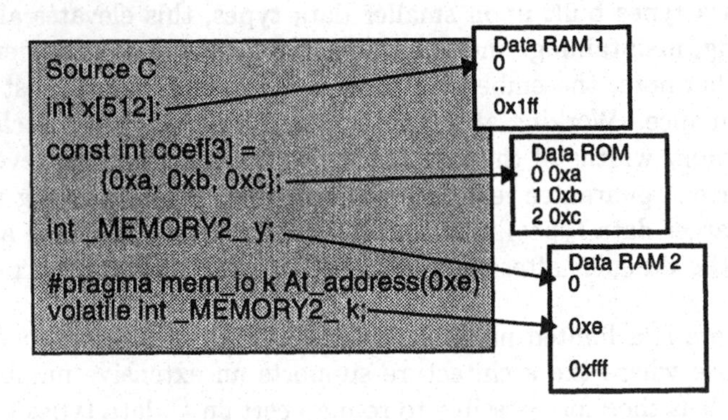


**Figure 5.11** Storage class specification and type qualification for multiple memories.

Extending the methodology further, storage class specifiers can also be applied to memory-mapped I/O (MMIO). However, in this case there are two further important characteristics of a memory-mapped I/O variable. First, as a specific address is required, the user must be able to force the variable into that location in memory, either through a #pragma or with an ANSI C extension. Secondly, the compiler must be made aware that it cannot remove accesses to the variable through optimization. The type qualifier *volatile* accurately defines this characteristic.

Any software support system for an embedded processor should support multiple coding levels which allow programmers to reach all the functionality of an architecture. If a designer cannot meet his performance objectives through the capabilities of a compiler, he/she must have the ability to reach the equivalent quality of hand-written assembly code, while maintaining the ability to use a high-level of programming for parts which do meet the objectives. For pragmatic reasons, it is essential that the bridge to higher levels of automation always be spanned as smoothly as possible.

**A built-in function** is a compiler-recognized function which is mapped directly onto a set of instructions of the processor. These allow the execution of operations which are not found in C, for example: interrupt instructions, hardware do-loops, wait mechanisms, hardware operators, co-processor directives, the setting of addressing modes (modulo, bit-reverse, etc.).

A retargetable compiler should allow coding on various levels of abstraction as well as the mixing of these levels. This is to allow the designer to reach all the functionality of the processor, at perhaps the expense of code portability. Our experience has shown that while the development effort on optimizations is important to achieve a more portable level of code, the effort in providing the handling of lower levels of coding is essential.

For C, we define four levels of coding styles:

1. **high level behavioral ANSI C** This level is characterized by the use of variables, arrays, structures and all the operations available in C.

2. **mid level** This level allows the use of built-in functions. Any arrays or structures that are declared in memory must be accessed by pointers. Variables and pointers may be allocated into extended storage classes and register sets.

3. **low level** This level allows the assignment of variables and pointers in specific registers.

4. **assembly level** This level allows the programmer to write in-line assembly directly in C code.

Level 1 is the goal for compiler technology. This is the level at which all optimization and retargeting capabilities should aim as it provides the most abstract and portable source descriptions. It is also the level at which a programmer can freely write algorithms without being concerned with the underlying hardware.

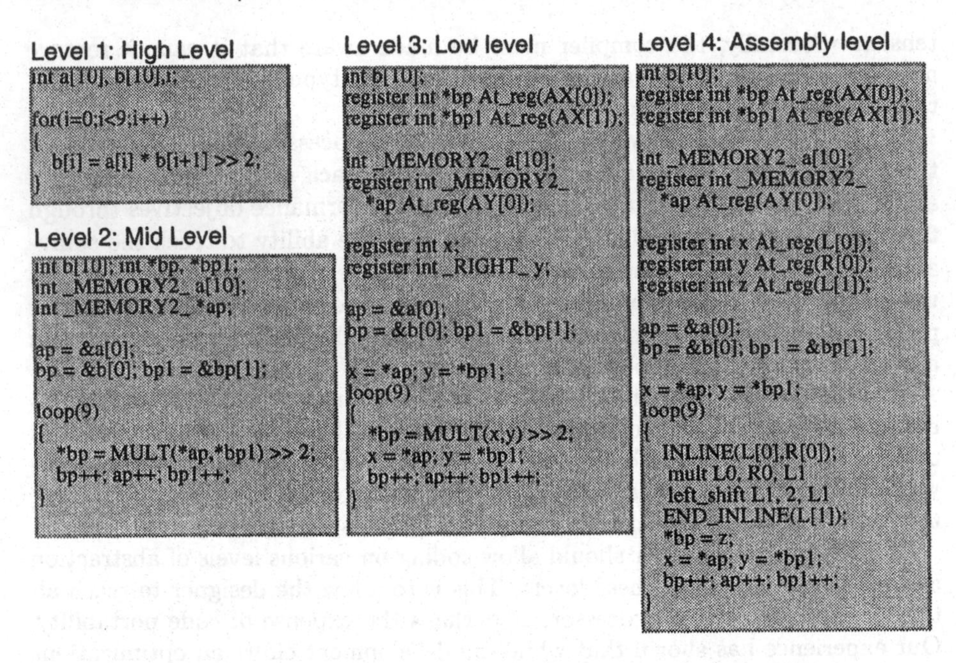


**Figure 5.12** Example of C code on different abstraction levels.

Examples for these four levels are shown in Figure 5.12. The high level example shows the use of behavioral C constructs like the for-loop and references to arrays. The mid level example shows the exclusion of array references replaced by pointers. It also shows the declaration of certain arrays into specific memories. As well, the built-in functions loop is used for a hardware do-loop, and MULT is used for a multiply operation. The low level example shows pointers assigned to specific registers. In addition, a manual loop pipelining operation has been done by specifying new variables allocated to register sets. Finally, the assembly level example shows two in-line assembly instructions specifying specific operations and registers.

### 5.6.1 *Compiler Validation*

The very definition of the term retargetable compiler suggests a countless number of targets and even possible targets. This places a huge importance on the validation methodology. The confidence that a compiler produces correct code is a significant factor that the embedded system designer cannot neglect.

Compiler validation is done today predominantly based upon simulation. For these schemes, the selection of a suitable test suite which covers possible faults is an issue which arises. Commercially available C test suites are available, such as Plum-Hall (see `http://www.plumhall.com`) Perenial, and MetaWare (see `http://www.metaware.com`). These suites are made up examples which test all the facilities of C in a thorough manner. Unfortunately the test suites are not directly applicable to embedded processors because an embedded processor compiler typically uses a subset of C. For example, only some data types and some operations may be supported. Moreover, any extensions to C are not tested.

There are numerous advantages to including a parallel compilation path on the host platform in addition to the target compilation path. The development and debugging environment of the host is generally available prior to the availability of the target development tools. This means that application development can begin before anything else is in place. Even when the target compilation environment is in place, a host equivalent execution will run much faster than any instruction-set simulation of the target processor.

With this methodology in mind, it is important to provide a bit-true library for: any built-in functions that are provided by the target compiler; any operators with data types differing from ANSI C; and any other operations that are implemented differently on the target hardware than on the host processor. The construction of the bit-true library typically involves careful handling of bit-widths with shifts and bit-masking.

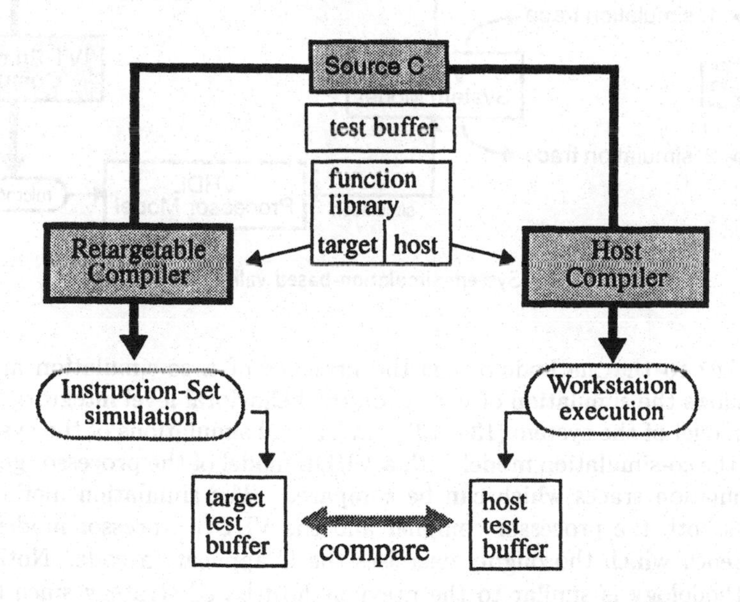


**Figure 5.13** Host-based compiler validation strategy.

Once a bit-true library is in place, a validation methodology as shown in Figure 5.13 is possible. The function library contains functions which allow writing values to a pre-defined test buffer. After compilation on both paths, this buffer is compared for any differences, which indicate a discrepancy in the retargetable compiler, instruction-set simulator, or the bit-true library. In most cases, the host compiler is assumed to produce correct code, as it usually must pass it's own validation stage.

A second validation strategy is demonstrated by its use for the ST Integrated Video Telephone described earlier. As the processors of the IVT are an integral part of the entire system, the most important aspect to verify is the function in the system. The validation of the each processor functionality was done as shown in Figure 5.14.

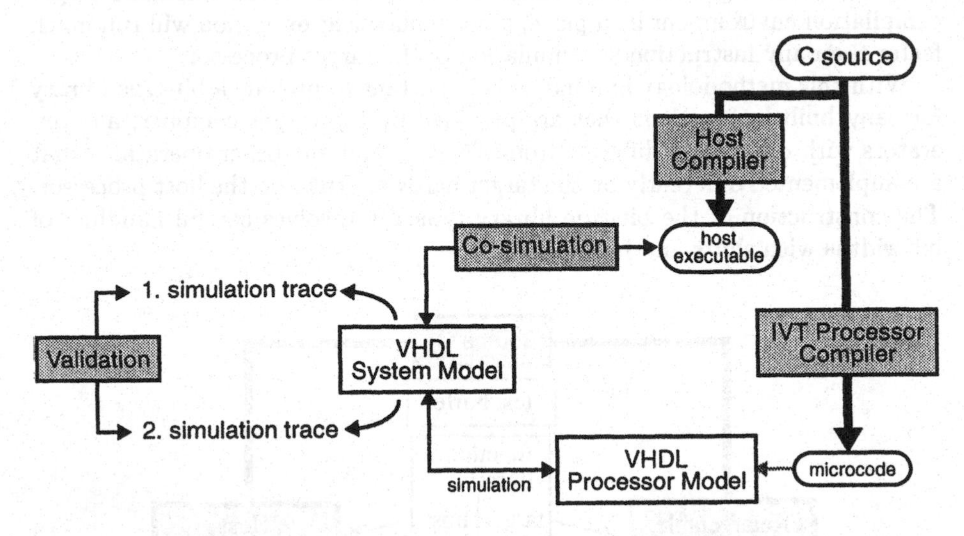


**Figure 5.14** System simulation-based validation strategy.

The key to the methodology is the presence of a co-simulation approach which allows the simulation of C code on the behavioral level integrated with a VHDL model of the system [134, 135]. Individual simulations of the system replacing the co-simulation model with a VHDL model of the processor generates two simulation traces which can be compared. The simulation methodology validates both the processor compiler and the VHDL processor model, given a test bench which thoroughly exercises the C application code. Notice that the methodology is similar to the previous host-based strategy since the two principal elements are present: the host compiler and the target compiler.

### *5.6.2 Source-Level Debugging*

A compilation path to the host workstation or PC allows standard source-level debugging tools to be used. An example public domain debugger is gdb, the GNU source-level debugger distributed by the Free Software Foundation [140], which has many user interfaces (e.g. xxgdb, Emacs, ddd). As the host compilation path is naturally the faster path and the tools are immediately available, functional validation and debugging of the source algorithms should ideally be done at this level.

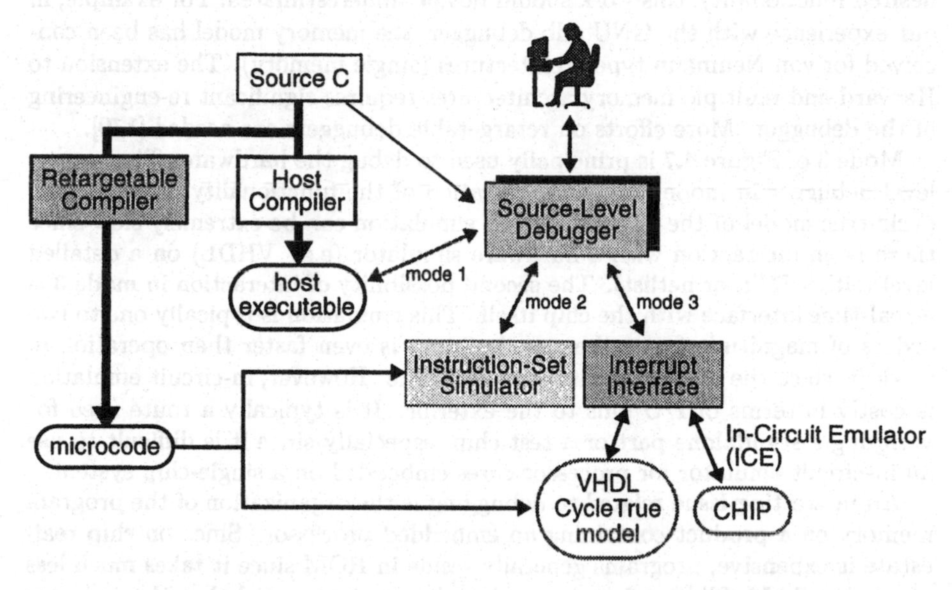


**Figure 5.15** Debugging an embedded processor using different modes.

Figure 5.15 illustrates a methodology whereby the host source-level debugging interface is reused in different modes. The interaction labelled mode 1 is the familiar host debugging mode; the interaction labelled mode 2 is the mode using the instruction-set simulator; and the interaction labelled mode 3 is the mode which interfaces with a cycle-true model of the processor or the chip itself through an in-circuit emulator (ICE) interface. Mode 2 of debugging is principally used for verifying the retargetable compiler. It is basically a debugging means for the validation strategy shown in Figure 5.13. Of course, bugs can occur in the instruction-set simulator or the bit-true library as well. In mode 2 of the debugging scenario, depending on the type of instruction-set simulator, the boundary of whether the debugging functions belong to the instruction-set simulator or the source-level debugger can become blurred. For example, general purpose DSPs and MCUs are usually distributed with an instruction-set simulator rich in debugging functions. In this case, the source-level debugger needs simply to provide a user interface to those functions. One method is to detach a user interface (e.g. xxgdb, ddd, Emacs GUD) from a debugger (e.g. gdb, dbx) and attach the user interface directly to the instruction-set simulator. This can often provide a simple route to basic debugging capabilities.

Some debugging functions can be quite complex, making it advantageous to interface the debugger to the instruction-set simulator. Depending on the desired functionality, this work should not be underestimated. For example, in our experience with the GNU gdb debugger, the memory model has been conceived for von Neumann type architectures (single memory). The extension to Harvard and multiple memory architectures requires significant re-engineering of the debugger. More efforts on retargetable debuggers are needed [170].

Mode 3 of Figure 4.7 is principally used to debug the hardware. The source-level debugger in mode 3 allows verification of the functionality of the VHDL cycle-true model of the processor. This simulation can be extremely slow since there is an interaction with a hardware simulator (e.g. VHDL) on a detailed level (either RTL or netlist). The second possibility of interaction in mode 3 is a real-time interface with the chip itself. This emulation is typically one to two orders of magnitude faster than mode 2 and is even faster than operation in mode 1, since the chip is operating in real-time. However, in-circuit emulation is costly in terms of I/O pins to the exterior. It is typically a route used for verifying a standalone part or a test chip, especially since it is difficult to use an in-circuit emulator for processor cores embedded on a single-chip system.

An interesting issue related to debugging is the organization of the program memory on a product containing an embedded processor. Since on-chip real-estate is expensive, programs generally reside in ROM since it takes much less area than RAM. This makes sense, since it is not expected that the program need to be changed in an embedded system. However, in test chips it is often the case that things go wrong; and therefore, designers may want to change the contents of the embedded program. One approach is a design to balance this trade-off. It is possible for a designer to enhance the program ROM with a small space of downloadable program RAM. Dedicated instructions may be included in the processor as a provision to execute this *patch* code when debugging the test chip. This method is beginning to show more frequently as a popular way to debug final systems. On the other hand, the decision on the sizes of ROM and RAM remains a difficult guessing game.

Although the mechanics of source-level debugging are well understood, practical implementation can still be a difficulty. The techniques of compilation for embedded processors produce optimized code which is specialized for the architecture. The symbolic debugging of optimized code is an enormous and complex problem [169]. Even the simplest of compilation tasks can produce a tangling mess for the debugging symbols. For example, a register assignment algorithm which assigns the same variable to ten different registers at different points of its lifetime means that the object code must carry ten times the symbolic information than previously. Furthermore, if an aggressive scheduling algorithm is used, operations can move to different points in the code, including into and out of loops. As well, unreachable code optimizations can make code disappear completely. In addition, older debug formats (e.g. COFF [171]) have no way of dealing with optimized code, which has given rise to company-specific variants (e.g. XCOFF, ECOFF, EXCOFF). Some efforts are underway for debugging standards which address some of these concerns (e.g. ELF, DWARF [172, 140]). However, for today's embedded processors, the debugging problem for optimized code remains largely unsolved, and users take what they can get. That is to say, source-level debugging can work well for some types of optimized code, and not so well for others.

### *5.6.3 Architecture and Algorithm Exploration*

While a retargetable compiler represents the principal implementation technology for the design of embedded systems, design exploration tools are also of great use for the development of a system. For example, a compiler does not provide many metrics to the designer of the processor to measure how well the instruction-set has been designed. Ideally, a maximum of feedback indicating the static and dynamic usage of instruction codes is the type of information a designer would like to see as he designs the processor, and most certainly, as the processor evolves and is reused.

When a product finally hits its target window and is on the market, a designer often ponders on how well he made the design decisions. In many cases, this is further motivated by a decision to redesign a low-cost version of the product which is already on the market. The designer is then faced with the challenge of better fitting the programmable processor to the existing application code. In any case, the design evolution of an instruction-set processor is a factor which cannot be overlooked. Designers need tools which provide feedback on how well a piece of application code fits on an architecture, such as statistics on resource usage and suggestions for changes to the program to improve its fit onto the architecture.

Figure 5.16 shows two examples of tools that fall into this category [117]. These are design aids which allow the exploration of the relationship between the instruction-set and the corresponding application code of custom embedded processors. After analyzing the instruction set and code, the designer can then use the set of editing functions to adjust the instruction set to the ap-

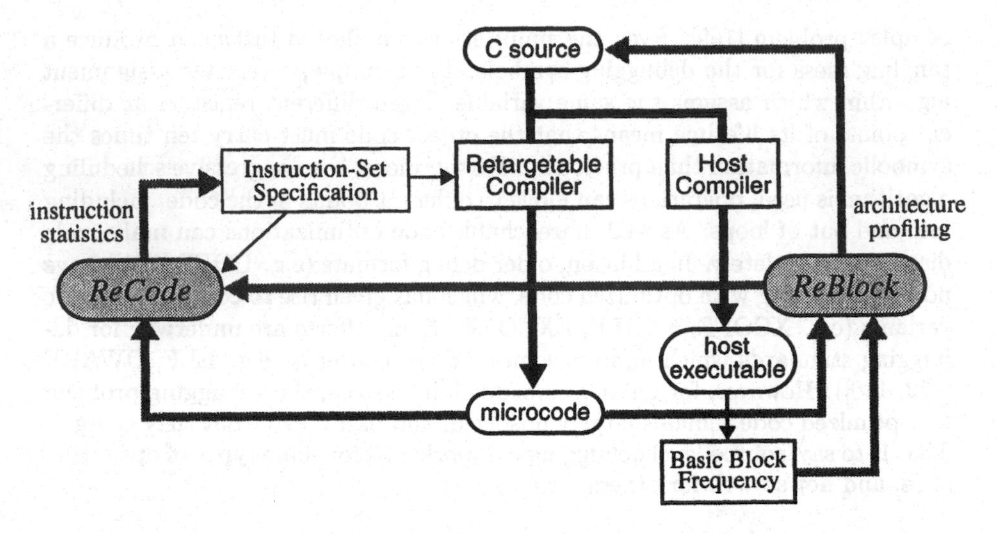


**Figure 5.16** ReCode and ReBlock: tools for algorithm and architecture exploration.

plication code. The designer can make gains by removing unused hardware, relieving bottlenecks in the hardware, and removing unused instruction codes. The tools work together with the instruction set specification used by a retargetable compiler and can automatically regenerate the specification changes.

The tools also support dynamic analysis. A standard profiler function is provided requiring only the retargetable compiler and a host compiler for frequency information. Links are automatically performed between the microcode and the basic block execution on the host. Functions are available which estimate real-time performance based on the host execution. Furthermore, the same information is reused for dynamic analysis of the instruction usage. This can be used for exploring changes in the architecture.

Other approaches include those of Huang [173] and Holmer [174] which determine the most useful instructions from a base formula including parameters such as the execution profiles of compiled benchmarks. A model of the architecture data-path and the performance metrics allows the system to suggest good instructions to keep and remove instructions of marginal benefit.

A similar approach has been proposed for the implementation of ASIP architectures in the PEAS system [175, 176]. This approach attempts to minimize both software and hardware costs based on the compilation of source algorithms. A set of primitive operations allow compilation by the GNU gcc compiler, while a basic and extended set of operations may be included, based on the performance measures. For an objective area and power constraint, the performance is maximized by the system. However, optimization of the architecture below the primitive set of operations is not possible.

The high expense of program memory for embedded processors has given rise to efforts of program width reduction such as instruction-word encoding. Some have approached the problem using a technique which reduces the width of the final program memory by exploiting redundancy [177]. Assuming all the application code is available, the entire program memory is divided into columns. Each instruction column which can possibly be generated from another instruction column is eliminated. This can be done through small hardware modifications such as exchanging multiplexer control lines and adding small pieces of logic. Nevertheless, this is a last ditch optimization that is not likely to allow the compiler to add any new software.

The ASIP methodology opens new doors to both architecture exploration. Drawing the line between hardware and software is often a balancing game between flexibility and performance.

## 5.7 CONCLUSIONS

Instruction-set processors will continue to implement an increasing fraction of the functionality of VLSI systems because they are flexible, simplify design reuse, and allow concurrent engineering. Modern CAD techniques will make it increasingly possible to develop custom ASIPs for more applications, allowing higher cost/performance ratios in a broader range of markets. The biggest obstacle to the adoption of ASIPs has been the need to develop compilers and other software support tools. Emerging techniques are drastically improving the quality of ASIP compilers and the speed with which those compilers can be developed, giving strong impetus to the increased use of ASIPs in system-on-silicon. Continued improvements in compilers and the complete software support suite for ASIPs should continue to increases the architectural options available to system designers.

# 6 DESIGN SPECIFICATION AND VERIFICATION

J. Staunstrup

Computer Systems Section
Department of Information Technology
Technical University of Denmark
Lyngby, Denmark

## 6.1 INTRODUCTION

This chapter describes high-level design techniques for developing hardware or software and combinations of the two. The chapter gives both an overview of the key concepts found in a range existing languages and tools and a specific proposal for modeling the abstract behavior of a design. The aim of such high-level design techniques is to reduce the design time or effort by shifting as many decisions and analysises as possible from low-level to high-level models.

In a large design project it is necessary to keep track of a large amount of detail; so much that no single person will ever have a complete grasp of all aspects at any one time. This is not a new situation and engineers have always been capable of building models that gives them sufficient overview. A circuit diagram is a good example, this is a model that provides an abstraction



*J. Staunstrup and W. Wolf (eds.), Hardware/Software Co-Design: Principles and Practice, 193-233.*

where certain aspects are ignored like the physical placement of components and wires. Instead the model allows the designer to focus on combining components yielding the intended computation (functionality). Likewise, a computer program is an abstract model of the computations done by the physical computer. The ever increasing complexity of the applications of electronics puts an increasing demand on the models (abstractions) needed to design and realize these applications. This has pushed the models used for circuit designs towards higher and higher levels and today it is not uncommon to model the computation of a circuit at a level of abstraction that is similar to the one found in software. Design of software and hardware is therefore becoming very similar and they certainly share one common difficulty: bridging the gap between the ***static description*** of a computation commonly found in high-level models and the ***dynamic behavior*** found in the executions evolving in the realization. Practice has shown that it can be very difficult to keep the description and the intended behavior consistent. One problem is that the nature of the two are very different, one is static and the other is dynamic. Another difficulty is to grasp and carefully consider the enormous space of possibilities that the realization may encounter. This chapter focuses on models that supports the designer in reducing both of these difficulties, i.e.,

- reduce the differences between the static description and the dynamic behavior,
- allow the designer to do exhaustive checking covering the entire space of possible behaviors.

The first (reducing the differences between the static description and the dynamic behavior) was originated in the pioneer work on **structured programming** [178]. The second is often referred to as **formal methods.**

### 6.1.1 Design

The generic term "**design**" is used throughout this chapter to cover computations realized in software or hardware and combinations of the two. As pointed out above, there is an increasing similarity between hardware and software development. Today, neither hardware nor software are unique easily identifiable areas, they both cover a whole range of technologies for realizing (executing) computations such as: interpreted code, compiled code, hardwired code, programmable hardware, microcode, and ASICs.

To specify a design, one needs an abstract model that is general enough to cover both hardware and software components, and that enables a designer to describe his design without making a commitment as to what should be realized as software and what should become hardware. This chapter motivates

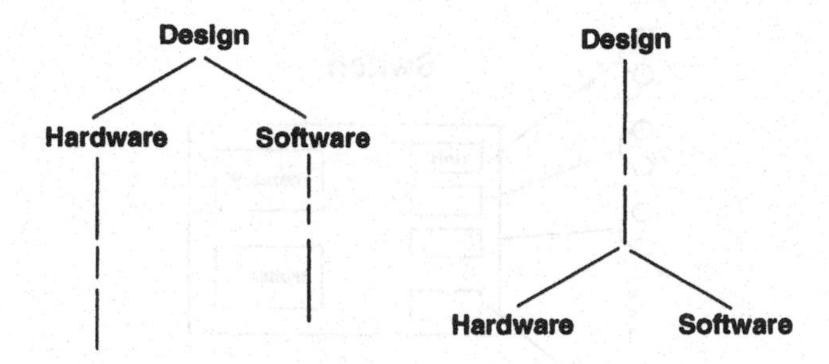


**Figure 6.1** Early vs. late binding of hardware and software.

and introduces such a model of a computation that can be realized in either hardware, software, or a mixture of the two.

Figure 6.1a) shows a typical situation where a product is designed by separating the hardware and software development at a very early phase of the design. If instead an abstract model covering both hardware and software is used, one can do as suggested in figure 6.1b), where a large part of the design is done without binding the components to a particular kind of realization. Co-design is a commonly used term for approaches aiming towards such a late binding of components to technologies that can be both hardware and software.

### *6.1.2 Co-design*

Co-design has applications in a wide variety of specialized electronic products, e.g., communications equipment, controllers, and instruments. These are often given the common label **embedded systems.** In this chapter a telephone switch is used to illustrate the concepts and models used for designing such embedded systems. The example is a simplified version of the TigerSwitch [179]. The focus is on the digital aspects of the switch such asd establishing a connection and transmitting data in digital form. The analog parts are not considered, e.g., conversion of sound to and from a digital form. Fig. 6.2 gives an overview of the **simplified telephone switch** that can handle a number of phones. The switch has a separate unit for handling each phone, this unit has physical parts like the connector for plugging in the telephone and abstract parts for doing the necessary computation. The switch has an arbiter that reserves the buffers needed for communication and resolves conflicts like two phones calling the same receiver. Finally, the memory contains the buffers needed for exchanging data.

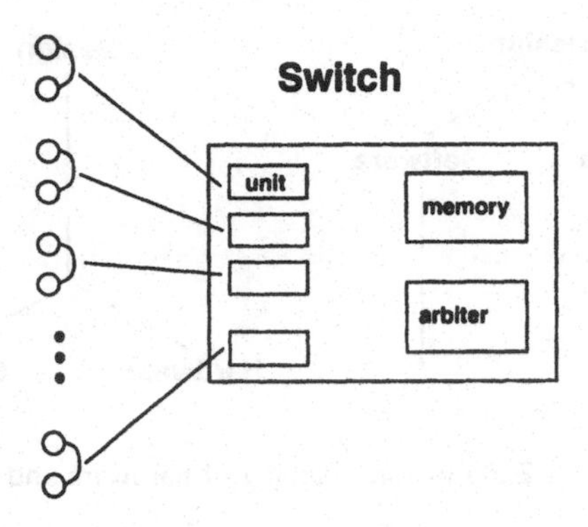


**Figure 6.2** Overview of the telephone exchange.

### *6.1.3 The Co-design Computational Model*

The chapter presents a number of key concepts for specifying and verifying high-level designs. These are introduced in a generic form that is independent of their appearance in the currently available technologies, notations, and tools. To illustrate the concepts and make it possible to describe examples, like the telephone switch, a specific computational model is introduced. This model has similarities with commercially supported languages like VHDL, C, Java, and Verilog, but whereas all these are primarily intended and used for either software or hardware, the **co-design computational model** introduced in this chapter is not biased towards a particular technology; it aims at specifying computations at a level where it can be realized in a range of hardware and software technologies. The *co-design computational model* focus on concurrency, modules, and communication. Although the computations are the most important aspect, it is necessary to have a notation for describing these, the syntactical aspects of this notation are of less importance and they are introduced through examples only. A more systematic presentation of the notation is given in [52].

The choice of a model for co-design (or any other complex task) is a delicate balance between abstract and concrete. If the model is too concrete, the designer is constrained by low level decisions even in the early phases of the design. On the other hand, if the model is too abstract, it may later become difficult to make an efficient realization. When doing co-design, the model should not favor a particular kind of realization, for example, a software realization.

There is a close relationship between the model and the notation/language/tools used for describing a design. For co-design one needs a model and a corresponding language that is abstract enough to allow descriptions of com-

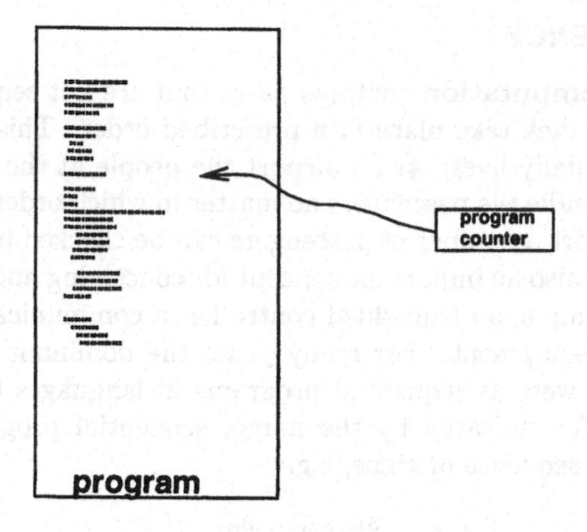


**Figure 6.3** Sequential program

putations in a range of technologies. However, almost all existing models and languages aim at either describing software (programming languages) or circuits (hardware description languages). Both kinds of language aim at describing a *computation*, but they are usually based on very different concepts. Almost all programming languages are based on the Von-Neumann model, where a computation is described by a *sequence* of instructions operating on data stored in a memory, see figure 6.3. An instruction counter keeps track of which instruction in the sequence to execute next. However, in most circuits there are many parts operating in parallel; so sequencing is at best an extra obstacle, and in some cases it might even preclude an efficient realization. For a hardware designer, the spatial relationship between different components is a primary concern; therefore, many hardware description languages emphasize structural descriptions where the physical organization of parts is explicitly specified.

Both programming and hardware description languages are used for describing computations, and this is also the main purpose of a co-design notation. However, the computation should neither be described sequentially nor structurally, because that would bias the realization. Section 6.2 shows that concurrency is a fundamental concept for describing an design at the abstract level. It is, therefore, obvious to base a co-design language on such a model. It has been demonstrated that a computation modeled with fine grain parallelism can be used both to develop software [180, 181] and to synthesize circuits [182, 52].

## 6.2 CONCURRENCY

A **concurrent computation** contains parts that are not sequenced, i.e., it is not required that they take place in a prescribed order. This is a well known concept from our daily lives. At an airport the people at the check-in counter are prepared to handle the passengers no matter in which order they arrive; and at any larger airport a number of passengers can be checked in simultaneously.

Concurrency is also an important concept for conceiving and describing computations, for example, an embedded controller, a communication device, or a piece of medical equipment. For many years the dominant ways of describing computations were as sequential programs in languages like, FORTRAN, C, and Pascal. As indicated by the name, sequential programs, describe a computation as a sequence of steps, e.g.:

$$s_1;\ s_2;\ \ldots s_n;$$

Although all programming languages have mechanisms for structuring a computation into loops, conditional parts, procedures etc. these constructs all define a sequencing of the computation. The first iteration of a loop happens before the second, a procedure call is completed before proceeding etc. Therefore, the computations described in these languages are sequential. This fits very well with the predominant computer architectures that are constructed for executing a sequence of instructions. Even todays high-end processors with caches and advanced execution units are fundamentally oriented towards executing a single sequence of instructions. However, this is not an adequate model of the embedded computations where the order of external events and computational steps cannot be prescribed prior to the computation. A concurrent model is a more appropriate model, because it explicitly opens for the possibility of not prescribing the order in which its steps are performed, e.g.:

$$s_1 \parallel s_2 \parallel \ldots s_n$$

This describes a computation where $s_1$, $s_2 \ldots s_n$ are concurrent, i.e., no order has been prescribed.

The concurrency abstraction covers a number of different situations, for example:

**nondeterminacy** The order of the steps of a computation are not known a priori, for example, where the arrival time of external events is not known.

**simultaneity** Steps of a computation happen simultaneously, for example, where external events must happen at the same time.

**multiprocessing** Steps of a computation are done in parallel on multiple distinct physical processors to speed up a computation.

**structuring** Steps of a computation are separated into different components, for example, handling of different external devices.

Note that an abstract concurrent model of a computation, for example, the airline passengers arriving at a check-in counter does not preclude that the actual computation is sequential, because only a single person at a time can be handled at the counter. However, since it cannot be known a priori in which order the passengers arrive, a concurrent model provides a way of abstracting away this detail in order to get a simple description of the computation. At other times the concurrency abstraction is used for describing a computation where the steps are done simultaneously, as for example, an airport with several check-in counters which enhances the performance.

### *6.2.1 Components*

An independent part of a computation done concurrently with other parts is often called a **process**. This is the term used in many programming languages and operating systems, for example, VHDL, OCCAM, and UNIX. One also finds terms such as **thread** (Java), **lightweight process**, and stream. In the ***codesign computational model***, used in this chapter, the term component is used, and the notation shown below is used for describing such a concurrent component.

*COMPONENT name*
  *declaration*
  *computation*

This describes a component called *name* containing computations that are concurrent with other computations of the design. The *declaration* describes local quantities of the component used in its *computation*.

The telephone switch mentioned in the introduction is a good example of a design exploiting concurrency. Each phone is a distinct physical object and the people using the phone system are independent. Hence, the switch must be capable of handling simultaneous actions from different phones. This is done by having a separate (concurrent) component called a *unit* for each physical phone

*COMPONENT unit*
  *p: phone_connector*
  *...(* computation *)*
*END unit*

A unit can set up and break down connections. It detects when the receiver is lifted of hook and transforms a sequence of touch-tone signals into an identification of the receiving unit. The unit also performs the steps needed for disconnecting two units when the caller puts the receiver back on hook.

Once a number has been dialed on the calling phone, the associated unit tries to set up a connection to the receiving unit. This requires that the receiver is not already engaged in another call and that there is free capacity, e.g., empty buffers for the data transmission in the switch. This is handled by the arbiter and in case the connection cannot be established then the caller must get a busy signal, otherwise buffers are reserved for the communication. The component *connect* handles this.

*COMPONENT connect*

The transmission of data (sound) is handled by the component, *transfer*, that transmits digitized sound to and from buffers in the common memory. When a connection between two phones has been established data (sound) is constantly transmitted in both directions.

*COMPONENT transfer*

The switch has a number of distinct units to handle the physically separated telephones, this illustrates handling of simultaneous events where it is the behavior of the environment that guides the designer into using a number of concurrent units. The separation of the setup and transfer into the components *connect* and *transfer* is an example of using concurrency for structuring a design. Finally, the component *transfer* provides an example of multiprocessing to enhance the performance, in this case by making it possible to overlap sending and receiving of data. **End of example**

### 6.2.2 Nondeterminism

This section shows the importance of **nondeterminism** in high-level design where it is essential to abstract away all details that are not essential (otherwise there would not be any point in making the high-level design). The difficulty is to find the right balance where the high-level design contains the essential overall decisions while leaving the low-level concerns open for later stages of the design process. One typical separation is to use abstract data types in the high-level design thus postponing decisions about representations, word-width, buffer-sizes etc. for later. The precise ordering of events is another detail that is often better left unspecified in a high-level design. Therefore, a high-level design is often nondeterministic. Like concurrency, considered in the previous section, nondeterminism may take many forms

**external nondeterminism** The order of external events cannot be specified.

**internal nondeterminism** The order of events should not be specified.

**randomization** The order of events must be unpredictable.

At a first glance, nondeterminism may appear to be the same as concurrency, both focus on *not* ordering events. However, there is an important difference, because nondeterministic events *do* happen in some order; but *which one is not prescribed.* With concurrent events it is acceptable that they happen simultaneously or partly overlapping.

Most programming and hardware description languages do not have explicit constructs for describing nondeterministic choice between different steps of a computation. In the notation used in these notes, it is simply specified by giving a list of alternative steps.

$$s_1 \mid s_2 \mid \ldots s_n$$

This specifies a computation where one of the $n$ alternative actions is done. This is very similar to the guarded commands proposed by E.W.D. Dijkstra [183] to handle nondeterminism.

The telephone switch illustrates the different forms of nondeterminism. While dialing a telephone number, it not known a priori (by the telephone) which digit is coming next, hence the unit collecting the digits must be prepared to accept any one of the 10 digits. This is an example of external nondeterminism. Note the difference to the concurrency caused by two distinct telephones on which numbers are being dialed at the same time. To specify a design with external nondeterministic choice between two buttons $A$ and $B$ one may write:

$\ll$ *button_A* $\rightarrow$ *action to handle A* $\gg$ |
$\ll$ *button_B* $\rightarrow$ *action to handle B* $\gg$

**End of example**

As mentioned several times already, a high-level specification abstracts away unnecessary details, for example, the exact choice of data representations. The general principle is that when several alternatives are acceptable, a specification should avoid stating which one is chosen. Otherwise, over-specification can prevent choosing a better (e.g., simpler, faster, or cheaper) implementation later in the design process. Often, the detailed ordering of the computational steps can be left unspecified.

The meeting scheduling problem and its solution are presented in the book *Parallel Program Design* by [180]. It is a nice illustration of how internal nondeterminism is used to obtain an elegant under-specified design. The problem is to find the earliest meeting time acceptable to every member of a group of people. Time is represented by an integer and is nonnegative. As a simple example, consider a group consisting of three people: $F, G$, and $H$. Associated with persons $F, G, H$ are functions $f, g, h$ (respectively), that map time to time; $f(t)$ is the earliest time at or after $t$ when person $F$ can meet. Hence, for any $t$, person $F$ can meet at time $f(t)$ and cannot meet at any time $u$ where $t \leq u < f(t)$. The meanings of $g$ and $h$ are analogous. For example:

| $t$ | 8 | 9 | 10 | 11 | 12 | 13 | 14 | 15 | 16 | 17 | 18 |
|---|---|---|---|---|---|---|---|---|---|---|---|
| $f(t)$ | 9 | 9 | 10 | 13 | 13 | 13 | 16 | 16 | 16 | 17 | 20 |

In this example, $F$ can meet at 9 or 10, but not at 11. The earliest $F$ can meet after 11 is 13. It is assumed that there is at least one possible meeting time for $F, G$, and $H$. Let $t$ denote an estimate of the meeting time, if $f(t) > t$ then $F$ cannot meet at $t$ and $f(t)$ indicates the earliest time after $t$ when $F$ can meet; hence the estimate of the meeting time should be updated. This is expressed more succinctly by the transition

$$\ll f(t) > t \rightarrow t := f(t) \gg$$

Similar transitions for $G$ and $H$ express their local updating of the estimate $t$. By combining the following three transitions, one obtains a computation that produces the earliest time at which $F, G$, and $H$ can meet (after or at the time given by the initial value of the state variable $t$):

$$\ll f(t) > t \rightarrow t := f(t) \gg \mid$$
$$\ll g(t) > t \rightarrow t := g(t) \gg \mid$$
$$\ll h(t) > t \rightarrow t := h(t) \gg$$

The state variable $t$ is repeatedly increased by one of the three transitions until $t$ gets a value such that

$$t = f(t) = g(t) = h(t)$$

which by definition of $f, g$, and $h$ is a possible meeting time. Note the nondeterministic nature of this computation: each person updates the meeting time independently of the others. However, no potential meeting time is missed by one person making an excessive update. It might happen that one person updates the meeting time in small increments, whereas one of the others could make a larger increment. There is no attempt to avoid such superfluous operations by sequencing the computation in a clever way. In many circuits, it would be less efficient to try to avoid these superfluous operations. This is not always the case, but then the optimization is a separate concern, left for a later refinement of the design. **End of example**

The third form of nondeterminism, called randomness above, is used for specifying computations where the ordering of steps vary from one computation to another, i.e., they appear in random order. An arbiter, such as the one used in the telephone switch, is a nice example. It is used to resolve a conflict where several clients such as the telephone units compete for a set of limited resources. In the telephone switch there is a limited number of buffers (communication channels) for transmitting data. The arbiter reserves buffers when a connection is established. In situations where the number of requests exceed the available resources, the arbiter must make a choice to honor the requests in some order. This order should not be fixed or predetermined, so nondeterminism is a good high-level abstraction of the arbiters choice.

Consider a simple arbiter that control the access to some common resource from two clients $A$ and $B$. Each client indicates that it wants to use the resource by issuing a request. The arbiter must not give priority to one of the two when resolving conflicting requests. At the very abstract level, the arbiter is specified as a simple nondeterministic computation as follows.

$$\ll \textit{request}_A \rightarrow \textit{grant access to } A \gg \mid$$
$$\ll \textit{request}_B \rightarrow \textit{grant access to } B \gg$$

The arbiter is discussed in further detail later in the chapter.

**End of example**

Randomization is also a powerful concept for designing efficient algorithms [184]. Hashing is a well-known example of a randomized algorithm that works very well on the average. More recently such randomized algorithms have been described for many time consuming tasks in for example cryptoanalysis and computational geometry.

### *6.2.3 Concurrency in Standard Languages*

This section gives an overview of concepts for describing concurrency and nondeterminism in a number of programming and hardware description languages in common use. Although these languages enable the designer to describe a computation as a collection of concurrent components, there are significant differences that influence the design style and ease of use.

**Concurrency in Programming Languages.** A number of concepts and constructs for concurrency have been proposed for a variety of programming languages. Among the first were PL360 [185] and Concurrent Pascal [186]. Later languages such as CSP/Occam [181] and ADA [187] have received considerable attention. Currently, Java [188] has been introduced and the concurrency concepts of Java, called multi-threading, are discussed several times in this chapter. It is maybe a little bit surprising that none of the dominant programming languages such as FORTRAN, Pascal, Lisp, C, C++, have explicit constructs for handling concurrency. The concurrent computations developed in these languages are based on non-standardized extensions such as real-time kernels and dialects such as Concurrent C.

Concurrent components in Java are called threads and the concurrency aspects of the language is often referred to as multi-threading. A single thread (concurrent component) is described by a **method** (the Java term for a procedure, function, or subroutine) that can then be executed concurrently with a number of other methods.

**Example: Scheduling a Meeting (continued).** Concurrency in Java can be illustrated by the algorithm for scheduling a meeting introduced above. Each person, $F, G$, and, $H$, are described by a method. For example, to describe F:

```
public void run() {
  do {
    s.t= f(s.t);
  } while (true);
}
```

As in the example above, each person $F, G$, and $H$ has a function (*f, g, h* respectively), that maps time to time; *f(t)* is the earliest time at or after $t$ when person $F$ can meet. The best current estimate of a possible meeting time is given by the variable $s.t$ which is accessible (shared) by all three threads $F, G$, and $H$. As above, the definition of the functions *f, g,* and *h* and termination are ignored at this point. The handling of shared variables, such as $s$ is discussed in section 6.3.

In Java, the multi-threading is realized by the standard class *Thread* in the Java library. This class offers a number of methods in addition to the *run* method which is the one providing the possibility of creating concurrent components. Among the methods offered in the threading class are ones for controlling the execution and priorities. It is, for example, possible to stop the execution of a thread temporarily by calling the methods *yield, sleep* or *wait.* This indicates that the concurrency concepts in Java are intended for multiprogramming, i.e., sharing a single processor between a number of concurrent activities. This is useful for handling concurrency needed to handle nondeterminacy or structuring (see page 198). However, the Java multi-threading as offered by the class *Thread* is not intended for multiprocessing or handling simultaneous operation of distinct hardware components.

Concurrency is a fundamental concept in hardware description language such as VHDL [189] and Verilog [190]. Syntactically there are several ways of expressing a concurrent computation, however, they are variations of the fundamental concept of a concurrent component (process) described as follows:

*name: PROCESS(parameters)*
  *declarations*
*BEGIN*
  *computation*
*end PROCESS name*

This describes a concurrent process called *name.* The computation is described by VHDL statements which are not discussed in further detail here. Several instances of a process may exist as illustrated by the units in the telephone switch. One of the strengths of VHDL is the ability to describe several "variations" of a process differing in their internal behavior but with the same appearance to the outside world.

External nondeterminism can be expresses in VHDL by having several processes handling each of the external events. However, it is not possible to express internal nondeterminism. This is a deliberate decision by the creators of the language, to make all computations reproducible, i.e., running a program twice on the same inputs will always yield exactly the same output. This simplifies testing significantly.

### *6.2.4 Synchronous and Asynchronous Computations*

Concurrent computations are often categorized as synchronous or asynchronous, for example, is VHDL sometimes referred to as synchronous language. This classification refers to the underlying computational model on which the language constructs are defined. As discussed above, a component (thread, process, stream, ...) is modeled as a sequence of separate steps. The exact nature of these steps is not important at this point, it depends on which level of abstraction one takes. At a low level, the steps corresponds to instructions of the underlying processor, and at a higher level, the steps corresponds to statements in a high-level language.

Synchronous and asynchronous models of concurrent computations differ in how steps from different components are combined:

**synchronous** A synchronous computation consists of a sequence of macro-steps, where each macro-step has a (micro)-step from each of the participating components.

**asynchronous** An asynchronous computation consists of a number of independent sequences consisting of steps from all the participating components.

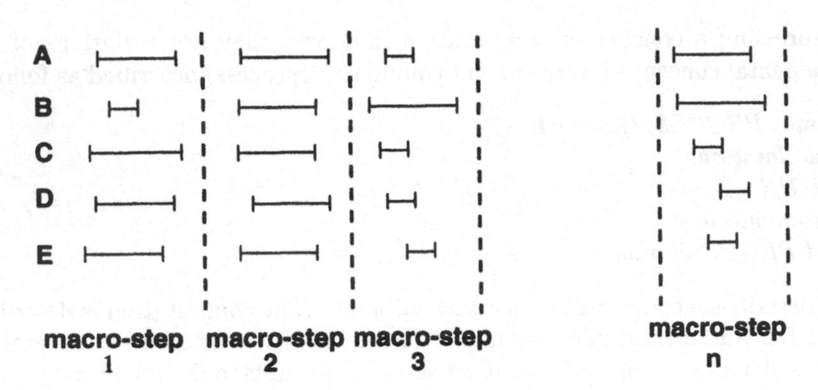


**Figure 6.4** Synchronous computation.

In both cases the concurrent computation is modeled by a sequence which might seem counter intuitive at a first glance, we will return to this point after a short elaboration on the differences between synchronous and asynchronous computations.

The key aspect of the synchronous model is the coupling of the steps from all participating components. Each macro-step has a step from *each* component. This means that they all proceed in synchrony, if one component takes $n$ steps to perform a sub-task, any other components will also have taken exactly $n$ steps in the meantime, see Figure 6.4.

It is the lock-step nature of the computation that is the key aspects of a synchronous model of a computation. This makes it possible for one component to predict the exact progress of other components. It is not necessary that all steps take exactly the same amount of time. Some components might complete their (micro-)step very quickly while others take longer. However, no component is allowed to proceed to the next step before all others have completed their current (micro-)step.

The synchronous model fits well with the behavior of traditional clocked circuitry where a global common clock signal is used to control the progress from one macro-step to the next. It is, however, not an inherent part of the synchronous model that all macro-steps have the same physical duration. This is just a convenient mechanism that has been used with great success at the circuit level. However, at higher levels, one might, for example, use a barrier synchronization mechanism to control the simultaneous progress of all components.

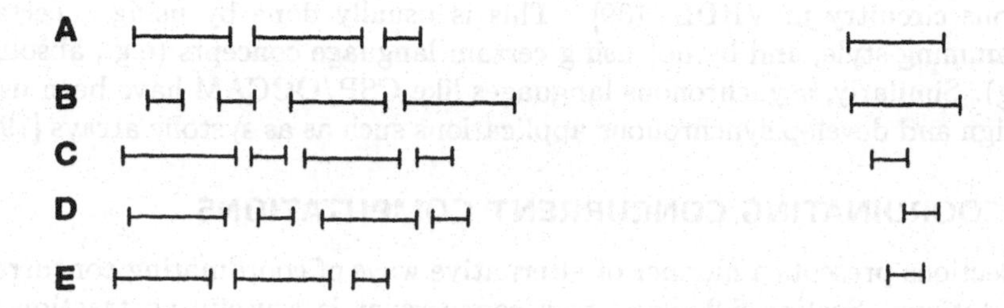


**Figure 6.5** Asynchronous computation.

An asynchronous model describes a computation where all components proceed independently and, therefore, it has a separate sequence for each component, see Figure 6.5.

There is no causal or timing relations between the steps from different sequences, and it is therefore not possible in one to component to rely on when other component reach particular steps of their computation, the components are asynchronous.

The asynchronous models fits well with the behavior of a distributed system with parts that are geographically spread. Although the majority of the computation of each component is done independently of others there is a need to occasionally coordinate them, e.g., to exchange data. Mechanisms for handling this are described in section 6.3. However, the asynchronous model is also applicable at the circuit level where it is used for designing self-timed circuits [191, 192].

### *6.2.5 Classification of High-Level Languages*

Both the synchronous and the asynchronous model has been used as a foundation for high-level design/programming languages. Most programming languages fall in the asynchronous category, e.g., Occam and Java. However, the French school of **synchronous languages** such as ESTEREL, LUSTRE and SIGNAL [193] are notable exceptions. VHDL is also an example of language based on the synchronous model. On top of the fundamentally synchronous model VHDL a number of additional concepts are found in VHDL such as exact timing.

One should not interpret this classification very dogmatically; first of all, it should not be interpreted as an evaluation in the sense that synchronous languages are good and asynchronous bad (or conversely). Furthermore, both kinds of languages are quite flexible, it is for example possible to design asynchronous circuitry in VHDL [189]. This is usually done by using a certain programming style, and by not using certain language concepts (e.g., absolute timing). Similarly, asynchronous languages like CSP/OCCAM have been used to design and develop synchronous applications such as as systolic arrays [194].

## 6.3 COORDINATING CONCURRENT COMPUTATIONS

This sections presents a number of alternative ways of coordinating concurrent computations. Section 6.2 shows how concurrency is a useful abstraction for specifying the behavior of a design. The concurrency makes it possible to design computations with independent (sub)components. Although the components are largely independent they must occasionally interact, e.g., to communicate or to synchronize their access to a common resource. For example, in the telephone switch the units communicate to send and receive (sound) data; the units must also synchronize their use of the limited number of buffers available in the memory.

**Coordination** is used here to encompass mechanisms that allows concurrent components to synchronize, communicate, or compete. Elsewhere one finds terms such as **synchronization primitives, communication mechanisms**, and **process interaction** to cover the same concepts. However, coordination is used here because it is a more general term covering all the aspects of interaction between concurrent computations.

Many concepts have been proposed to handle the coordination of concurrent computations, surveys can be found in [195]. This sections describes a few key concepts that are common to almost all the proposed coordination mechanisms. A simple version of the arbiter in the telephone switch is used to illustrate and compare the different approaches. The purpose of the arbiter is to administer the use of memory buffers which are a common and limited resource. Before using a memory buffer, permission is requested from the arbiter, and after use of the buffer, it is explicitly released. Therefore, the computations of all units in the telephone exchange follow this pattern:

```
...
request
   use buffer
release
...
```

The steps *request* and *release* involve the arbiter and provides the needed coordination with other units to ensure that memory buffers are used correctly, for example, to avoid that the same buffer is allocated to two competing units. Furthermore, the arbiter must control the scheduling of competing components, i.e., the order in which they get to use the buffers. It is the designer who should control the order in which buffers are allocated to waiting units, this is called medium term scheduling.

One of the first mechanisms introduced for coordinating concurrent processes was the semaphore concept [196]. A **semaphore**, $s$, has two atomic operations *P(s)* and *V(s)*. The **atomicity** ensures that if several processes attempt to operate on the semaphore concurrently, these operations are done in some sequence, one at a time. The algorithm determining the order is called **short-term scheduling**. This algorithm is fixed in the implementation of the semaphore. The semaphore is initialized with some integer value, $k > 1$. After initialization, the first $k$ P-operations may complete immediately. However, the $k + 1$st ($k + 2$nd) cannot be completed until a V-operation is done by some other process. When several processes are waiting to complete a P-operation, each V-operation enables one of them to continue. During the computation, the number of completed P-operations can never exceed the number of V-operation by more than $k$. If on the other hand the difference between the number of completed P- and V-operations is less than $k$, the next P-operation will complete immediately. The special case of a semaphore where $k = 1$ is called a binary semaphore. When using a binary semaphore the number of P-operations can never exceed the number of V-operations by more than 1.

**Example: Simple Arbiter Using Semaphores.** It is possible to make a very simple arbiter using a binary semaphore where the number of P-operations can never exceed the number of V-operations by more than 1. Therefore, if a request is realized with a P-operation and a release with a V-operation it is ensured that at most one process has completed a request without having done the corresponding release

...
***P(s) - request***
  ***use buffer***
***V(s) - release***
...

This realization ensures that the buffer is allocated to at most one process at a time. However, in this realization the designer is not able to control the medium-term scheduling. Consider the situation shown in figure 6.6. A number of processes, $A, B, C, D, E, \ldots$ make requests very close to each other. $A$ is allocated the buffer first and before releasing it, both $B$, $C$, $D$ and $E$ also make requests. At the point where $A$ releases the buffer, it is the short-term scheduling algorithm that determines to which unit the buffer is allocated next. Since this algorithm is a fixed part of the semaphore implementation it is not possible for the designer to provide an alternative algorithm, hence, it is not

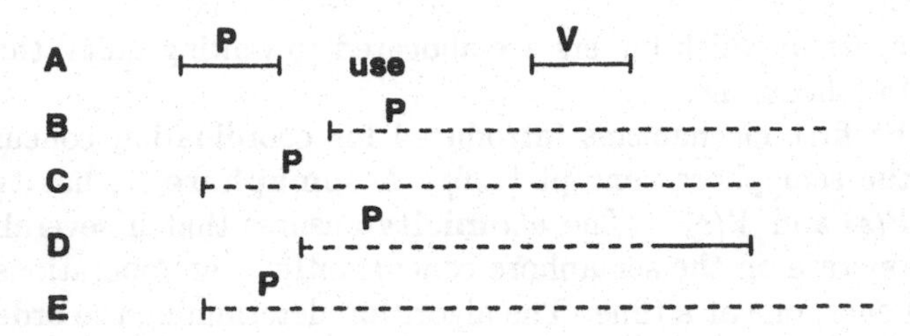


Figure 6.6 Scheduling of competing request.

possible to control the medium-term scheduling. All the arbiters shown below makes it possible for the designer to decide which **medium-term scheduling** should be used.

### *6.3.1 Classification*

This section presents three orthogonal dimensions for classifying coordination mechanisms. The classification is used for analyzing and describing some principal characteristics of the many existing coordination mechanisms while avoiding syntactical issues of less importance. The three dimensions are:

- shared state *versus* messages;
- open *versus* closed operations; and
- blocking *versus* nonblocking.

Each of the dimensions is explained in more detail below. Almost all of the coordination paradigms found in systems, programming, and hardware description languages can be classified along these three dimensions. For example, Verilog [190] has shared state, nonblocking and open operations whereas LOTOS [197] uses messages and fixed blocking operations.

The three orthogonal characterizations are shown in Fig. 6.7

### *6.3.2 Shared State Versus Messages*

Concurrent processes may coordinate their computation by sharing part of their state space. For example, by having one a more shared variables that they can all read and write. Alternatively, they could coordinate their activities by exchanging messages (values) in which case their states spaces are separated completely. To illustrate the two alternatives, consider two concurrent processes, $P$ and $Q$, sharing a variable called *data*.

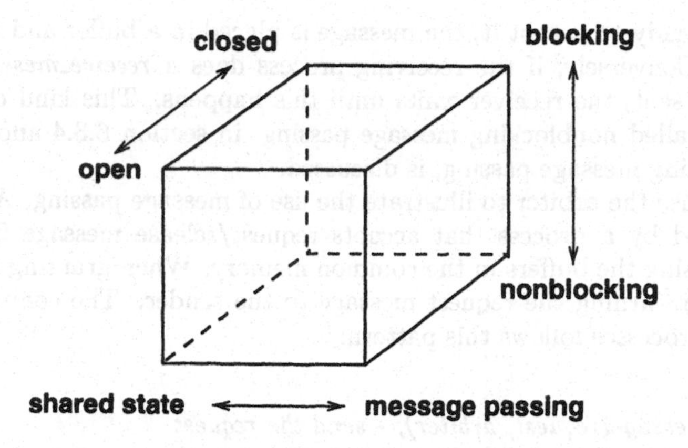


**Figure 6.7** Characterization of coordination paradigms.

```
        PROCESS P:                              PROCESS Q:
          data:= F(...)
          ...                                     ldots
                                                  v:= data
```

The process $P$ produces some data, illustrated by the function $F$, and assigns it to the shared variable *data*. The other process $Q$ reads the data and copies it to $v$ for further processing.

The same is easily achieved by message passing.

```
PROCESS P:                              PROCESS Q:
  send_message(F(...))
  ...                                     ...
                                          receive_message(v)
```

Here the data is again produced with the function $F$ and then communicated as a message that is received by the process $Q$. The important distinction is that in the first case (using the shared variable *data*) there is no restriction on the use of this variable by either process, hence $Q$ may assign values to it, and $P$ may also assign a new value before $Q$ has finished processing the current value. When using message passing, the processes are protected from each other, and they may only interact through the message passing primitives. Many different variations of message passing have been suggested, but fundamentally it is based on only two operations: *send_message* and *receive_message*. As indicated by the names the former is used to send a message that can be received with the latter. If the sending process sends the message before the receiver is ready to accept it, the message is placed in a buffer and the sender proceeds. Conversely, if the receiving process does a *receive_message* before it has been sent, the receiver waits until this happens. This kind of message passing is called nonblocking message passing, in section 6.3.4 another kind, called blocking message passing, is discussed.

We will use the arbiter to illustrate the use of message passing. Arbitration is performed by a process that accepts request/release message from client processes using the buffers in the common memory. When granting access this is done by returning the request message to the sender. The computation of the client processes follows this pattern:

```
client: ...
  send_message(request, arbiter); - send the request
  wait_message(mes, arbiter); - wait for the grant
    use buffer;
  send_message(release, arbiter); - send the release
  ldots
```

A request by the client is done by sending a message containing the request and waiting for the arbiter to return this message. When the client has finished its use of the resource it sends a release message to the arbiter.

The arbiter must always be able to accept both request and release messages. If the resource is in use the arbiter stores requests in a local queue. The ordering of this queue determines the medium-term scheduling, it is just assumed that the following operations are available on the queue:

- *insert* a new element in the queue;
- *remove* the first element of the queue; and
- *empty* which is true if there are no elements in the queue.

There are a number of different scheduling algorithms and these can be based on many parameters such as waiting times, priorities, and resource utilization. We will not go into further details in these notes.

The arbiter gives a grant to a client by returning the request-message as follows *send_message(r, r.sender)*. The arbiter repeats the following loop forever

```
arbiter: PROCESS
  wait_message(r, any process);
  IF r=request THEN
    IF free THEN free:= false; send_message(r, r.sender) ELSE insert(r, q)
  IF r=release THEN
    IF ¬empty(q) THEN r:= remove(q); send_message(r, r.sender)
```

*ELSE free:= true*

**End of example**

Message passing is found in a number of widely available packages and languages. It is also supported by most commercial multiprogramming/real-time kernels. The first widely available and published implementation was the multiprogramming kernel for the RC-4000 [198]. It was partly written in a dialect of ALGOL with message-passing primitives. It is a concept that has been reinvented numerous times leading to a large number of variations of the fundamental concept illustrated in this section.

### *6.3.3 Open Versus Closed Operations*

The coordination mechanism is embedded in the notation for specifying the computation. This is often done by a number of predefined operations, e.g., ***send_message*** and ***receive_message***, which are then the ***only*** way of coordinating the concurrent processes. Alternatively, the set of operations can be left open to be specified by the designer (programmer). The monitor concept [199, 200] is a good illustration of this approach. A monitor is defined by the designer, and it provides a number of operations (procedure calls) for coordinating the concurrent processes.

To illustrate the monitor concept, yet another version of the arbiter is shown below. As it is the case with message passing, there are many variants of the monitor concept, the example here is based on monitors as they are found in the language Java.

A **monitor** consists of a datastructure (set of variables) and a number of operations (procedures), in Java these are called methods. The datastructure can only be accessed through the procedures, so direct access is not possible. Execution of the operations is atomic (critical regions) so access to the monitors datastructure that is a shared variable ***can only be done*** in an indivisible operation. In Java this is specified by prefixing the method (procedure) heading with the keyword "synchronized". The datastructure of the arbiter consists of a boolean indicating whether the resource is free and a queue of waiting processes in case the resource is not free.

The clients using the buffers in the common memory follow this pattern:

***client:*** . . .
    ***request;***
        ***use buffer;***
    ***release;***
    . . .

The arbiter gives a grant to a client by returning from the request procedure. The arbiter monitor is shown below.

```
arbiter: MONITOR
  SYNCHRONIZED PROCEDURE request
    WHILE NOT free DO wait();
    free:= false;

  SYNCHRONIZED PROCEDURE release
    free:= true;
    notify();
```

The operations "wait" and "notify" is used to temporarily stop an indivisible operation and to later restart it. When a "notify" is executed, one of the waiting components is resumed so it can attempt to regain indivisible access.

**End of example**

### 6.3.4 *Blocking Versus Nonblocking Operations*

Concurrent processes are independent which means that they do most of their computations isolated from the other processes, the exceptions are those steps involving coordination with other processes. The coordination itself can be done in two very different ways: blocking or nonblocking. Blocking means that the coordination steps in all the involved processes is viewed as a common step, e.g., that it takes place at the same time. On the other hand nonblocking coordination does not require the processes to take their coordination steps commonly. As an example, consider again the message passing operations *send_message* and *receive_message.* In a blocking interpretation, the sending and the receiving process wait for each other, until both are ready to exchange the message, this is then done as a common step after which the processes are again independent. Alternatively, in a nonblocking interpretation, the sending process delivers the message when it is ready and goes on, independently of when the receiving process chooses to accept the message. If it tries to receive the message before it is sent, the receiving process waits. Nonblocking message passing requires a buffering mechanism for holding messages sent, but not yet received. The languages CSP [181] and LOTOS [197] are based on blocking message passing while nonblocking message passing is used in many commercial multiprogramming kernels [198].

**Terminology.** In this chapter the terms blocking and nonblocking are used, however, blocking is also often called **synchronous communication** [181] or **rendezvous** [187]. Similarly, one may find nonblocking communication referred to as asynchronous.

Once more, the arbiter is used as an illustration, the version shown below uses blocking message passing. It is quite similar to the one using nonblocking message passing, but there are also some important differences. The client computations follow exactly the same pattern:

*client: ...*
*send_message(request, arbiter); - send the request*
*wait_message(mes, arbiter); - wait for the grant*
*use resource;*
*send_message(release, arbiter); - send the release*
*ldots*

The arbiter process maintains a queue of waiting requests as in the other version shown in this section.

*arbiter:*
*WHEN*
*wait_message(request, any process) THEN*
*IF free THEN free:= false;send_message(r,r.sender) ELSE insert(r,q)*

*wait_message(release, any process) THEN*
*IF ¬empty(q) THEN r:= remove(q); send_message(r, r.sender)*
*ELSE free:= true*

Blocking message passing is based on a handshake between sender and receiver, so the operation *wait_message(-, any process)* in the client and the operation *send_message(-, arbiter)* in the arbiter are done simultaneously.

**End of example**

Although the three versions of the arbiter are rather similar, there are also some significant differences that are typical of the three approaches. The most important is the handling of **external nondeterminism** (see also section 6.2.2). In this example, the arbiter does not know the order in which request and release operations are performed. Hence, it must at all times be prepared to accept both requests and releases from any client. This external nondeterminism is reflected in the monitor version by the two procedures that can both be called at any time. Similarly, in the blocking message passing version where the nondeterministic choice is represented with the when construct:

*WHEN*
*wait_message(request, any process) THEN ...*
*wait_message(release, any process) THEN ...*

All the languages based on blocking message passing, e.g., CSP, OCCAM, and LOTOS, have such a construct for handling external nondeterminism.

When using nonblocking message passing, the external choice must be programmed explicitly. In the arbiter with an *IF THEN ELSE*. It is not very elegant, and in larger cases this solution becomes quite clumsy. The when-construct cannot be used for nonblocking message passing where the semantics of *wait_message* implies waiting in case the message has not yet arrived.

However, blocking message passing also has some limitations in handling external nondeterminism. At first it seems obvious to also allow *send_message* in a when construct, e.g.,

*WHEN*
  *wait_message(a) THEN ...*
  *wait_message(b) THEN ...*
  *send_message(c) THEN ...*

The handshake semantics implies that a communication is done with either *a*, *b*, or *c* depending on which of the three processes are prepared to communicate. This powerful construct provides quite elegant specifications. However, it turns out to be difficult to find a simple and efficient implementation of such a general when-construct, and therefore, most of the programming languages, e.g., CSP and OCCAM, does not allow send operations to appear in their when-construct. It is possible to implement the construct however, the implementation is quite complex and so far most programming languages with blocking message passing allow only restricted forms of external selection.

### 6.3.5 Remote Procedure Calls

The **remote procedure call** is a generalization of the monitor concept, often used for coordinating concurrent computations implemented on a network of computers. As indicated by the name, the coordination of two processes is done by one of the two calling a (remote) procedure in the other. The main difference to the monitor concept is that the called part is not just passively waiting for one of its procedures to be called, it may contain a local computation determining making it possible to accept different procedure calls at different points of the computation.

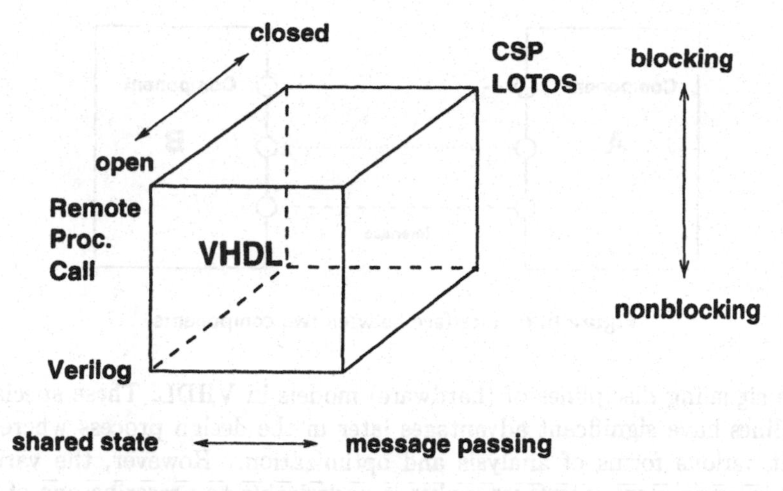


**Figure 6.8** Characterization of coordination paradigms.

### *6.3.6 Classification of Coordination Paradigms*

Almost all existing languages and tools for developing concurrent computations use one the paradigms described in this section, it has for example been mentioned several times that the specification language LOTOS [197] is based on blocking message passing. Figure 6.8 shows the classification of a number of existing languages.

The next section continues the description of a ***codesign computational model*** that was initiated in Section 6.2.1. The (concurrent) components introduced there are the building blocks of the model. The next section describes interfaces that are used for gluing together components to form a complete design.

## 6.4 INTERFACING COMPONENTS

This section introduces a model for interfacing the (concurrent) components of a design. The **interface** determines the coordination of the components including their data transfer and synchronization. As shown in section 6.3 many coordination mechanism have been proposed, however, co-design demands a model that allows different components to be realized in different technologies such as software (for a general purpose computer), software (for a specialized controller), various programmable hardware technologies such as FPGA, special purpose dedicated processors, synthesizable circuitry, and hand-crafted ASICs.

Therefore, the **interface model** does not prescribe a particular coordination mechanism, favoring a particular technology, e.g., blocking message passing dictated by (software) models such as CSP [181], CCS [201], LOTOS [197]

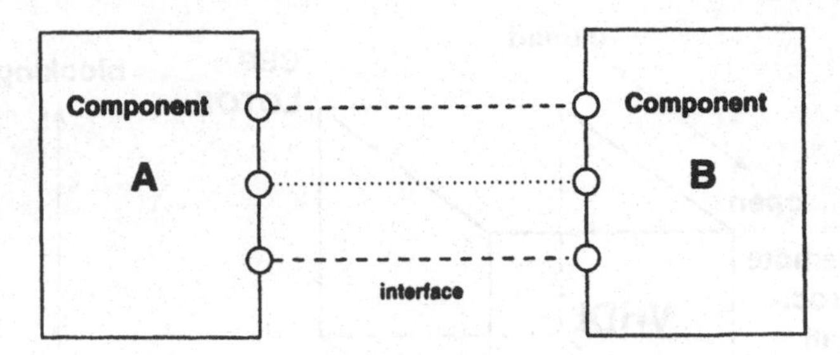


**Figure 6.9** Interface between two components

or the signaling disciplines of (hardware) models in VHDL. These specialized disciplines have significant advantages later in the design process where they permit various forms of analysis and optimization. However, the variation found in component interfaces makes it undesirable to prescribe one of these rigid discipline when doing high-level design.

The interface model allows components to ***share one or more state variables.*** Such shared state variables may contain a simple boolean, a composite value like an integer, a buffer, or any other data type. Sharing implies that the value of the state variables in the interface may be changed by several different components.

Undisciplined use of shared variables can easily lead to time dependent design errors which can be difficult to locate. We will later describe how shared variables can be tamed to help designs avoid such unpleasant errors.

The telephone switch illustrates how interfaces based on shared variables are defined. The design is divided into five kinds of components, see figure 6.10. One part of the interface between the dial component and the telephone receiver is the indication of whether the receiver is lifted or not. This is specified with the boolean variable *on_hook*. The interface to the dial component has other elements that we will return to later, they are just indicated with ... below.

*COMPONENT dial(on_hook: BOOLEAN; ...)*

Inside a dial component, the shared (interface) variable *on_hook* is used like any other state variable, e.g., in expressions and assignments. It may, for example, appear in a guard controlling some action to be taken when the receiver goes off hook.

$\ll \neg$ *on_hook* $\rightarrow$ *accept dialing sequence* $\gg$

After having received and decoded a dialing sequence the dial component has the number of the until to be called, this number is exchanged with the transfer component and is therefore also part of the interface.

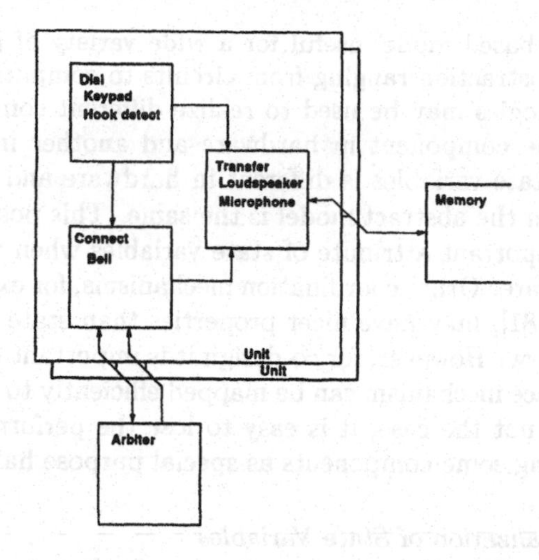


**Figure 6.10** Design of the telephone switch.

*COMPONENT dial(on_hook: BOOLEAN; rec_number: number; ...)*

where number is a type ranging over all possible telephone numbers used in the switch. The use of types restricting the range of possible values is an important part of the interface. Section 6.5.4 describes techniques for restricting the use of the shared variables in the interface and for verifying that these restrictions are imposed consistently in different component. Type checking is a first step in this direction.

Together the *on_hook* and *rec_number* makes up the interface between the dial and connect components of one unit, they are therefore grouped into a record.

*TYPE InterfaceDialConn = (* Interface: dial - connect component *)*
*RECORD*
*on_hook: BOOLEAN (* Indicates receiver on hook *)*
*rec_number: number (* Dialled phone number *)*
*END*

Using this type definition the interface of the dial component becomes

*COMPONENT dial(ifdc: InterfaceDialConn; ...)*

The other parts of the interface to a dial component indicated with ... are not described in further detail in this section.

A state based model is used here in contrast to the message passing approach found in CSP [181] and CCS [201], but in agreement with UNITY [180]. We have found a state based model useful for a wide variety of purposes [52] on different levels of abstraction ranging from circuits to computer networks.

Different technologies may be used to realize different components, for example, realizing one component in hardware and another in software. The realization of the state variables is different in hardware and in software; but the interpretation in the abstract model is the same. This possibility of a dual realization is an important attribute of state variables when used to interface hardware and software. Other coordination mechanisms, for example, messages as found in CSP [181], may have nicer properties than state variables from a software point of view. However, for co-design it is important that the abstract model of the interface mechanism can be mapped efficiently to a variety of technologies. If this is not the case, it is easy to lose the performance advantage obtained by realizing some components as special purpose hardware.

### *6.4.1 Physical Realization of State Variables*

In a co-design different components may be realized in different technologies, for example, by realizing one in hardware and another in software. It is, therefore, important that there are efficient realizations of state variables making up the interfaces in *both* hardware and software. In software, state variables are realized as program variables represented in memory. In hardware there are several alternatives depending on the technology; one possibility is to represent state variables with wires connecting the sub-circuits using the variables. These wires always contain the current value of the corresponding state variables, and they are connected with the inputs of sub-circuits reading them. Similarly, they are connected to the outputs of sub-circuits writing them. Associated with each wire is some refresh-mechanism maintaining the current value.

We can sketch how the model introduced above can be generalized to allow the interface to contain continuously changing (analog) variables. From a descriptive viewpoint this is quite straightforward. The physical realization can also be quite simple if one uses the approach sketched above where a shared variable is realized by a physical wire. It is not always that simple, but we will not go deeper into that here.

A state variable $x$ is interpreted as a function from *Time* to its range of values, e.g., if $x$ is boolean, we have

$$\mathcal{M}_b \quad x : Time \rightarrow B$$

This may seem like an overly generalized model, however, it opens some very interesting and useful perspectives, for example, describing variable changes by a continuous rather than a discrete function. It is, however, also possible to use a much simpler notion of time, e.g., one with discrete steps where each steps corresponds to an atomic update of one or more state variables. In such a model the computation can be viewed as a sequence

$$s_1; s_2; s_3; \ldots s_k; s_{k+1}; \ldots$$

Each $s_i$ represents a time unit (step) where all state variables have well defined values in the ranges determined by their type. Hence, the mapping $\mathcal{M}$ takes the form

$$\mathcal{M}_b \quad x : Nat \rightarrow B$$

The domain is the natural number determining in which step of the computation the variable is inspected. In Section 6.5.4, a special case of this model is considered where it is possible to form expressions relating the current (denoted *x.pre*) and previous (denoted *x.post*) values of a variable $x$.

This is a general definition of the mapping $\mathcal{M}$

$$\mathcal{M} \quad x : Time \rightarrow Range$$

As noted above *Time* can be continuous and this makes it possible to model interfaces bridging components without a discrete notion of time, for example, nonblocking communication links. Going further the *Range* can be any kind of set including infinite and non-countable ones.

As an example, consider the analog signal coming from the microphone of a telephone. This signal can be represented by an (interface) state variable

$$\mathcal{M}_v \quad voice : Time \rightarrow Frequency$$

where both *Time* and *Frequency* may be non-discrete sets. Let this type be denoted *analog_frequency*. The interface of the A/D converter transforming the analog input to discrete 8-bit samples is described as follows:

*COMPONENT adconvert(in: analog_frequency; out: [0..255])*

Syntactically, this is very similar to the way digital interfaces are described. However, at the present stage there are very few tools that are capable of handling such specifications. It is, for example, not even clear what is the correct notion of type checking when the types involve sets that are not discrete.

The first steps in this direction have however been taken with the analog extensions of both Verilog and VHDL. The theoretical foundation is currently a very active research area and is usually referred to as "Hybrid Systems".

The shared state variable is a very general mechanism for coordinating concurrent computations. However, the generality may also cause problems. Take, for example, a shared state variable that may vary as a continuous function over time, this is almost useless as a coordination mechanism. At the very lowest level one easily runs into the meta-stability problem [202]. At higher levels unrestricted use of shared variables may also cause problems, consider for example a simple counter, $x$, serving as the interface between two concurrent components:

$x := x+1 \parallel x := x-1$

Assume that $x$ has the value $k$ at some point of the computation. If both of the two components shown above attempt to update $x$ it is unpredictable what value $x$ gets next, $k$, $k+1$, and $k-1$ are all possible (together with many others depending on lower level details such as the number representation used for integers). If, for example, both components do their updates almost simultaneously they will both read out the old value of $x$. One of them increments the value to $k+1$ and the other decrements it to $k-1$ and they will then both attempt to store their new value in $x$. The result of this is unpredictable. These problems with shared variables have been recognized for a long time, see for example [196] and [203].

In software it is common to require that operations on shared variables are atomic (indivisible) which means that an operation is completed once it is started and with the same result as if all other components were momentarily stopped. In hardware the synchronous clock periods serve the same purpose of avoiding races between concurrent updates.

## 6.5 VERIFICATION

**Verification** is an important part of any non-trivial design project. Ideally one would like to do an exhaustive check, where all behaviors of the design are exercised. However, this is seldomly possible in practice, and only a small sample of the actual behaviors are checked by executing/simulating them. Recently, advances in algorithms, data structures and design languages have provided exhaustive verification techniques which are powerful enough to handle some significant practical examples [204, 205]. In order to use these, both the intended and actual behavior must be expressed in formal notation, e.g., as a program in a programming language or as a logic formula.

The rigor provided by formalism is useful for handling the complicated parts of a design. However, there are also parts where formal methods have no role to play. One should not insist that everything is done formally, but leave it to the designer's good taste to decide when formal analysis is appropriate and when more informal approaches are adequate. Formality is a possibility, but should not be mandatory.

This section describes three kinds of formal (exhaustive) verification used at different stages of the design process.

**interface verification** Separate components of a design interact through an interface that consists of both (physical) means for interacting, for example, a coordination mechanism, and conventions (**protocols**) determining how to perform and interpret the interaction. The primary reason for separating a design into distinct components is to enable the designer(s) to separate their concerns and work on one component at a time. This separation creates the possibility that an interface is not treated consistently in all components communicating through the interface. When working on one component, the designer might treat the signaling on a communication line in one way and later when working on another component the signaling is treated differently. This is a very common source of errors in particular in large designs done by a large group of designers. What often happens is that the designers initially have an informal agreement about the conventions for using the interface. Later in the design process the details of this agreement are then interpreted differently in different components leading to an inconsistency. Interface verification consists of verifying that there are no such inconsistencies. It is possible to do this before fixing the internal details of the different components. Interface verification is described in further detail in section 6.5.4.

**design verification** Design verification consists of verifying selected key requirements of incomplete (high-level) models. Consider again the arbiter where it is important that at most one device has access to a common resource (mutual exclusion). Once the interface to the bus has been designed, it is possible to verify that it does not violate the mutual exclusion property, even if other components of the design are still missing.

**implementation verification** In order to construct an efficient product, it is usually necessary to refine the initial (abstract) design into a concrete realization. This refinement typically includes a number of restrictions enforced by physical limitations and other practical constraints. It is, for example, necessary to restrict integer values to a certain range in order to represent the values in a fixed number of bits. Implementation verification, demonstrates that a concrete realization is a correct implementation of an abstract specification. This clearly requires that the concrete realization has been done and hence implementation verification is mainly relevant rather late in the design process. Implementation verification is becoming available in commercial design systems. The survey paper [206] contains a nice overview of the various approaches that has been used for circuit verification.

Next, we consider a simplified version of the arbiter. It has two clients: l(eft) and r(ight). A client issues a request by setting its request variable true (*reql* and *reqr* respectively). The arbiter grants access to a client by setting its grant variable (*grl* and *grr*) to true. When a client has finished its use of the common memory, it sets its request to false and the arbiter then resets the associated grant variable to false also.

```
COMPONENT arbiter(reql, grl, reqr, grr: BOOLEAN)
  INITIALLY grl = FALSE grr = FALSE
BEGIN
  ≪ reql ∧ ¬ grr → grl:= TRUE ≫ ||
  ≪ ¬ reql ∧ grl → grl:= FALSE ≫ ||
  ≪ reqr ∧ ¬ grl → grr:= TRUE ≫ ||
  ≪ ¬ reqr ∧ grr → grr:= FALSE ≫
END arbiter
```

An example of design verification would be to check that *grr* and *grl* are never both true. This requirement, called **mutual exclusion**, is expressed by the assertion: $\neg(grl \wedge grr)$. In section 6.5.1 it is described how this is verified formally/exhaustively.

The arbiter is by no means representative of the complexity found in realistic designs or the capabilities of currently available techniques and tools. However, the arbiter provides a simple illustration of the different approaches to verification described in this section. **End of example**

A high-level description of a design is an abstraction of its physical behavior. It is important to realize that formal verification deals with the abstraction and *not* with the physical realization. Despite the exhaustiveness provided by formal methods they are at most a guarantee of the *consistency between two symbolic models*, for example, the assertion that *grl* and *grr* are not true at the same time and the model of the computations of the arbiter such as the one given above or in other high-level notations (VHDL, C, Verilog, ...). A formal verification of a requirement, like the mutual exclusion of the arbiter, is not an absolute guarantee against malfunctioning, for example, errors caused by a a power failure. If this happens, the formal description is no longer a model of the physical realization, and hence, properties verified from the description may no longer hold. The same applies to simulation which is also based on a model of the physical reality. If the model is not an adequate abstraction, a simulation, no matter how exhaustive, does not provide any guarantee against malfunctioning. *Verification ensures consistency between two descriptions*, for example, one modeling the behavior of a design and another modeling a requirement. It is outside the realm of formal methods to ensure that a model adequately reflects physical phenomena such as light or changing voltages. Despite these reservations, formal verification can be a very powerful tool, but it is important to realize that *it can never become* an absolute guarantee.

### *6.5.1 Design Verification*

Design verification establishes consistency between a precisely described model, the design, and a rigorous formulation of selected key requirements. In the arbiter example, the mutually exclusive access is expressed with the assertion: $\neg(grl \wedge grr)$.

Requirements of a design are formalized as predicates constraining the computation, for example, that two grants must never be given simultaneously. To illustrate design verification we consider the special case of verifying invariants. An **invariant** defines a subset of the state space containing the initial state. Furthermore, there must not be any transitions from a state within the subset to a state outside. Hence, invariants describe properties that hold throughout the computation, because no transition will go to a state violating it. An invariant is written as a predicate, $I(S)$, on a state $S$.

Assume that $I$ is an invariant and that $t$ is a transition of a design, then $t$ is said to maintain the invariant if,

$$I(pre) \wedge t(pre, post) \Rightarrow I(post)$$

i.e., if the invariant holds in the pre-state then it is shown to hold in the post-state. By showing that the invariant holds in the initial state and by showing the implication for *each transition*, $t$, of the design one may conclude that the invariant holds throughout the computation. This verification technique is really an induction proof [207] (over the computations of the design) where the implication shown above corresponds to the induction step.

To show that the assertion $\neg(grr \wedge grl)$ is an invariant for the arbiter four implications (one for each transition) must be shown. For example:

$$reql \wedge \neg grr \wedge grl \Rightarrow \neg(grr \wedge grl)$$

There are many CAD tools available to help designers master the complexity of large designs. Most of these tools focus on later stager of the design process, for example, layout, code-generators, and synthesis. There is a similar need for tools aiming at the early design stages, for example, to support verification. Prototypes of such tools have been available for some time and they are now gradually becoming available in the commercial CAD packages. Section 6.5.3 gives a short survey.

### *6.5.2 Implementation Verification*

In order to construct an efficient product, it is usually necessary to refine the initial (abstract) design into a concrete realization. This refinement typically includes a number of restrictions enforced by physical limitations and other practical constraints. It is, for example, necessary to restrict integer values to a certain range in order to represent the values in a fixed number of bits. This section describes a formal approach to such refinements enabling a designer to rigorously verify that a concrete realization correctly implements a given abstract model. The approach requires that both the abstraction and the realization are described formally and they are called the **abstract design** and the **concrete design**, respectively. Once more the arbiter is used as an illustration.

An abstract design describes the computation with as few constraints as possible. This gives the implementor maximal freedom in choosing an efficient realization. In the arbiter example, the abstract design allows for the widest possible choice of scheduling algorithm. In the concrete design one of the possible realizations is chosen. Therefore, a concrete design may not exhibit all the behaviors of the abstract design. Informally, the behavior of a design is the computations that are externally visible which means that changes to local (internal) state variables are not directly reflected in the behavior of a component. A more rigorous definition of behavior is given in [208].

The abstract design of the arbiter serves as a good example. This design allows the arbiter to make an arbitrary choice between several requesting clients.

```
COMPONENT abstract_arbiter(reql, grl, reqr, grr: BOOLEAN)
  INITIALLY grl = FALSE grr = FALSE
  INVARIANT ¬ (grl ∧grr)
BEGIN
  ≪ reql ∧¬ grr → grl:= TRUE ≫ ||
  ≪ ¬ reql ∧grl → grl:= FALSE ≫ ||
  ≪ reqr ∧¬ grl → grr:= TRUE ≫ ||
  ≪ ¬ reqr ∧grr → grr:= FALSE ≫
END abstract_arbiter
```

**Example: Concrete Design of the Arbiter.** This section describes one of the many possible realizations of the abstract arbiter.

```
COMPONENT concrete_arbiter(reql, grl, reqr, grr: BOOLEAN)
  STATE lastl: BOOLEAN
  INITIALLY grl = FALSE grr = FALSE
  INVARIANT ¬ (grl ∧grr)
BEGIN
  ≪ reql ∧¬ grr ∧(¬ reqr ∨ ¬ lastl) → grl:= TRUE ≫ ||
```

$\ll \neg$ *reql* $\wedge$ *grl* $\rightarrow$ *grl, lastl:= FALSE, TRUE* $\gg$ ||
$\ll$ *reqr* $\wedge \neg$ *grl* $\wedge (\neg$ *reql* $\vee$ *lastl)* $\rightarrow$ *grr:= TRUE* $\gg$ ||
$\ll \neg$ *reqr* $\wedge$ *grr* $\rightarrow$ *grr, lastl:= FALSE, FALSE* $\gg$
*END abstract_arbiter*

This is a correct realizations in the sense that it only exhibits behaviors that could also be exhibited by the abstract design. The next section describes a technique for verifying (formally) that this is the case.

We can formalize our notion of refinement. Based on this formalism, we can develop a technique for verifying that one design is a refinement of another. Informally, the definition of refinement requires that any behavior exhibited by the concrete design could also be exhibited by the abstract. Note that it is not necessary that it can exhibit *all* behaviors.

*A concrete design is a* **refinement** *of an abstract design if the set of computations of the concrete design is a subset of the computations of the abstract design.*

Using this definition implies that the concrete design can be substituted for the abstract in any environment where the abstract is used, see also [52, 209]. However, the definition does not directly suggest a useful way of verifying refinement because it requires that all computations of both the abstract and concrete designs are described in a form that makes it possible to show that one is a subset of the other. Below a more practical verification technique for showing refinement is presented; as is the case with the other verification techniques described in this chapter, there is a compromise between power and practical feasibility.

To be a refinement, the concrete design must resemble the abstract one, yet there must be significant differences between the two. Typically, the representation of data is different, the abstract design could for example be expressed using integers, whereas the concrete design uses bit-vectors. This relationship is captured by an abstraction function.

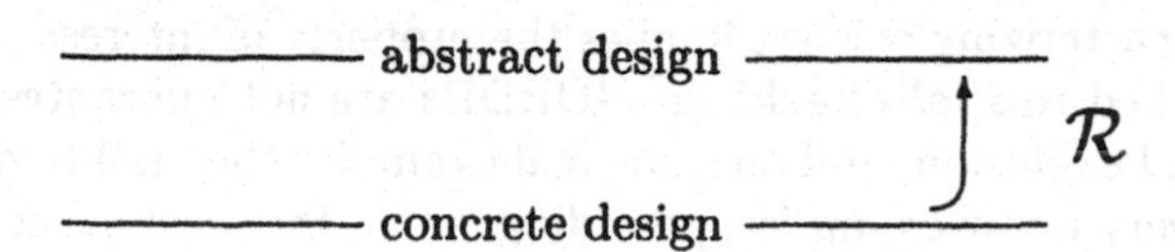


Formally, an abstraction function is a mapping from the concrete state space, $S_c$, to the abstract state space, $S_a$.

In the arbiter example, the refinement mapping maps the concrete instances of *reql, reqr, grl,* and *grr* to the corresponding abstract variables.

To show that a concrete design is a refinement, it must be shown that any behavior exhibited by the concrete design is also a possible behavior of the abstract design. This is ensured if the initial state of the concrete design is mapped to an initial state of the abstract, and if every execution of a concrete transition, $t_c$, either has no visible effect or there is a corresponding abstract transition, $t_a$, with a similar effect.

Typically, a concrete design includes details that are not in the abstract design, for example, extra internal state variables such as the carry bits in an adder or the internal state of a finite state machine. Changes in these are not reflected at the abstract level, which means that it must be possible for the concrete design to do transition that are made invisible by the abstraction function, these are often called **stuttering steps**.

### 6.5.3 *Verification Tools*

During the past 10 years, a number of powerful formal **verification tools** have evolved [204, 205]. The majority of these focus on implementation verification, the survey [206] presents and compares a number of these tools. A more recent comparison can be found in [210]. As mentioned above, formal verification corresponds to an exhaustive check of the state space. Using an explicit representation very quickly leads to a combinatorial explosion of the number of states generated, resulting in poor performance (the **state explosion**). However, ***implicit*** representations of state sets with clever datastructures can in many real examples overcome the problem. One of the most significant implicit representation is the **Reduced Ordered Binary Decision Diagrams** [211], ROBDDs. They provide compact representations of boolean functions. All the standard boolean operations are reflected by ROBDD-operations that are implemented as efficient algorithms on the underlying datastructure. Representing sets of states by their characteristic boolean functions provides the needed representation.

Using ROBDDs, the initial states, the set of transitions, and the reachable states are all represented as boolean functions. After computing the set of reachable states, the verification task is reduced to checking that the boolean function characterizing this set implies the property of interest. This approach is usually called **model checking**. ROBDDs are not guaranteed to avoid the combinatorial explosion, and on some real examples they fail to do so [212]; but they do on very many examples, providing one of the most successful heuristics currently known.

General purpose theorem provers [213, 214, 215] have also been used successfully to verify a large number of designs. The general purpose nature of these often means that they are less efficient than the ROBDD-based tools when these work. On the other hand, the general purpose tools have been used to verify designs with unbounded state spaces where the ROBDD-based tools are often not applicable.

### 6.5.4 Interface Verification

A common source of errors and delays in design and development projects is misunderstanding caused by inconsistent views on common interfaces. This section gives a brief description of how it is possible to verify that different components have a consistent view of their interface. It is important to allow them to have different views as long as these are not in conflict. To illustrate this consider a packet in a communication protocol. One component may treat this is an uninterpreted collection on bits to be transmitted whereas another component may impose a structure on the packet with different fields indicating addresses, control, and checksum.

In the *codesign computational model* used in this chapter the interface of a component consists of

- a set of state variables and
- **a protocol.**

The state variables are shared with other components and the protocol describes constraints on the use of the state variables. The sharing implies that the value of the state variables in the interface may be changed both by local computations and computations by other modules.

**A Simple Protocol.** This section illustrates interface verification on simple arbiter. In the next section, it is shown how to do interface verification on the somewhat more complicated arbiter used in the telephone switch.

To operate correctly, the arbiter may assume that all request/grant pairs follow the four-phase protocol: A client requests the privilege by setting *req* to true. When *gr* becomes true, the client may enter its critical region. When leaving it, *req* is set to false which is followed by *gr* becoming false. This can be expressed formally as follows:

$$\mathit{req.post} \neq \mathit{req.pre} \Rightarrow \mathit{grant.post} \neq \mathit{req.post} \wedge$$
$$\mathit{grant.post} \neq \mathit{grant.pre} \Rightarrow \mathit{grant.post} = \mathit{req.post}$$

Both the *arbiter* and the *client* may assume that the other components follow this protocol. To verify that the changes made to *gr* by the *arbiter* follow the protocol, it must be shown that all its transitions obey:

$$\mathit{gr.post} \neq \mathit{gr.pre} \Rightarrow \mathit{req.post} = \mathit{gr.post}$$

Similarly, all transitions of the client must obey:

$$req.post \neq req.pre \Rightarrow req.post \neq gr.post$$

To ensure interface consistency, it must be shown that these protocols do not contradict each other. In this simple example it is not very difficult.

The exact proof obligations needed to formally verify interface consistency are given in [52, 216]. It amounts to showing two logical implications for each (instance) of a component, i.e., four implications in this case.

The telephone switch contains a number of interfaces that are all a bit more complex than the simple four-phase protocol used for the arbiter shown above.

Consider first the interface between the connect and arbiter components. It is based on the following datastructure

```
TYPE
  connreq = (none, half, alloc, dealloc)
  unitno = 0..no_of_units (* Unit number; 0 = invalid. *)
  bufno = 0..no_of_buffers (* Buffer number; 0 = invalid. *)
  InterfaceConnArb = (* Interface connect <-> arbiter component. *)
    RECORD
      req: connreq (* -> Request to arbiter. *)
      caller: unitno (* -> Caller unit number. *)
      called: unitno (* -> Called unit number. *)
      reply: BOOLEAN (* <- Indicates arbiter reply. *)
      buf: bufno (* <- Granted buffer number, or 0. *)
  END
```

Informally, the interface is specified as follows:

> Assume that a unit (the caller) wants to make a call to another (the called). Before starting, *reply* must be *FALSE* to indicate that the arbiter is idle. The caller then sets *req* = *alloc* and presents the unit numbers of the caller and the called units.
>
> If a buffer is available and the called unit is not busy, the arbiter gives a grant by setting *reply TRUE*, and *buf* is set to the allocated buffer number. Then *both* of the two units acknowledge receipt of the buffer number. First, the called unit sets *req* = *half*, and thereafter the caller sets *req* = *none*. Finally, the arbiter sets *reply* to *FALSE*.
>
> If no buffer is free, the arbiter will respond with *reply* = *TRUE*, *buf* = *0* and the called and the caller will have to acknowledge before the arbiter can remove the answer.
>
> When the caller wants to release its buffer (hang up), again the 5-phase protocol applies, but the caller uses *req* = *dealloc*, and the arbiter responds with the buffer number being released.

This is formalized in the following protocol:

```
STATE
  ifca1: InterfaceConnArb1 (* Interface unit -> arbiter. *)
  ifca2: InterfaceConnArb2 (* Interface unit <- arbiter. *)
PROTOCOL
  (¬ ifca2.reply.pre ∧ ¬ SAME(ifca2.reply) ⇒
  (ifca1.req.pre=alloc ∨ ifca1.req.pre=dealloc) ∧ SAME(ifca1.req)) ∧
  (ifca2.reply.pre ∧ ¬ SAME(ifca2.reply) ⇒
    ifca1.req.pre=none ∧ SAME(ifca1.req)) ∧
  (ifca1.req.pre=none ∧ ¬ SAME(ifca1.req) ⇒
    ¬ ifca2.reply.pre ∧ SAME(ifca2.reply)) ∧
  ((ifca1.req.pre=alloc ∨ ifca1.req.pre=dealloc) ∧ ¬ SAME(ifca1.req) ⇒
  ifca1.req.post=half) ∧
  (ifca1.req.pre=half ∧ ¬ SAME(ifca1.req) ⇒ ifca1.req.post=none)
```

As it was the case with the simple arbiter and the four-phase protocol, each of the componets have a somewhat simpler version of the protocol reflecting their views.

```
COMPONENT arbiter(
  ifca1: InterfaceConnArb1; (* Interface connect -> arbiter. *)
  ifca2: InterfaceConnArb2 (* Interface connect <- arbiter. *) )
(* Constraints on changes of reply. (Arbiter output behaviour.) *)
  PROTOCOL
    ¬ ifca2.reply.pre ∧ ¬ SAME(ifca2.reply) ⇒
      (ifca1.req.pre=alloc ∨ ifca1.req.pre=dealloc) ∧ SAME(ifca1.req) ∧
    ifca2.reply.pre ∧ ¬ SAME(ifca2.reply) ⇒
      ifca1.req.pre=none ∧ SAME(ifca1.req) ∧
    (* Constraints on changes of req. *)
    ifca1.req.pre=none ∧ ¬ SAME(ifca1.req) ⇒
     ifca2.reply.pre=FALSE ∧ SAME(ifca2.reply)∧
    (ifca1.req.pre=alloc ∨ ifca1.req.pre=dealloc) ∧ ¬ SAME(ifca1.req) ⇒
     ifca1.req.post=half ∧
    ifca1.req.pre=half ∧ ¬ SAME(ifca1.req) ⇒ ifca1.req.post=none
COMPONENT unit(
  ifca1: InterfaceConnArb1; (* Interface unit -> arbiter. *)
  ifca2: InterfaceConnArb2; (* Interface unit <- arbiter. *)
... )
  PROTOCOL
    (ifca1.req.pre=none ∧ ¬ SAME(ifca1.req) ⇒
      ¬ ifca2.reply.pre ∧ SAME(ifca2.reply)) ∧
    ((ifca1.req.pre=alloc ∨ ifca1.req.pre=dealloc) ∧ ¬ SAME(ifca1.req) ⇒
```

*ifca1.req.post=half)* $\wedge$
*(ifca1.req.pre=half* $\wedge \neg$ *SAME(ifca1.req)* $\Rightarrow$ *ifca1.req.post=none)*

Again two implications must be shown for each instance of a component. One to demonstrate that the environment meets all the expectations of the component and another to ensure that the components satisfy all requirements set by the environment. The technique is described in further detail in [216].

The importance of interface consistency and verification is receiving increasing attention. At least three levels can be identified

**static** It is common practice in many languages to type check separately compiled components, e.g. by requiring a rudimentary specification of externally used components. In VHDL one must give an ENTITY declaration specifying the procedure headings of another component in order to use it. However, no dynamic information is provided and therefore only a static consistency check is possible.

**safety** The approach sketched above allows one to specify constraints on the use of an interface (a protocol). These constraints usually express dynamic properties and hence they can not be checked statically at compile time, instead a verification tool is required. This kind of verification is sometimes called **functional verification**.

**timing** It is very common that an interface contains timing assumptions, a good example is the timing diagrams found on most data sheets of hardware components. There are several approaches, e.g., [217] to specifying such timing constraints in a symbolic manner that allows a formal consistency check.

## 6.6 CONCLUSIONS

High-level specification and verification have for a long time been research areas with a promising *future*. However, in these years they seem to be finding their way into practical application. There are, in my view, several recent developments that have made this possible. The most important is undoubtedly the appearance of powerful verification tools (such as the ones described in section 6.5.3). However, another important development is the growing realization that formal verification is not an all-or-nothing proposition. It can be applied in a rather pragmatic way allowing the designer to decide what is important enough to warrant the extra effort needed to obtain the certainty offered by the exhaustive checking of a formal verification technique.

**Acknowledgements.** This work has been supported by the Danish Technical Research Council's Co-design project.

# 7 LANGUAGES FOR SYSTEM-LEVEL SPECIFICATION AND DESIGN

A.A. Jerraya, M. Romdhani, C.A. Valderrama, Ph. Le Marrec, F. Hessel, G.F. Marchioro, and J.M. Daveau

System-Level Synthesis Group
TIMA/INPG
Grenoble, France

## 7.1 INTRODUCTION

Chapter 6 introduced concepts and abstract models for designing at the system level. This chapter discusses concrete specification languages and intermediate models used for transforming such system-level designs into realizations.

As discussed in Chapter 1, specification is an essential task in system design, and languages for doing this plays an essential role. Selecting a language is generally a tradeoff between several criteria: the expressive power of the language, the automation capabilities provided by the model underlying the language, and the availability of tools and methods supporting the language. Additionally, for some applications several languages are needed to do different modules, as for example, designs where different parts belong to different application classes, e.g., control/data or continuous/discrete.



*J. Staunstrup and W. Wolf (eds.), Hardware/Software Co-Design: Principles and Practice, 235-262.*

Intermediate forms are used by the system design tools for the refinement of the specification into an architecture. Although intermediate forms are generally not accessible to the users of such tools, understanding them help to better understand how the tool works.

There two major types of intermediate forms: language-oriented and architecture-oriented. Language-oriented intermediate forms use a graph representation. The main employed representations are the **data flow graph** (DFG), **control flow graph** (CFG), and a combination of the DFG and CFG known as the **control-data flow graph** (CDFG). The architecture-oriented intermediate forms use an FSM/datapath representation that is closer to the final realization. Intermediate forms may be based on simple or multiple threads (Section 6.2 in Chapter 6).

The use of language-oriented intermediate forms makes it easier to apply system-level transformations. However, the use of architecture-oriented models allows one to handle more sophisticated architectures with features such as synchronization and specific communication buses. Such intermediate representations will be detailed in Section 7.3.5.

The next section describes three system-level approaches to introduce homogeneous and heterogeneous specification for co-design. Each of the specification strategies implies a different organization of the co-design environment. Section 7.4 introduces several languages and outlines a comparative study of these languages. Finally, section 7.5 deals with multi-language specification and co-simulation.

## 7.2 SYSTEM-LEVEL SPECIFICATION

The system-level specification of a mixed hardware/software application may follow one of two schemes [218]:

- **homogeneous** A single language is used for the specification of the overall system including hardware parts and software parts.
- **heterogeneous** Different languages are used for hardware parts and software parts; a typical example is the mixed C-VHDL model.

Each of the two above specification strategies implies a different organization of the co-design environment.

### *7.2.1 Homogeneous Specification*

Homogeneous specification implies the use of a single specification language for the specification of the overall system. A generic co-design environment based on homogeneous specification is shown in Figure 7.1. Co-design starts with a global specification given in a single language. This specification may be independent of the future implementation and the partitioning of the system into hardware parts and to software parts. In this case co-design includes a partitioning steps aimed to split this initial specification into hardware and software. The outcome is an architecture made of hardware processors and software processors. This is generally called a **virtual prototype** and may be given in a single language or different languages (e.g. C for software and VHDL for hardware).

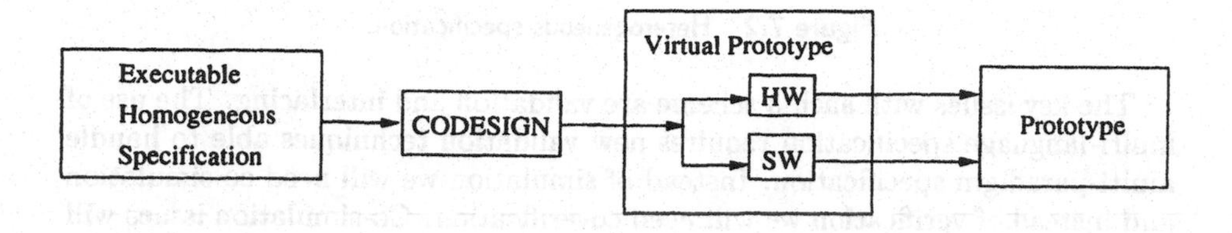


**Figure 7.1** Homogeneous specification.

The key issue with such co-design environments is the correspondence between the concepts used in the initial specification and the concepts provided by the target model, which is also known as a **virtual prototype**. For instance the mapping of the system specification language including high level concepts such as distributed control and abstract communication onto low level languages such as C and VHDL is a non trivial task [219, 220].

Several co-design environments follow this scheme. In order to reduce the gap between the specification model and the virtual prototype these tools start with a low level specification model. Cosyma (see chapter 8) starts with a C-like model called $C^x$ [221, 222]. Vulcan starts with another C-like language called Hardware C [223]. LYCOS (see chapter 9) and Castle start with C. Several co-design tools start with VHDL [9]. Only few tools tried to start from a high level specification. These include Polis [74] that starts with an Esterel specification [224, 225], SpecSyn [8, 226] that starts from SpecCharts [227, 228, 229] and [220] that starts from LOTOS [230]. Chapter 10 details Cosmos, a co-design tool that starts from SDL.

### 7.2.2 *Heterogeneous Specification*

Heterogeneous specification allows the use of specific languages for the hardware and software parts. A generic co-design environment based on a heterogeneous specification is given in Figure 7.2. Co-design starts with a virtual prototype when the hardware/software partitioning is already made. In this case, co-design is a simple mapping of the software parts and the hardware parts on dedicated processors.

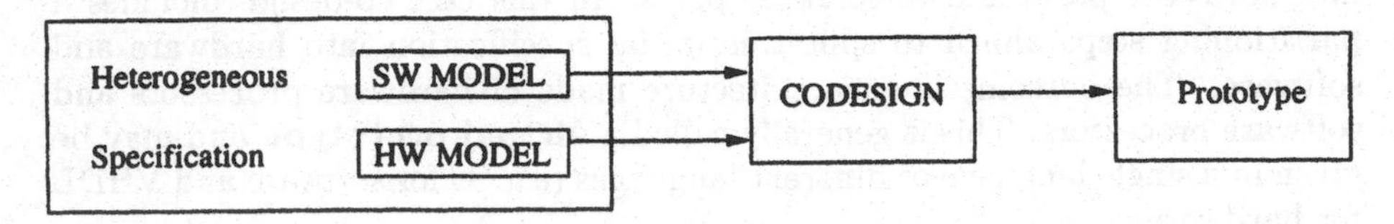


**Figure 7.2** Heterogeneous specification.

The key issues with such a scheme are validation and interfacing. The use of multi-language specification requires new validation techniques able to handle multi-paradigm specification. Instead of simulation we will need co-simulation and instead of verification we will need coverification. Co-simulation issues will be addressed in Section 7.5. Additionally, multi-language specification brings the issue of interfacing sub-systems which are described in different languages. These interfaces need to be refined when the initial specification is mapped onto a prototype.

Coware [231] and Seamless [232] are typical environments supporting such co-design scheme. They start from mixed description given in VHDL or VERILOG for hardware and C for software. All of them allow for co-simulation. However, only few of these system allows for interface synthesis [233]. This kind of co-design model will be detailed in Section 7.5.

## 7.3 DESIGN REPRESENTATION FOR SYSTEM LEVEL SYNTHESIS

This section deals with the design specifications used in co-design. The goal is to focus on the computational models underlying specification languages and architectures.

### *7.3.1 Synthesis Intermediate Forms*

Most co-design tools make use of an internal representation for the refinement of the input specification into architectures. The input specification is generally given in a human readable format that may be a textual language (C, VHDL, SDL, Java, etc.) or a graphical representation (Statechart, SpecChart, etc.). The architecture is generally made of a set of processors. The composition scheme of these processors depends on the computational model underlying the architecture. The refinement of the input specification into architecture is generally performed into several refinement steps using an internal representation also called intermediate form or internal data structure. The different steps of the refinement process can be explained as a transformation of this internal model.

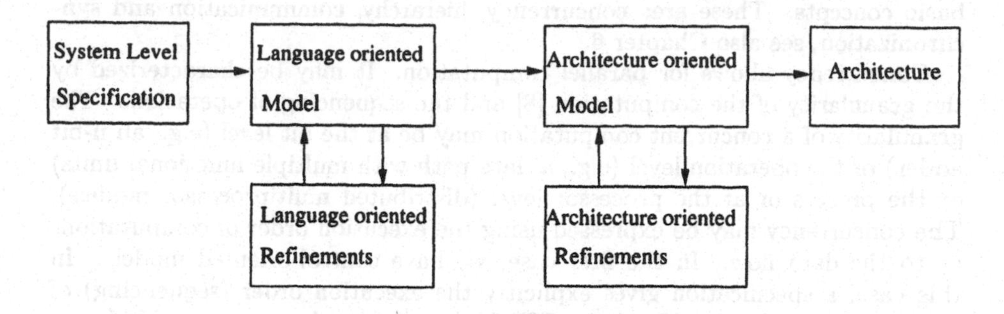


**Figure 7.3** Language-oriented vs architecture-oriented intermediate forms.

There are mainly two kinds of intermediate forms: the first is language-oriented and the latter is architecture-oriented. Both may be used for system representation and transformations. Language-oriented intermediate forms are generally based on graph representation. It models well the concepts handled in specification languages. Architecture-oriented intermediate forms are generally based on FSM models. These are closed to the concepts needed to represent architectures.

Figure 7.3 shows a generic co-design model that combines both models. In this case, the first co-design step translates the initial system-level specification into a language-oriented intermediate form. A set of refinement steps may be applied on this representation. These may correspond to high level transformations. A specific refinement step is used to map the resulting model into architecture-oriented intermediate form. This will be refined into an architecture.

Although general, this model is difficult to implement in practice. In fact, most existing tools make use of a unique intermediate form. The kind of intermediate form selected will restrict the efficiency of the resulting co-design tool. Language-oriented intermediate form make easier the handling of high level concepts (e.g. abstract communication, abstract data types) and high level transformations. But, it makes difficult the architecture generation step and the specification of partial architectures and physical constraints. On the other side, an architecture-oriented intermediate form makes difficult the translation of high level concepts. However, it makes easier the specification of architecture related concepts such as specific communication and synchronization. This kind of intermediate form generally produces more efficient design.

### *7.3.2 Basic Concepts and Computational Models*

Besides classic programming concepts, system specification is based on four basic concepts. These are: concurrency, hierarchy, communication and synchronization, see also Chapter 6.

Concurrency allows for parallel computation. It may be characterized by the granularity of the computation [8] and the sequencing of operations. The granularity of a concurrent computation may be at the bit level (e.g., an n-bit adder) or the operation level (e.g., a data path with multiple functional units) or the process or at the processor level (distributed multiprocessor models). The concurrency may be expressed using the execution order of computations or to the data flow. In the first case, we have control-oriented models. In this case, a specification gives explicitly the execution order (sequencing) of the element of the specification. CSP-like models and concurrent FSMs are typical examples of control-oriented concurrency. In the second case we have data-oriented concurrency. The execution order of the operations fixed by the data dependency. Dataflow graphs and architectures are typical data-oriented models.

Hierarchy is required to master the complexity. Large systems are decomposed into smaller pieces which are easier to handle. There are two kinds of hierarchies. Behavioral hierarchy allows constructs to hide sub-behaviors [226]. Procedure and substates are typical forms of behavioral hierarchies. Structural hierarchy allows to decompose a system into a set of interacting subsystems. Each component is defined with well defined boundaries. The interaction between subsystem may be specified at the signal (wire) level or using abstract channels which hide complex protocols.

The computational model of a given specification can be seen as the combination of two orthogonal concepts: (1) the **synchronization model** and (2) the **control model**. As discussed in chapter 6 the synchronization can be classified as blocking or non-blocking. The control model can be classified into control-flow or data-flow. As explained above, this gives the execution order of operations within one process. The control-oriented model focuses on the control sequences rather than the processing itself. The data-oriented model focus expresses the inxbehavior as a set of data transformations.

According to this classification we obtain mainly four computational models that may be defined according to concurrency and synchronization, as illustrated in Figure 7.4. Most of the languages mentioned in the figure will be discussed in the next sections.

| Concurrency \ Synchronization Model | Synchronous | Asynchronous |
|---|---|---|
| Control-driven | SCCS, Statechart<br>Esterel, SML<br>VHDL | CCS, CSP<br>OCCAM, SDL |
| Data driven | SILAGE<br>LUSTRE<br>SIGNAL | Z, B |

**Figure 7.4** Computational models of specification languages.

### *7.3.3 Language Oriented Intermediate Forms*

Various representations have appeared in the literature [234, 235, 236, 237], mostly based on flow graphs. The main kinds are **data, control** and **control-data flow** representations as introduced above.

**Data Flow Graph (DFG).** The data flow graph is the most popular representation of a program in high level synthesis. Nodes represent the operators of the program, edges represent values. The function of node is to generate a new value on its outputs depending on its inputs.

A data flow graph example is given in Figure 7.5, representing the computation *e:= (a+c)*(b-d)*. This graph is composed of three nodes $v_1$ representing the operation +, $v_2$ representing - and $v_3$ representing *. Both data produced by $v_1$ and $v_2$ are consumed by $v_3$. At the system level a node may hide complex computation or a processor.

In the **synchronous data flow** model we assume that an edge may hold at most one value. Then, we assume that all operators consume their inputs before new values are produced on their inputs edges.

In the **asynchronous data flow** model, we assume that each edge may hold an infinite set of values stored in an input queue. In this model we assume that inputs arrivals and the computations are performed at different and independent throughputs. This model is powerful for the representation of computation. However it is restricted for the representation of control structures.

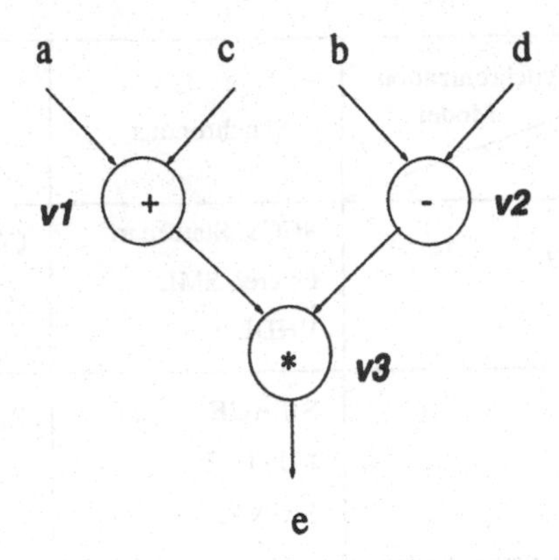


Figure 7.5 Example of a simple data flow graph.

**Control Flow Graph (CFG).** Control flow graphs are the most suited representation to model control design. These may contain many (possibly nested) loops, global exceptions, synchronization and procedure calls; in other words, features that reflect the inherent properties of controllers. In a CFG, nodes represent operations and edges represent sequencing relations.

While this kind of graph models well control structure including generalized nested loops combined with explicit synchronization statement (`wait`), control statements (`if`, `case`), and exceptions (`EXIT`), it provides restricted facilities for data flow analysis and transformations.

**Control-Data Flow Graph (CDFG).** This model extends DFG with control nodes (`if`, `case`, `loop`) [238]. This model is very suited for the representation of data flow oriented applications. Several co-design tools use CDFG as intermediate form [221].

### *7.3.4 Architecture Oriented Intermediate Forms*

This kind of intermediate form is closer to the architecture of the realization than to the initial input description. The data path and the controller may be represented explicitly when using this model. The controller is generally abstracted as an FSM. The data path is modeled as a set of assignment and expressions including operation on data. The main kind of architecture oriented forms are FSM with data path model FSMD defined by Gajski [63] and the FSM with co-processors (FSMC) [239].

**The FSMD representation.** The FSMD was introduced by Gajski [63] as a universal model that represents all hardware design. An FSMD is an FSM extended with operations on data.

In this model, the classic FSM is extended with variables. An FSMD may have internal variables and a transition may include arithmetic and logic operation on these variables. The FSMD adds a data path including variables, operators on communication to the classic FSM.

The FSMD computes new values for variables stored in the data path and produces outputs. Figure 7.6 shows a simplified FSMD with two states *Si* and *Sj* and two transitions. Each transition is defined with a condition and a set of actions that have to be executed in parallel when the transition is fixed.

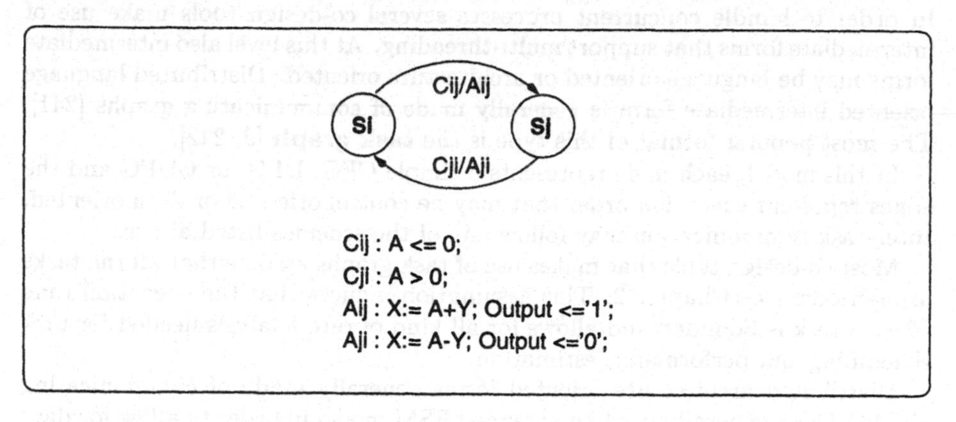


**Figure 7.6** The FSMD model.

**The FSM with Co-Processors Model (FSMC).** An FSMC is an FSMD with operations executed on co-processors. The expressions may include complex operations executed on specific calculation units called co-processors. An FMSC is defined as an FSMD plus a set of N co-processors C. Each co-processor *Ci* is also defined as an FSMC. FSMC models hierarchical architecture made of a top controller and a set of data path that may include FSMC components as shown in Figure 7.7.

Co-processors may have their local controller, inputs and outputs. They are used by the top-level controllers to execute specific operations (expressions of the behavioral description). Several co-design tools produce an FSMC based architecture. These include Cosyma [222], Vulcan [240] and the LYCOS system of Chapter 9. In these co-design tools, the top controller is made of a software processor and the co-processor are hardware accelerators.

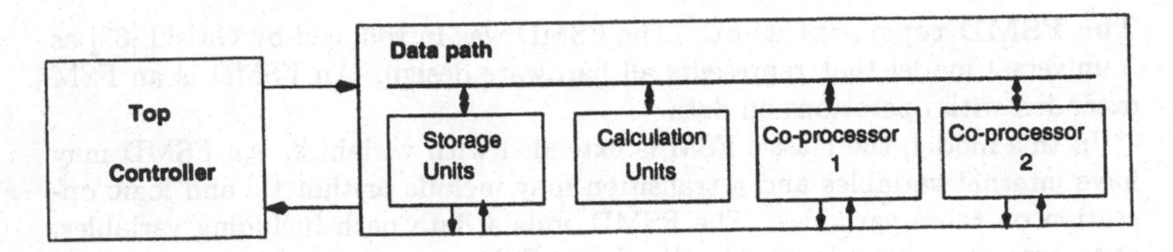


**Figure 7.7** The FSMC architecture model.

### *7.3.5 Distributed Intermediate Forms*

In order to handle concurrent processes several co-design tools make use of intermediate forms that support multi-threading. At this level also intermediate forms may be language oriented or architecture oriented. Distributed language oriented intermediate form is generally made of communicating graphs [241]. The most popular format of this type is the **task graph** [9, 242].

In this model, each node represents a simple CFG, DFG, or CDFG and the edges represent execution order that may be control oriented or data oriented. Inter-task communication may follow any of the schemes listed above.

Most co-design tools that makes use of task graphs assume that all the tasks are periodic (see Chapter 2. This assumption induces that the execution time of each task is bounded and allows for all kind of rate analysis needed for task scheduling and performance estimation.

Distributed architecture oriented forms generally made of communicating FSMs. These generally used an extended FSM model in order to allow for data computation within the FSMs. Only few co-design tools support communicating FSMs. SpecSyn [8] and Polis [74] make use of interconnected FSMs. The model used in Polis is based on a specific kind of FSMDs, called Co-design FSMs (CFSM). In this model the communication is made at the signal level and a system is organized a DAG where each node is a CFSM. The Cosmos system is based on a communicating FSMC model called SOLAR[243]. This model allows for a generalized composition model and makes use of RPC [244] forinter-modules communication.

## 7.4 SYSTEM LEVEL SPECIFICATION LANGUAGES

### *7.4.1 The Plethora of System Specification Languages*

Specification languages were first introduced in software engineering in order to support the early steps of the software development [245]. The high cost of the software development and maintenance raised the need for concentrating on the specification and the requirements analysis steps. The software quality and productivity is expected to be improved due to formal verification and gradual refinement of software starting from higher level specification.

The first hardware description languages were aimed to the specification of instruction-set computers. Von Neumann used an ad hoc hardware description language for the description of its machine. DDL [246], PMS, and ISP are typical examples introduced in the late 60's and the early 70's [8] for hardware specification. Since that time, a plethora of languages is presented in the literature. These are the result of several fields of research. The most productive areas are:

1. **VLSI system design** Research in this area modern hardware description language (HDL). ISPS [247], CONLAN [248] and more recently Hardware C [223] , Spec-Charts [228, 229] and VHDL [249] are typical HDLs. These languages try to deal with the specific characteristics of hardware design such as abstraction levels, timing and data flow computation.

2. **protocol specification** Several languages have been created for protocol specification. In order to allow for protocol verification, these languages are based of what is called formal description technique (FDT) [250]. SDL [251], LOTOS [230] and ESTELLE [252, 253] are the main languages in this area.

   LOTOS [230] (LOgical Temporal Ordring Specification) is a formal specification languages for protocols and distributed systems. It is an ISO standard. The LOTOS specification is composed of two parts:

   - A behavioral part based on the theory of process algebra.

   - A definitional part for data definition based on abstract data types.

   The formal basis of LOTOS are well defined. The specification approach consists in producing a first executable specification, then validate it, and derive an implementation.

   The SDL language [251] was designed in order to specify telecommunication systems. The language is standardized by the ITU (International Telecommunication Unions) as 2.100. SDL is particularly suitable for systems where it is possible to represent the inxbehavior by extended finite state machines.

An SDL specification can be seen as a set of abstract machines. The whole specification includes the definition of the global system structure, the dynamic behavior of each machine and the communication in between.

SDL offers two forms of representation: A graphical representation named SDL-GR (SDL Graphical Representation) and a textual representation named SDL-PR (SDL Phrase Representation).

ESTELLE [253] is an ISO standardized language for the specification of protocol and their implementation. The specifications are procedural having a Pascal-like constructions. ESTELLE is rather a programming language than a specification language. In fact, the ESTELLE specification includes implementation details. These details are generally not necessary during the specification phase.

ESTELLE adopts an non-blocking model for the communication based on message passing. It presents several limitations in the specification of the parallelism between concurrent processes.

3. **reactive system design** reactive systems are real time application with fast reaction to the environment. Esterel [224], LUSTRE [254] and Signal [255] are typical languages for the specification of reactive systems. Petri nets [256] may also be included in this area.

   Esterel [224] is an imperative and parallel language which has well defined formal basis and a complete implementation.

   The basic concept of Esterel is the event. An event corresponds to the sending or receiving of signals that convey data. Esterel is based on a synchronous model of computation. This synchronism simplifies reasoning about time and ensures determinism.

   LUSTRE [254] particularly suits to the specification of programmable automata. Some real-time aspects have been added to the language in order to manage the temporal constraints between internal events.

   The SIGNAL language [255] differs from LUSTRE in the possibility of using different clocks in the same program. The clocks can be combined through temporal operators, allowing flexible specification.

   Statecharts [257] is visual formalism for the specification of complex reactive systems created by D. Harel. Statecharts describes the behavior of those systems using a state-based model. Statecharts extend the classical finite state machine model with hierarchy, parallelism and communication.

   The behavior is described in terms of hierarchical states and transitions in-between. Transitions are triggered by events and conditions. The communication model is based on broadcasting, the execution model is synchronous.

Petri nets [256] are tools that enable to represent discrete event systems. They do enable describing the structure of the data used, they describe the control aspects and the behavior of the system. The specification is composed of a set of transitions and places. Transitions correspond to events; places correspond to activities and waiting states.

Petri nets enable the specification of the evolution of a system and its behavior. The global state of a system corresponds to a labelling that associates for each place a value. Each event is associated to a transition, firable when the entry places are labelled.

4. **programming languages** Most (if not all) programming languages have been used for hardware description. These include Fortran, C,Pascal, Ada and more recently, C++ and Java. Although these languages provide nice facilities for the specification hardware systems, these generally lack feature such as timing or concurrency specification. Lots of research have tried to extend the programming languages for hardware specification with limited results because the extended language is no more a standard.

5. **parallel programming languages** parallel programs are very closed to hardware specification because of the concurrency. However, they generally lack timing concepts and provide dynamic aspects which are difficult to implementing hardware, see also Chapter 6.

6. **functional programming and algebraic notation** Several attempts were made to use functional programming and algebraic notations for hardware specification. VDM, Z and B are examples of such formats.

   VDM [258] (Vienna Development Method) is based of the set theory and predicate logic. It has the advantage of being an ISO standard. The weakness of VDM are mainly the non support concurrency and its verbosity.

   Z [259] is a predicative language similar to VDM. It is based on the set theory. Z specification language enables to divide the specification in little modules, named "schemes". These modules describe at the same time static and dynamic aspects of a system.

   B [260] is composed of a method and an environment. It was developed by J.R. Abrial, who participated to the definition of the Z language. B is completely formalized. Its semantic is complete. Besides, B integrates the two tasks of specification and design.

### *7.4.2 Comparing Specification Languages*

There is not a unique universal specification language to support all kinds of applications (controller, heavy computation, DSP, ...). A specification language is generally selected according to the application at hand and to the designer culture. This section provides 3 criteria that may guide the selection specification languages. These are [261]:

- **Expressiveness** Expressive power is determined by the computational model (see 7.3.2). The expressive power of a language fixes the difficulty or the ease when describing a given behavior

- **Analytical power** This is related to the analysis, the transformation and the verification of the format. It is mainly related to tool building.

- **Usability** This criterion is composed of several debatable aspects such as clarity of the model, related existing tools, standardization efforts, etc.

The main components of the expressive power of a given language are: Concurrency, communication, synchronization, data description and timing models. See Chapter 6 for more details.

The analytical power is strongly related to the formal definition of the language. Some languages have a formal semantics. These include Z, B, SCCS, temporal logic, etc. In this case, mathematical reasoning can be applied to transform, analyze or prove properties of the system specification. The formal description techniques provide only a formal interpretation: these may be translation using a well-defined method, to another language that have a formal semantics. For instance, the language Chill [262] is used for the interpretation of SDL. Finally, most existing language have a formal syntax which is the weakest kind of formal description.

The existence of formal semantics allows an easier analysis of specification. This also make easier the proof of properties such as coherence, consistency, equivalence, deadlocks, etc.

The analytical power includes also facilities to build tools around the language. This aspect is more related to the underlying computational model of the language. As stated earlier, graph based models make easier the automation of language-oriented analysis and transformation tools. On the other side, architecture-oriented models (e.g. FSMs) makes easier the automation of lower level transformation which are more related to architecture.

The usability power include aspects such as standardization, readability and tool support. The readability of a specification plays an important role in the efficiency of its exploitation. A graphical specification may be quoted more readable and reviewable than a textual specification by some designers. However, some of the designers may prefer textual specification. The graphical and textual specifications are complementary.

The availability of tools support around a given specification language is important in order to take best benefit from the expressiveness of this language. Support tools include editors, simulation tools, provers, debuggers, prototypers, etc.

The above mentioned criteria shows clearly the difficulty of comparing different specification languages. Figure 7.8 shows an attempt for the classification of few languages. The first four lines of the table shows the expressive power of the languages, the fifth line summarizes the analytical power and the three last lines give the usability criteria.

| | LOTOS | SDL | Esterel | StateChart | SpecChart | Petri Net | VHDL | C | JAVA |
|---|---|---|---|---|---|---|---|---|---|
| Concurrency | *** | ** | * | *** | *** | *** | ** | X | *** |
| Data Structure | *** | *** | * | ** | * | X | * | * | *** |
| Communication | ** | ** | * | * | ** | * | * | X | *** |
| Time | X | * | *** | ** | *** | *** | *** | X | * |
| Analysis | * | ** | *** | ** | *** | *** | ** | ** | ** |
| Standard | ** ISO | *** ITU | * | ** | X | * | *** IEEE | *** | ?? |
| Lisibility | * T | ** T+G | * T | *** G | *** | *** G | * T | * T | ** |
| Availability | * | ** | * | * | * | * | *** | *** | * ?? |

Figure 7.8 Summary of some specification languages.

Each column summarizes the characteristic of a specific language:

*** : the language is excellent for the corresponding criteria

** : the language provides acceptable facilities

* : the language provides a little help

X : the language lacks the concepts

T : the language is textual

**G** : the language is graphical

**ISO** : the language is an ISO standard

**ITU** : the language is an ITU standard

**IEEE** : the language is an IEEE standard

? : non proved star

**C** provides a high usability power but it fails to provide general basic concepts for system specification such as concurrency and timing aspects.

**LOTOS** provides one of the highest expressive power for concurrency data structure (ADT) and communication (multiple Rendez-Vous). However it fails to provide real time concepts. Recent work introduced RT-LOTOS which extend LOTOS with real time aspects [220].

**Esterel** provide powerful concepts for expressions time and has a large analytical power. However its communication model (broadcasting) is restricted to the specification of synchronous systems.

**Petri nets** provide also a great deal of analytical power. However, they generally lack concepts for the specification of data structures. One should note that there is no standard. The literature include all kind of extension to the original Petri nets (colored, marked, timed, ...). When associated with data structure, Petri nets lose most of their analytical power.

**VHDL** provides an excellent usability. However, its expressive power is quite low for communication and data structures.

**Statecharts** provide high usability. However, like Esterel, it has a restricted communication model. The expression power is enhanced by the existence of different kinds of hierarchies (activities, states, modules).

**SpecCharts** is a promising language in terms of expression power and analytical power, however it lacks standardization.

**SDL** may be a realistic choice for system specification. Although it got only few "***", SDL provides acceptable facilities for most criteria. The weakest point of SDL is timing specification where only the concept of timer is provided. Additionally, SDL restrict the communication to the asynchronous model. The recent version of SDL introduces more generic communication model based on RPC.

**Java** is generating lots of hope for system-level specification. If we exclude the dynamic features of Java such as garbage collection, we obtain a model with a very high expression power. Although Java includes no explicit time concepts, these can be modeled as timers using the RPC concept. The usability power of Java is still unknown. Although the language is generating lots of books and tools, it is no clear if there will be a unique standard.

## 7.5 HETEROGENEOUS SPECIFICATION AND MULTI-LANGUAGE CO-SIMULATION

Experiments with system specification languages [263] show that there is not a unique universal specification language to support the whole life cycle (specification, design, implementation) for all kind of applications. The design of a complex system may require the cooperation of several teams belonging to different cultures and using different languages. New specification and design methods are needed to handle these cases where different languages and methods need to be used within the same design. These are multi-language specification design and verification methods. The rest of this section provides the main concepts related to the use of multi-language and two examples of multi-language co-design approaches. The first starts with a heterogeneous model of the architecture given in C and VHDL. The second makes use of several system-level specification languages and is oriented towards very large system design.

### *7.5.1 Basic Concepts for Multi-Language Design*

The concept of multi-language specification aims at coordinating the different cultures through the unification of the languages, formalism, and notations. In fact, the use of more than one language corresponds to an actual need in embedded systems design. In most system houses software design groups are separated from hardware design groups. In this case we find generally a third group in charge of the specification of the overall system. The design of large systems, like the electronic parts of an airplane or a car, may require the participation of several groups belonging to different companies and using different design methods, languages and tools. Besides, multi-language specification is driven by the need of modular and evolutive design. This is due to the increasing complexity of designs. Modularity helps in mastering this complexity, promotes for design re-use and more generally encourages concurrent engineering development.

There are two main approaches for multi-language validation: **compositional and co-simulation based.**

The compositional approach (cf. Figure 7.9) aims at integrating the partial specification of sub-systems into a unified representation which is used for the verification of the global behavior. This allows to operate full coherence and consistency checking, to identify requirements trace-ability links, and to facilitate the integration of new specification languages [245].

Several approaches have been proposed in order to compose partial programming and/or specification languages. Zave and Jackson 's approach [264] is based on the predicate logic semantic domain. Partial specifications are assigned semantics in this domain, and their composition is the conjunction of all partial specification. Wile's approach [265] to composition uses a common syntactic framework defined in terms of grammars ad transformations. The Garden project [266] provides multi- formalisms specification by means of a common operational semantics. These approaches are globally intended to facilitate the proofs of concurrent systems properties.

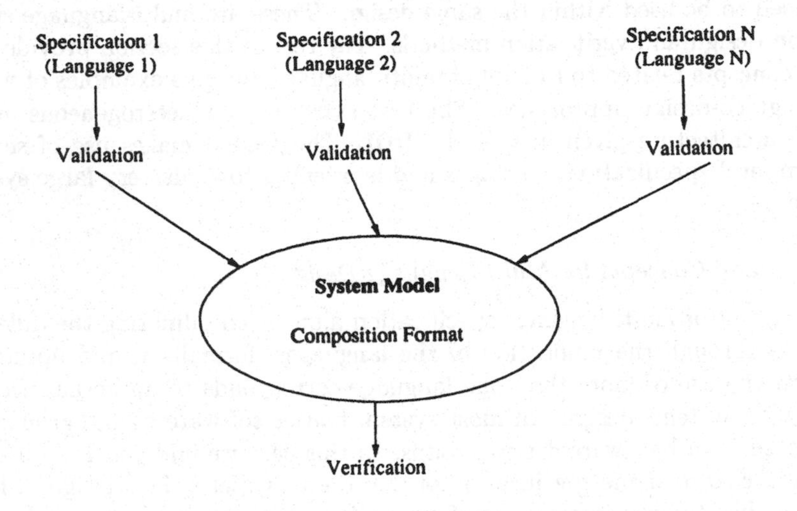


**Figure 7.9** Validation through composition.

The co-simulation based approach (cf. Figure 7.10) consists in interconnecting the simulation environments associated to each of the partial specification. Compared with the deep specification integration accomplished by the compositional approaches, co-simulation is an engineering solution to multi-language validation that performs just a shallow integration of the partial specifications.

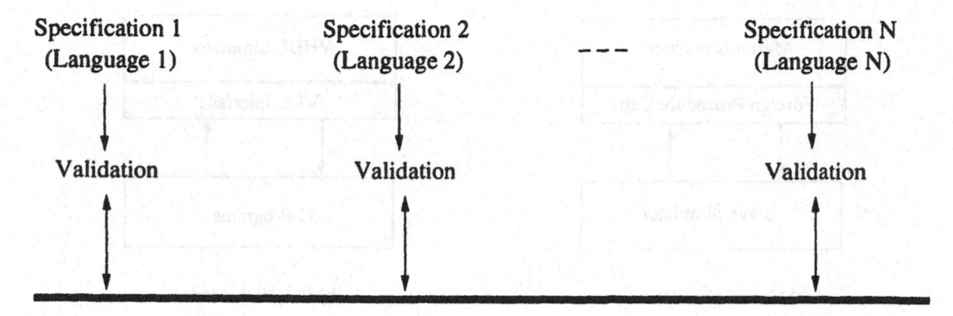


**Figure 7.10** Validation through co-simulation.

### *7.5.2 Co-simulation Models*

Co-simulation aims at using several simulators concurrently. The key issue for the definition of a co-simulation environment is the communication between the different simulators. There are mainly two simulation modes: the **master-slave** mode and the **distributed** mode.

With the master-slave mode, co-simulation involve one master simulator and one or several slave simulators. In this case, the slave simulators are invoked using procedure-call-like techniques and can be implemented in either of two ways:

1. calling foreign procedures (e.g. C programs) from the master simulator; or

2. encapsulating the slave simulator within a procedure call.

Figure 7.11 shows a typical master-slave co-simulation model.

This scheme is detailed using the co-simulation of mixed C-VHDL models. Most commercial VHDL simulators provide a basic means to invoke C functions during VHDL simulation following a master-slave model as shown in Figure 7.11b. This access is possible by using the foreign VHDL attribute within an associated VHDL architecture. The foreign attribute allows parts of the code to be written in languages other than VHDL (e.g., C procedures). Although useful, this scheme presents a fundamental constraint: a C program need to be sliced into a set of C procedures executed in one shot and activated through a procedure call. In fact, this requires a significant style change in the C flow, specially for control-oriented applications which needs multiple interaction points with the rest of the hardware parts [267]. By using this model, the

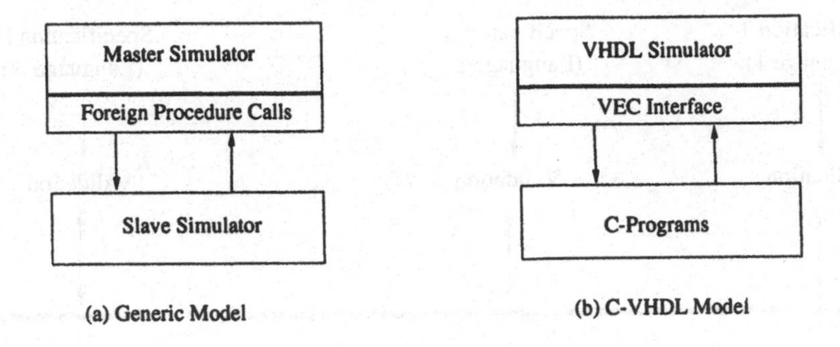


**Figure 7.11** Master-slave co-simulation.

software part for control-oriented applications requires a sophisticated scheme where the procedure saves the exit point of the sliced-C program for future invocations of the procedure. Moreover, the model does not allow true concurrency: when the C procedure is been executed, the simulator is idle.

The distributed co-simulation model overcomes these restrictions. This approach is based on a communication network protocol which is used as a software bus. Figure 7.12 shows a generic distributed co-simulation model. Each simulator communicates with the co-simulation bus through calls to foreign procedures.

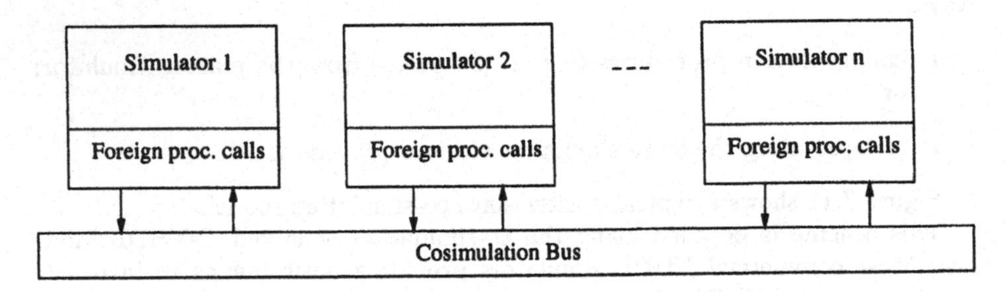


**Figure 7.12** The distributed co-simulation model.

The co-simulation bus is in charge of transferring data between the different simulators. It acts as a communication server accessed through procedure calls. The implementation of the co-simulation bus may be based on standard system facilities such as Unix IPC or sockets. It may also be implemented as an ad hoc simulation backplane [268].

Figure 7.13 shows a distributed C-VHDL model using the UNIX IPC as the co-simulation bus.

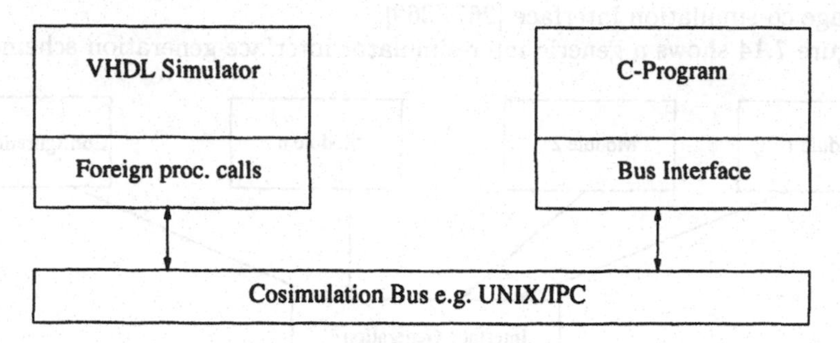


**Figure 7.13** Distributed C-VHDL co-simulation.

In this model, the communication may also be achieved through procedure calls. In practice, the capability to call a C procedure from VHDL is still in use, but the procedure contains only the protocol to address the software bus—that is, interprocess communication (IPC). In order to communicate, the VHDL simulator need to call a procedure that will send/receive information from the software bus (e.g. IPC channel). This solution allows the designer to keep the C application code in its original form. In addition, the VHDL simulator and the C program may run concurrently. It is important to note that the co-simulation models are independent of the communication mechanism used during co-simulation. This means that co-simulation can use other communication mechanisms than IPC (for example, Berkeley sockets or any other ad hoc protocol).

The distributed co-simulation model brings several advantages over the master slave co-simulation model:

**modularity** The different modules can be designed concurrently using different tools and design methods

**flexibility** Different modules can be simulated at different abstraction level during the design process

### *7.5.3 Automatic Generation of Co-Simulation Interfaces*

When dealing with co-simulation, the most tedious and error-prone procedure is the generation of the link between different simulation environments. In order to overcome this problem automatic co-simulation interface generation tools are needed. This kind of tools take as input a user defined configuration file, which specifies the desired configuration characteristics (I/O interface and synchronization between debugging tools) and produces a ready-to-use multi-language co-simulation interface [267, 269].

Figure 7.14 shows a generic inter-simulator interface generation scheme.

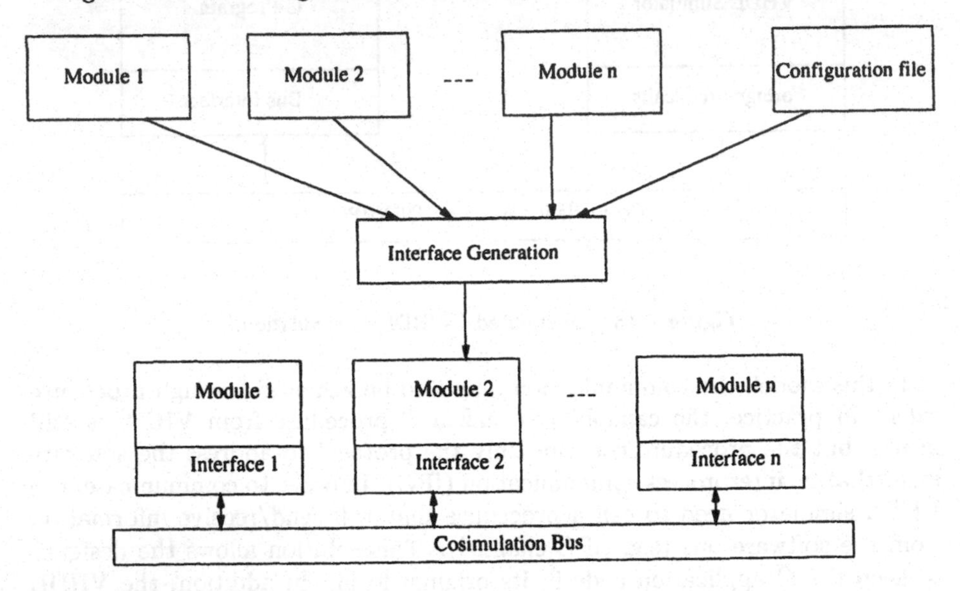


**Figure 7.14** Inter-simulator interface generation.

Based on the information provided by the configuration file, the interface generation process produces automatically the interface between the modules and the co-simulation bus.

The configuration files specify the inter-module interactions. This may be specified at different levels ranging from the implementation level where communication is performed through wires to the application level where communication is performed through high level primitives independent from the implementation. Figure 7.15 shows three levels of inter-module communication [270].

The complexity of the interface generation process depends on the level of the inter-module communication.

At the application level, communication is carried out through high level primitives such as send, receive and wait. At this level the communication protocol may be hidden by the communication procedures. This allows to specify a module regardless to the communication protocol that will be used later. Automatic generation of inter simulator interfaces for this level of communication

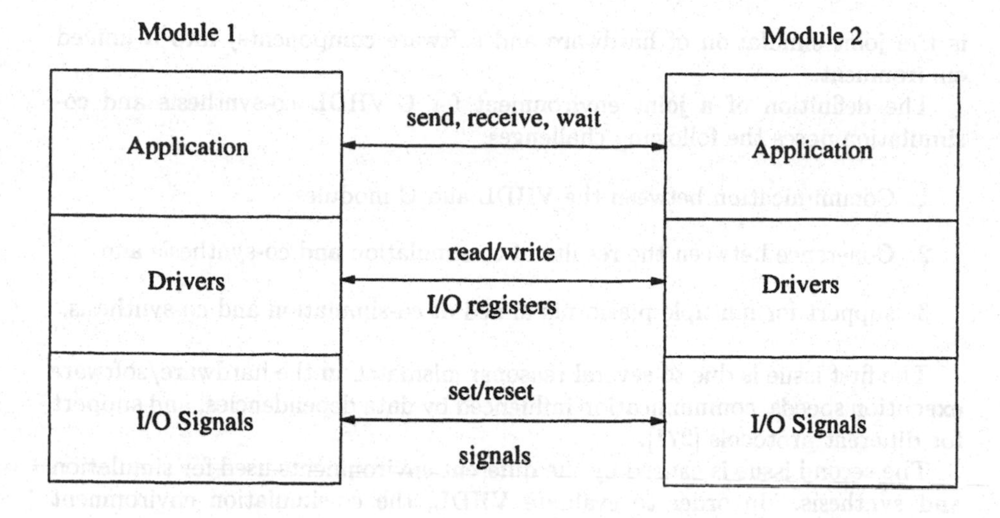


**Figure 7.15** Abstraction levels of inter-module communication

requires an intelligent process able to select a communication protocol. This process is generally called communication synthesis [9, 219] (see Section 2.5.6).

The next level may be called the driver level. Communication is performed through read/write operation on I/O registers. These operations may hide physical address decoding and interrupt management. Automatic generation of inter-simulator interfaces selects an implementation for the I/O operations. Coware is a typical co-simulation environment acting at this level [231].

At the lowest level all the implementation details of the protocol are known and communication is performed using operation on simple wires (e.g. VHDL signals). The inter-simulator interface generation is simpler for this level. Chapter 10 details VCI, a C-VHDL co-simulation method acting at the signal level.

### *7.5.4 Application: C-VHDL Specification and Co-Simulation*

This section deals with a specific case of multi-language approach based on C and VHDL. We assume that functions have already been allocated to hardware and software. The co-design process starts with a virtual prototype, an heterogeneous architecture composed of a set of distributed modules, represented in VHDL for hardware elements and in C for software elements. The goal is to use the virtual prototype for both co-synthesis (mapping hardware and software modules onto an architectural platform) and co-simulation (that is the joint simulation of hardware and software components) into a unified environment.

The definition of a joint environment for C-VHDL co-synthesis and co-simulation poses the following challenges:

1. Communication between the VHDL and C modules.

2. Coherence between the results of co-simulation and co-synthesis and

3. support for multiple platforms aimed at co-simulation and co-synthesis.

The first issue is due to several reasons: mismatch in the hardware/software execution speeds, communication influenced by data dependencies, and support for different protocols [271].

The second issue is caused by the different environments used for simulation and synthesis. In order to evaluate VHDL, the co-simulation environment generally uses a co-simulation library that provides means for communication between VHDL and C. On the other hand, the co-synthesis produces code and/or hardware that will execute on a real architecture. If enough care is not taken, this could result in two different descriptions for co-simulation and co-synthesis.

The third issue is imposed by the target architecture. In general, the co-design is mapping of a system specification onto a hardware/software platform that includes a processor to execute the software and a set of ASICs to realize the hardware. In such a platform (ex. a standard PC with an extended FPGA card), the communication model is generally fixed. Of course, the goal is to be able to support as many different platforms as possible.

C-VHDL environments differ by the abstraction level and communication model used.

The design of system on chip applications generally combine programmable processors executing software and application specific hardware components. The use of C-VHDL based co-design techniques enable the co-specification, co-simulation and co-synthesis of all the system. Experiments [272] show that these new co-design approaches are much more flexible and efficient than traditional method where the design of the software and the hardware were completely separated. More details about the design of embedded cores can be found in Chapter 5.

Figure 7.16 shows a typical C-VHDL based co-design method [267, 231]. The design starts with a mixed C-VHDL specification and handle three abstraction levels: the functional level, the cycle-accurate level and the gate level.

At the functional level, the software is described as a C program and the hardware as a behavioral VHDL model. At this level the hardware-software

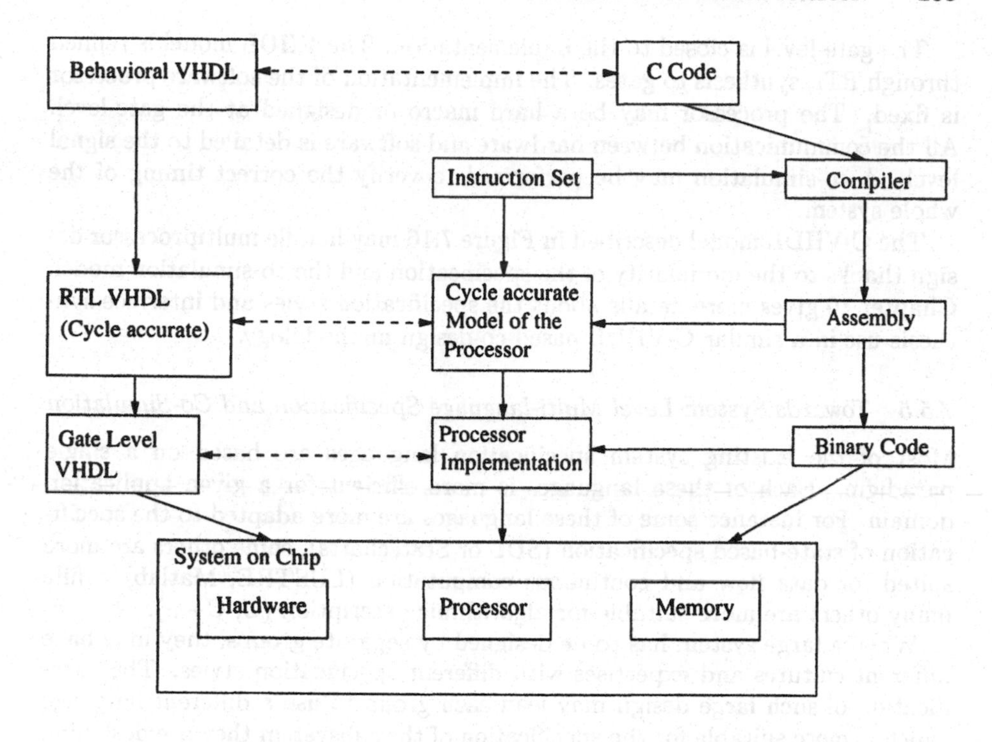


**Figure 7.16** C-VHDL based co-design for embedded processors.

communication may be described at the application level. A C-VHDL co-simulation may be performed for the verification of the initial specification. The C code is executed at the workstation and the VHDL is simulated at the behavioral level. Only functional verification can be performed at this level, accurate timing verification can be performed with the two next levels.

At the cycle accurate level, the hardware is described at the RT-levelregister-transfer design and the software at the assembly-level. A co-simulation may be performed for the verification of the behavior of the system at the cycle-level. The C code is executed on a cycle accurate model of the processor. This may be in VHDL or another software model such as C. At this level, the hardware/software communication can be checked at the behavioral cycle-level.

The cycle accurate model may be obtained automatically or manually starting from the functional level. The RTL VHDL model may be produced using behavioral synthesis. The assembler is produced by simple compilation as explained in Chapter 5; the main difficulty is the interface refinement [231].

The gate-level is closed to the implementation. The VHDL model is refined through RTL synthesis to gates. The implementation of the software processor is fixed. The processor may be a hard macro or designed at the gate-level. All the communication between hardware and software is detailed to the signal level. A co-simulation may be performed to verify the correct timing of the whole system.

The C-VHDL model described in Figure 7.16 may handle multiprocessor design thanks to the modularity of the specification and the co-simulation model. Chapter 10 gives more details about the specification styles and interface synthesis use in a similar C-VHDL based co-design methodology.

### *7.5.5 Towards System Level Multi-language Specification and Co-Simulation*

Most of the existing system specification languages are based on a single paradigm. Each of these languages is more efficient for a given application domain. For instance some of these languages are more adapted to the specification of state-based specification (SDL or Statecharts), some others are more suited for data flow and continuous computation (LUSTRE, Matlab), while many others are more suitable for algorithmic description (C, C++).

When a large system has to be designed by separate groups, they may have different cultures and expertises with different specification styles. The specification of such large design may lead each group to use a different language which is more suitable for the specification of the subsystem they are designing according to its application domain and to their culture.

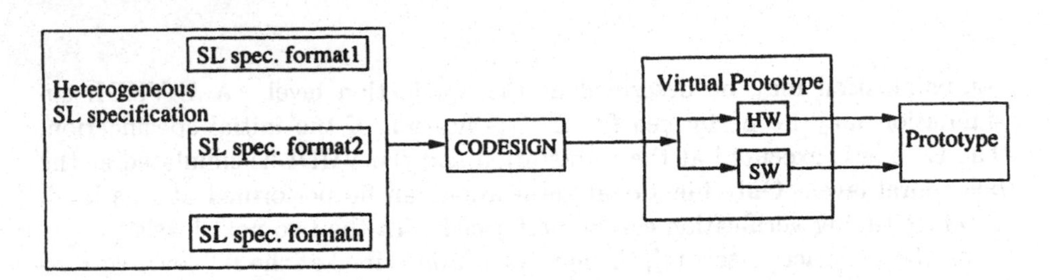


**Figure 7.17** System level heterogeneous specification.

Figure 7.17 shows a generic flow for co-design starting from multi-level specification. Each of the subsystems of the initial specification may need to be decomposed into hardware and software parts. Moreover, we may need to compose some of these subsystems in order to perform global hardware/software subsystems. In other words, partitioning may be local to a given subsystem or a global to several subsystems. The co-design process also needs to tackle the refinement of interfaces and communication between subsystems.

As in the case of the heterogeneous specification for system architecture, the problems of interfacing and multi-language validation need to be solved. In addition, this model brings another difficult issue: language composition. In fact, in the case, where a global partitioning is needed, the different subsystems need to be mapped onto a homogeneous model in order to be decomposed. This operation would need a composition format able to accommodate the concepts used for the specification of the different subsystems and their interconnect.

Only few systems in the literature allow such co-design models. These include RAPID [273] and the work described in [263]. Both systems provide a composition format able to accommodate several specification languages. This co-design model will be detailed in Section 7.5.

## 7.6 CONCLUSIONS

System-level specification is an important aspect in the evolution of the emerging system-level design methods and tools. This includes system-level specification languages which target designers and internal format which are used by tools.

Most hardware/software co-design tools make use of an internal representation for the refinement of the input specification an architectures. A intermediate form may be language-oriented or architecture-oriented. The first kind makes easier high-level transformations and the latter makes easier handling sophisticated architectures.

The intermediate form fixes the underlying design model of the co-design process. When the target model is a multi-processor architecture, the intermediate model needs to handle multi-thread computation.

A plethora of specification languages exists. These come from different research areas including VLSI design, protocols, reactive systems design, programming languages, parallel programming, functional programming and algebraic notations. Each of these languages excels within a restricted application domains. For a given usage, languages may be compared according to their ability to express behaviors (expressive power), their suitability for building tools (analytical power) and their availability.

Experiments with system specification languages have shown that there is not a unique universal specification language to support the whole design process for all kinds of applications. The design of heterogeneous systems may require the combination of several specification languages for the design of different parts of the system. The key issues in this case are multi-language validation, co-simulation and interfacing (simulator coupling).

The multi-language approach is an already proven method with the emergence co-design approaches based on C and VHDL. It is expected that for the future, multi-language design will develop and cover multi-system specification languages allowing for the design of large heterogeneous designs. In this case, different languages will be used for the specification of the different subsystems.

## Acknowledgments

This work was supported by France-Telecom/CNET under Grant 941B113, SGS-Thomson, Aerospatiale, PSA, ESPRIT programme under project COMITY 23015 and MEDEA programme under Project SMT AT-403.

# 8 THE COSYMA SYSTEM

Achim Österling, Thomas Benner, Rolf Ernst, Dirk Herrmann, Thomas Scholz, and Wei Ye

Technische Universität Braunschweig
Institut für Datenverarbeitungsanlagen
Braunschweig, Germany

## 8.1 OVERVIEW

This chapter gives an overview on Cosyma and is a complement to the existing literature in that it gives more tool and system specific information which gives a better impression of the whole system and is extremely helpful when using the system. More information, examples, the complete Cosyma system, a user manual, and a small library can be obtained from

ftp://ftp.ida.ing.tu-bs.de/pub/cosyma

This chapter also summarizes the Cosyma extensions which will have been released by the time of book publication.



*J. Staunstrup and W. Wolf (eds.), Hardware/Software Co-Design: Principles and Practice*, 263-282.

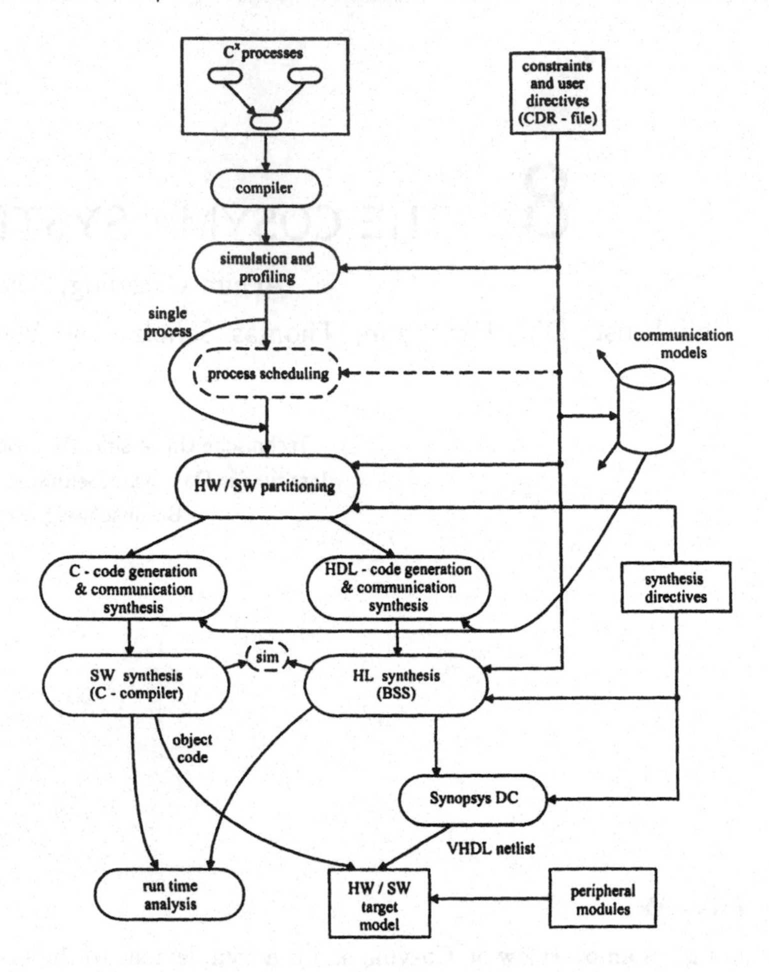


**Figure 8.1** The Cosyma design flow.

## 8.2 COSYMA ARCHITECTURE AND INPUT LANGUAGES

Cosyma (**COSY**nthesis for e**M**bedded micro **A**rchitectures) is a platform for design space exploration during hardware/software co-design.

The Cosyma target architecture consists of a standard **RISC processor** core (we provide a SPARC architecture model with 33 MHz clock and floating point coprocessor with Cosyma), a fast **RAM** for program and data with single clock cycle access time,and an automatically generated application specific

**coprocessor**. Peripheral units must be inserted by the designer. Processor and coprocessor communicate via **shared memory** in mutual exclusion.

The input description may consist of several communicating processes with timing requirements. The description consists of a function description in $C^x$, which is a minimum extension of the C programming language by a process statement allowing parallel processes. Process communication uses predefined C functions accessing abstract communication channels that are later mapped to physical channels (or are removed by optimization). It is similar to the approach used in Cosmos [46] except that both blocking and non blocking communication are supported [274]. Peripheral devices must be modeled as $C^x$ processes for simulation. $C^x$ processes are also used for stimuli generation. Both must be removed for process scheduling and partitioning.

The function description in $C^x$ is strictly separated from constraints and implementation directives to minimize C language extensions and to improve language portability. A *constraint and user directives file* contains time constraints which refer to labels in the $C^x$ processes, as well as channel mapping directives, partitioning directives, and component selection [274].

Constraints and global user directives, again, are separated from tool specific user control for a compact input description and to improve tool independence.

### *8.2.1 Cosyma design flow and user interaction*

Fig. 8.1 shows the Cosyma design flow. The input description is translated to a syntax graph using a customized version of the *Stanford SUIF compiler* [275]. This syntax graph is extended to an **Extended Syntax Graph (ESG)** by annotating the local data flow and the global data flow, as far as it can be analyzed. As an advantage, the program structure is conserved and the data annotations do not need to be complete, such as in a **data flow graph (DFG)** or in a **control-data flow graph (CDFG)**. Instead, data flow is analyzed as well as is possible while, for all other data dependences, default worst case assumptions can easily be derived from the syntax graph. This implementation strategy allowed us to extend data flow analysis with each version of Cosyma while maintaining correctness.

Next, the $C^x$ processes are simulated on a **RT-level** model of the target processor to obtain **profiling and software timing information**. Since *simulation* uses abstract channel functions for process communication, the result is an approximation of the final system behavior. In newer versions of Cosyma, profiling is replaced by a **symbolic analysis** approach, which is outlined later. Software timing data for each potential target processor is derived with simulation or symbolic analysis. This is a time consuming but necessary approach to obtain correct timing data for the later steps.

Single process systems directly proceed to hardware/software partitioning. For multiple process systems, Cosyma uses different approaches to scheduling. The first one uses **scalable performance scheduling (SPS)** [276] which finds a feasible and minimal serialization of processes which is then treated as a single process. In this case, *process scheduling* precedes hardware/software partitioning. This approach can only handle rate constraints on processes, while input/output timing constraints can only be regarded in a few special cases. A second, newer approach combines partitioning and scheduling and departs from mutual exclusive execution [277].

*Hardware/software partitioning* is the next step. Input to partitioning are the *ESG with profiling (or control flow analysis) information*, the *CDR-file* and *synthesis directives*. These synthesis directives include the number and type of functional units provided for the coprocessor implementation and they are needed to estimate the potential hardware performance. This parameter is controlled by user interaction.

Partitioning works at the basic block level. Since communication between basic blocks of a process is implicit, partitioning requires *communication analysis* and, thereafter, *communication synthesis*. This is a major difference to all approaches which partition on the level of processes using explicit communication described by the user. Communication is inserted when the internal ESG representation is translated back to C for software synthesis and to a hardware description language for high-level synthesis.

For *high-level synthesis*, the **Braunschweig Synthesis System (BSS)** is used. BSS creates a diagram showing the scheduling steps and function unit and memory utilization which allows to identify system bottlenecks. An RT-level synthesis tool, the *Synopsys Design Compiler*, generates the final netlist. For software synthesis, a standard C compiler is used. There is also a tool which allows to co-simulate the object code running on the target processor and an RT-level hardware description provided by BSS. This tool is mainly used for Cosyma software test.

*Run time analysis* is the last step which uses software simulation and hardware scheduling results in formal analysis for timing constraint validation, assuming worst-case communication channel timing. This saves an extra co-simulation step and has turned out be sufficiently accurate [278]. The netlist is finally completed by VHDL models for the peripheral modules which were excluded from the synthesis process.

The intermediate Cosyma results can be examined in many forms:

- a process schedule;
- a Gantt diagram and process graph for hardware/software partitioning;

- a marked list of hardware and software blocks for communication synthesis;
- communication statements in the synthesized C code;
- and scheduling diagram showing operations and memory access per control step.

## 8.3 HARDWARE/SOFTWARE PARTITIONING

Partitioning in Cosyma starts with an all software solution and tries to extract hardware components iteratively until all timing constraints are met. The partitioning goals are in order of decreasing importance:

- meet real-time constraints;
- minimize hardware costs;
- minimize the CAD system response time and, thus, allow the user to investigate the influence of system modifications on systems costs and performance.

While the first goal is a hard constraint because it is a necessary condition, the other ones are soft optimization goals. Hardware/software partitioning can be considered as a scheduling problem which is known to be NP-hard. Given a user specified maximum amount of hardware components (i.e. ALU, multiplier, ...), our task is to find a minimum set of code segments that lead to a sufficient speedup when implemented in hardware. We selected **simulated annealing**, a stochastic optimization algorithm.

The elements in the partitioning process are basic blocks which seems to be a manageable compromise between statement level (**fine-grain**) on the one side and function or process level partitioning (**coarse-grain**) on the other side. The total estimated costs of moving a single basic block $b$ from software to hardware amounts to:

$$\begin{aligned}\Delta c(b) &= w \cdot \Delta t(b) \\ &= w \cdot [(t_{HW}(b) - t_{SW}(b) + t_{com}(Z) - t_{com}(Z \cup b)) \cdot It(b)]\end{aligned}$$

where $\Delta c(b)$ is the estimated cost increment, $w$ is a weight factor to control simulated annealing, $\Delta t(b)$ is the estimated execution time increment, $t_{HW}(b)$ is the estimated coprocessor execution time of $b$, $t_{SW}(b)$ is the estimated execution time of b on the processor and $t_{com}$(Z) is the estimated processor-coprocessor communication time, given the current set $Z$ of basic blocks on the coprocessor. $It(b)$ is the number of iterations as determined by profiling/tracing on $b$. All estimations are required previous to partitioning because simulated annealing needs a fast cost function computation.

The execution time of a basic block on the processor and the coprocessor depends on the global optimization of the compiler and the synthesis tool. Each move of a block from software to hardware and vice versa changes the optimization potential. Since Simulated Annealing typically tries several hundreds of thousands of hardware/software configurations, off-line estimation techniques with fast incremental update after each move are necessary.

- $t_{SW}(b)$ is estimated with a local source code timing estimation based on simulation data. Estimation inaccuracy results from data dependent instruction execution times (SPARC: e.g. mult, div), optimization and register allocation in the compiler. Alternatively, trace data from processor simulation could be used [279].

- $t_{HW}(b)$ is estimated with a **list scheduler** (fast) or a **path-based scheduler** (more accurate) [280] on $b$ using the execution time (number of clock cycles) for each operator in high-level synthesis, given the user specified number of functional units used.

- $t_{com}(Z \cup b)$ is estimated by **data flow analysis**. For shared memory, communication costs are proportional to the number of variables to be communicated.

Since we have to handle hard constraints and soft optimization goals a special treatment of the applied Simulated Annealing is necessary. The complete cost function can be found in [281].

## 8.4 HARDWARE AND SOFTWARE SYNTHESIS

BSS (Braunschweig Synthesis System) is a high-level synthesis system developed specifically for fast coprocessor design. During partitioning, those basic blocks and functions which shall be implemented in hardware are marked in the ESG. In the first step, they are translated to the internal format of BSS which is a hierarchical CDFG. The marked basic blocks typically form sequences which are grouped as *segments*. When the *coprocessor* is activated by the *processor*, the processor sends a *segment identifier* to a reserved address, which activates the coprocessor and lets it execute the corresponding segment (see fig. 8.2). If we regard a segment as a complex instruction, we could consider the coprocessor to be extremely vertically microcoded. The data are passed in shared memory. Array variable accesses are split into index computation and memory read or write access.

Pointer variables are treated similarly. Passing array pointers instead of scalar variables drastically speeds up communication. The current version of

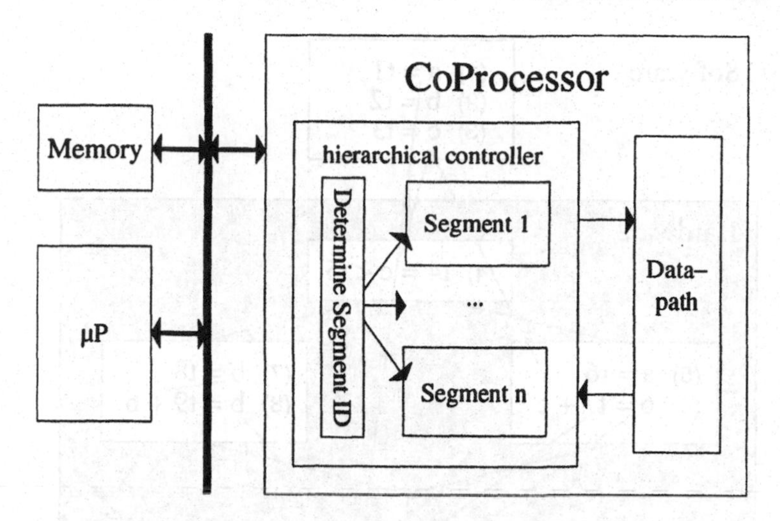


**Figure 8.2** Outline of the Cosyma hardware architecture.

BSS cannot handle functions, so they are inlined at translation time. So, the generated CDFG contains the segment identifier recognition, reading and writing operations to the shared memory for variable passing and the translated ESG segment. This segment can contain variables which are not in shared memory. Finally, the profiling and tracing information is included for synthesis and timing optimization.

The unmarked software part of the ESG is translated to C including segment identifier and parameter passing to the coprocessor. Suitable type definition and additional statements make sure that shared variables are not register allocated at the time of coprocessor activity. The GNU C compiler *gcc*, run at the maximum optimization level, then generates object code for the processor.

## 8.5 COMMUNICATION ESTIMATION AND CODE OPTIMIZATION

Communication estimation and code optimization is handled by an approach based on Aho, Sethi, and Ullman's algorithms on data flow analysis [138]. The underlying data structure is a variation of the **flow graph** (which is part of the ESG), presented at the same place.

A flow graph contains control structures like loops and conditional branches, as well as basic blocks. A basic block has exactly one entry point, one exit point at the end, and a strictly linear control flow. All calculations are contained in basic blocks in the shape of **three-address expressions**, where a target variable receives its value from an operation of at most two source operands (as

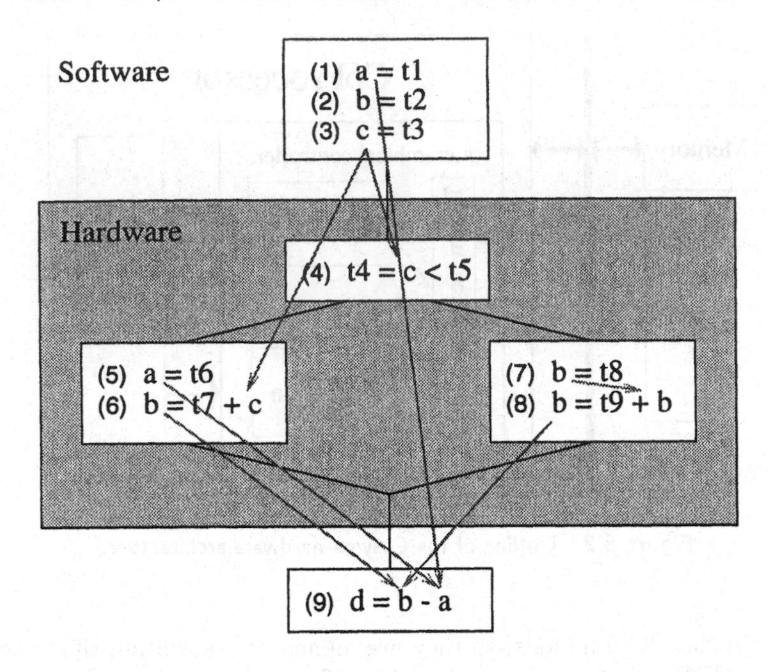


**Figure 8.3** Data flow analysis of an `if...then...else` structure.

an exception, function calls can have more parameters). By introducing temporary variables, every complex instruction can be broken down to a sequence of three-address expressions.

Data flow analysis, which is described in more detail in Chapter 5, determines possibly active expressions in the program. The results can be used to guide optimizations like **constant propagation, arithmetic optimization, tree height reduction,** and **common subexpression elimination.** Transformations on the higher level, like **loop unrolling,** need similar techniques to observe certain variables.

Cosyma estimates communication overhead $t_{com}$ for a given hardware/ software partition. For the moment, Cosyma does not support burst communication optimization. Therefore, communication overhead is proportional to the number of variables to be communicated. That number can be calculated using data flow analysis of a given flow graph. Every basic block of the flow graph is mapped to a certain module (e.g. hardware or software) of the co-design.

As an example, the shaded area of the graph in Figure 8.3 is mapped to hardware while the rest shall be realized in software. Every variable with arcs, starting and ending in different modules has to be communicated from the module containing the definition to the module containing the consumption. Thus, variable c of the example has to be transmitted from software to hardware and b has to take the opposite way. Things are a little more complicated for variable a. It obviously must be communicated from hardware to software, because a change may have happened in the left branch. Since there is exactly one exit path from the hardware module, that communication is done unconditionally. This implies the need to also communicate it to hardware to provide the correct definition (*1*) when the right branch is taken, albeit there is no explicit consumption in this module. This behavior is modeled by running data flow analysis a second time after inserting pseudo-code for all communications in the first run.

This method of communication estimation is not fixed to bi-partitioning. It can be used without changes for any number of modules. It is possible to use different communication schemes also, as long as communication is restricted to scalar variables. Currently, arrays have to be allocated to shared memory. Our future work will deal with array and pointer usage in distributed memory architectures.

## 8.6 THE SCALABLE-PERFORMANCE SCHEDULER

Starting from a description consisting of parallel processes, the scheduler tries to transform that description to one that can be executed in a single processor environment, thereby taking into account user supplied constraints [276].

To find a schedule close to the optimum, a binary search algorithm is put on top of the scheduling algorithm, i.e., for each stage of the binary search the scheduler tries to find a valid schedule. Binary search (and thus the scheduler) are controlled by the so-called **speedup** factor. The speedup factor tells the user how much the target architecture must be accelerated compared to the reference processor the run-time analysis was performed on. The calculated speedup gives the user a clue of how fast the target architecture approximately must be. This information is given in an early design stage prior to target architecture definition. Thus, the user can explore the design space because the automatic design support makes changes less expensive. Besides, the schedule and the speedup may help the user to identify and fine tune or remove emerged bottlenecks from the algorithmic description.

After having applied the run-time analysis to the serialized program the scheduler can be invoked to perform the scheduling. The first task is to calculate a **macro period** from the iteration rates (**rate constraints**) of the various processes. Assume we have three processes with rates 100, 150, and 200 time units. The **GCD** is 50, so the ratio of the rates is 2:3:4. The macro period amounts to the **LCM** of the ratio values multiplied by the GCD, in our case 12·50=600 time units, where the first process is executed 6 times, the second process 4 times, and the third process 3 times. Obviously, the smaller the LCM the shorter the macro period and the shorter the resulting serialized program.

Now, the scheduler serializes the processes as many times as determined by the macro period, thereby taking into account the user-supplied constraints. As mentioned above, each schedule pass is controlled by the speedup factor, i.e., all the instructions of the program are accelerated by that speedup factor. Obviously, the objective is to keep the speedup factor as small as possible because that would require either a less powerful target architecture or fewer segments mapped to hardware and, therewith, less area for the controller, otherwise the target architecture may become oversized.

If a valid solution has been found the speedup factor is reduced otherwise increased and a new schedule pass is initiated unless any end criterion of binary search is met.

### *8.6.1 An Example*

A model train control serves as an example. The train is controlled by a personal computer, from which 35 bit values are transferred over the rails. The model train has a rather powerful motion speed regulation that moves the train close to original large trains. A *decoder* scans the pulse duration on the rails with a rate of 313 kHz. Due to serious noise, a low pass filter and a 3 bit burst error correction code are used. The corrected and decoded value is passed to a regulator, that receives the actual velocity from a speedometer. Finally, a motor controller generates pulses for the motor electronics.

The $C^x$ description consists of five processes. The first process is the decoder that has a cycle time of $3.2\mu s$ and an execution time of $5.5\mu s$. This process alone cannot be implemented on a SPARC, but a speedup of 1.7 would be required. All the other processes are executed with a cycle time of $51.2\mu s$. The processor utilization is the sum of the quotients of the execution time and cycle time of each process (non scaled):

$$
\begin{aligned}
U_{ns} &= \sum_{i=1}^{n} \frac{T_i}{C_i} = \frac{5.5\mu s}{3.2\mu s} + \frac{18.0\mu s}{51.2\mu s} + \frac{16.7\mu s}{51.2\mu s} \\
&\quad + \frac{3.7\mu s}{51.2\mu s} + \frac{2.0\mu s}{51.2\mu s} = 2.51.
\end{aligned}
$$

This means, a speedup of 2.51 would be necessary in order to run the example on a SPARC processor without context switch and if all input signals would be available at the beginning of the scheduling period.

To include **context switching time**, the heuristic factor $a_{dep}$ is introduced. This factor is adapted in an iterative process starting with a heuristic value of 2.2. Figure 8.4 shows the Gantt diagrams of three valid schedules during the iterative process. A value of 2.20 (Figure 8.4a) for $a_{dep}$ leads to a scaling factor of 5.52. The processes are split into basic blocks. One block in the diagram represents the activation of at least one basic block of a process. The decoder process is activated every $3.2\mu s$. All the other processes are active only at the beginning of the period. In the interval from $16\mu s$ until $52\mu s$ only the decoder process is active. This leads to a scaled processor utilization of $U_s = 47.49\%$ ($CPU_{used}$ in fig. 8.4).

In the next iteration, $a_{dep}$ is set to 1.6 (not shown). After that, $a_{dep}$ is decreased to 1.3, as can be seen in Figure 8.4b. This adaption influences the order of the processes. Some activations of the decoder are brought forward. The processor utilization increases to 79.32%. As a result, the schedule becomes more compact.

One more valid schedule is found for $a_{dep}$=1.15. Since the difference of $a_{dep}$ of the last valid schedule (1.15) and the current $a_{dep}$ (1.075) is less than the supplied minimal interval width (0.1) the scheduler stops returning the best $a_{dep} = 1.15$.

As a consequence a speedup of 2.89 is required in order to implement the design on the SPARC. Most of this speedup is required to implement the decoder as seen in the Gantt diagram.

## 8.7 SYSTEM OPTIMIZATION EXAMPLE

### *8.7.1 Architecture Template*

Fig. 8.5 shows an architecture template for Cosyma applications in the **digital signal processing** domain. Input and output buffers are inserted to decouple the processor-coprocessor system from the strictly periodic I/O data stream thus widening the design space. The internal system is defined by a C description with a single rate constraint.

### *8.7.2 Example: Smooth Image Filter*

In the following we will show how to optimize an **FIR filter** implementation with Cosyma based on this template. Figure 8.6 shows the C description of the computation intensive part of the code. The image dimensions, filter window size and filter coefficients are fully parameterized. Even if the set of parameters

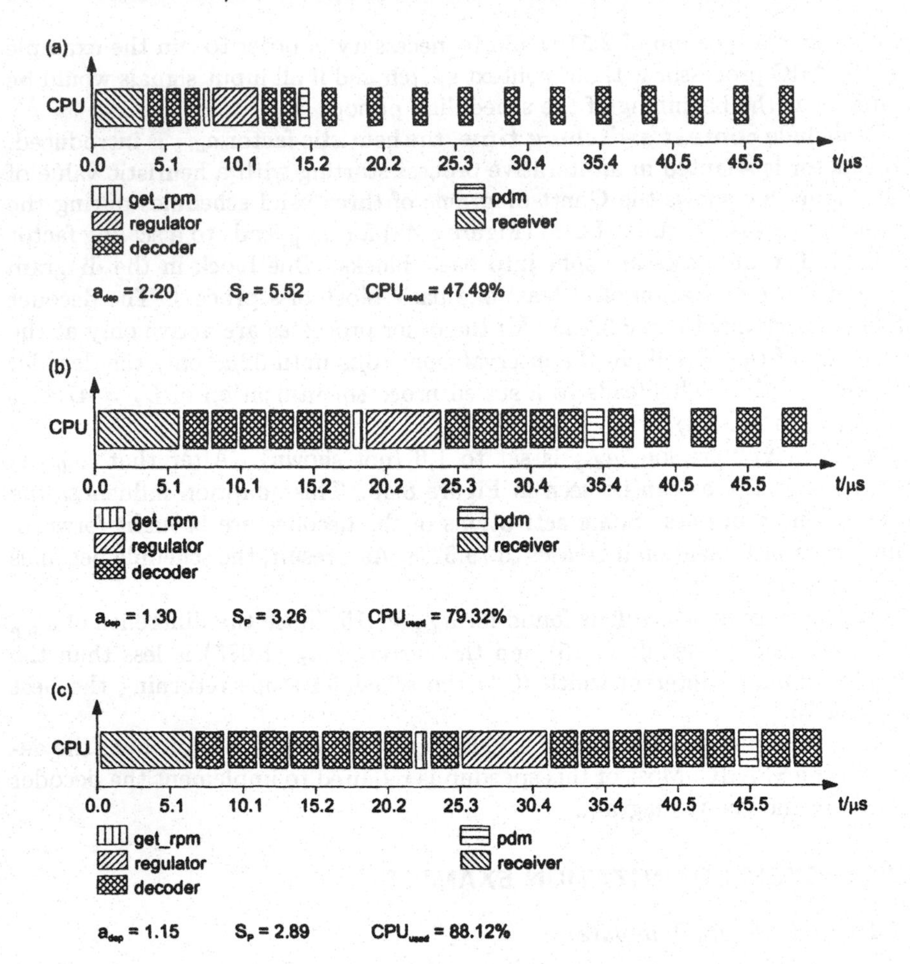


**Figure 8.4** Adaptation of the scaling factor.

would be fixed in the target system, it can be advantageous to realize a generalized algorithm because this increases the possibility of reusing the system later.

**Optimizing the Algorithm.** A first execution of Cosyma's run time estimation tool delivers an initial clock cycle count of 5348184 cycles for a pure software solution on a SPARC processor. In the sequel we will demonstrate how the execution time of the system can be reduced iteratively by transformations

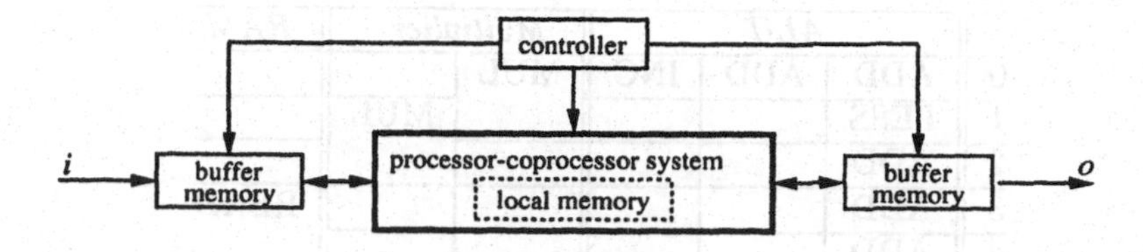


**Figure 8.5** A system architecture template.

```
for (row = 0; row < rows - wrows + 1; row++) {
    for (col = 0; col < cols - wcols + 1; col++) {
        sumval = 0;
        for (wrow = 0; wrow < wrows; wrow++) {
            for (wcol = 0; wcol < wcols; wcol++) {
                sumval += IN[(row + wrow) * cols + (col + wcol)]
                         * FIL[wrow * wcols + wcol];
            }
        }
        sumval = (sumval * scale) >> 8;
        OUT[(row + offset_rows) * cols + col + offset_cols] = sumval;
    }
}
```

**Figure 8.6** Filter C description.

on the input description. Cosyma and BSS are used to get an indication of which transformations would be beneficial, and to observe the improvements achieved after each transformation is applied. In table 8.2 the intermediate results are shown for a pure software solution as well as for a hardware/software system where the code of the three inner loops is executed in hardware by a coprocessor. The last column gives the speedup of the HW/SW system compared to the software execution time of the original solution. BSS is granted 4 ALUs and 2 multipliers as the maximum number of resources.

Trying to reduce the initial execution time by hardware/software partitioning shows that the highest reachable speedup for the fully parameterized code is 12.28 which is reached when all of the code is executed in hardware. A closer look at the hardware schedule in table 8.1 generated by BSS shows a poor usage of the given hardware components in the inner loops, which indicates that there is too little **instruction level parallelism** in the original version of the code.

To increase instruction-level parallelism, it is possible to execute several iterations of the innermost loop concurrently. By unrolling the inner loop we

| | *ALU* | | | *Multiplier* | | *RAM* |
|---|---|---|---|---|---|---|
| 0 | ADD | ADD | INC | MUL | | |
| 1 | LESS | | | | MUL | |
| 2 | ADD | | | | | |
| 3 | ADD | | | | | READ |
| 4 | ADD | | | | | |
| 5 | | | | | | READ |
| 6 | | | | | | |
| 7 | | | | MUL | | |
| 8 | | | | | | |
| 9 | ADD | | | | | |

**Table 8.1** Innermost loop schedule

can provide a larger loop body which gives more freedom to the BSS scheduler. Since in our example there are no data dependences between subsequent iterations of the inner loop, we can expect a major performance increase. At this point it turns out to be problematic that the dimensions of the filter window are parameterized and thus the iteration count of the inner loop is unknown: When unrolling the loop $n$ times, we have to take precautions for the cases that the iteration count of the inner loop is not a multiple of $n$. To keep matters simple, in our example we just fix the size of the filter window to three columns and unroll the inner loop completely. Thus we get a better optimized solution for the price of less flexibility. Increasing performance or optimization potential by decreasing generality is a typical tradeoff in co-design.

After this transformation is applied, the execution times of both the pure software and the hardware/software solution are reduced to 55% of the original execution times (see table 8.2). Repeating this transformation with the second inner loop leads to further comparable speedup.

Again, the schedule generated by the high-level synthesis tool indicates a problem: Reading pixel luminances and filter coefficients inside the unrolled loops requires 18 memory accesses taking two cycles each. Therefore, ALUs and multipliers are idle, waiting for data from RAM most of the time. It would be possible to widen the bottleneck by allowing more than one RAM interface, but we should try to omit unnecessary memory accesses first. We observe that the coefficients of the filter matrix are read from an array for every pixel. By preloading these values into scalar variables we can enforce buffering in registers at the source level.

Looking at the effects of preloading in Table 8.2 we notice that the execution time of the pure software solution is reduced to 92% compared to the software execution time of version 3, while the execution time of the hardware/software solution is reduced to 68% of the former hardware/software execution time. This relationship indicates that the number of memory accesses is a major obstructing influence against good exploitation of instruction level parallelism.

Further memory read operations can be saved by optimizing access to image data: Since the windows of pixel luminance data overlap in successive iterations, the number memory accesses per pixel can be decreased from 9 to 3 by introducing **buffers** for the overlapping regions. Only a single column of the height of the filter window has to be reread for the processing of every pixel.

| *Version* | *SW Cycl.* | *HW/SW Cycl.* | *Cycl./Pixl.* | *Speedup* |
|---|---|---|---|---|
| 1: original | 5 348 184 | 440 712 | 9 × 10 | 12.1 |
| 2: inner loop unrolled | 3 086 796 | 244 170 | 3 × 18 | 22.8 |
| 3: 2nd loop unrolled | 2 383 337 | 173 973 | 44 | 30.6 |
| 4: coefficients preloaded | 2 185 503 | 118 119 | 29 | 45.1 |
| 5: final | 1 938 169 | 71 991 | 17 | 74.0 |

**Table 8.2** Results for the smooth benchmark.

Applying these transformations, we end up with a final speedup of 74 compared to the software execution time of the original version of the algorithm. There is still a lot of optimization potential: Up to now no loop pipelining was applied for the generation of the hardware schedule, we could select faster memories or even allocate the data to different memories that could be accessed concurrently, restructure the algorithm further to use additional line buffers, perform unrolling again, ....

### *8.7.3 Real World Example*

The technique of incremental improvement of the input description and during the high-level synthesis phase was used in a case study [282], where a system was designed which executed two different professional studio video algorithms. The result of the project was a very competitive design of a **high performance, low-cost video system**. The project showed the massive impact of high-level transformations, memory architecture and memory access optimization to the performance of the target system, as should have become evident from our filter example as well.

The **flexibility issue**, which was mentioned in the previous section, played a major role in our case study. We observed a tremendous impact of flexibility requirements on the design process, the solution space, and on the target architecture. Only those pure software solutions, which did not have any remaining parameters at all, came close to the required performance. In the hardware/software solution the requirements were met, although still most parameters were kept. The case study showed, that hardware and software use different approaches to flexibility and that software is not flexible by definition, as well as hardware is not completely inflexible.

## 8.8 NEW APPROACHES IN COSYMA

### *8.8.1 Multi-Way Partitioning for Heterogeneous Systems*

In the original approach, Cosyma was restricted to a target architecture consisting of a single processor and one coprocessor with shared memory communication. A new version of Cosyma with a **multi-way partitioning algorithm** handles systems with **multiple heterogeneous processors** and **coprocessors** running in parallel and communicating over a **shared memory** or **point-to-point interconnect**.

Many applications, such as mobile communication, show a mixture of both control and data dominated applications on different levels of a design often included in a single task, such as a wireless communication channel. Data dominated applications, in particular periodic applications in signal processing and control engineering applications make extensive use of **buffering** to increase the design space and allow **high performance pipelining** on various levels of granularity.

On the other hand, the current control dominated co-synthesis approaches which focus on relatively small systems with processor-coprocessor architectures make little or no use of buffering and pipelining. [33] covers more complex heterogeneous multiprocessor architectures. They have presented several heuristic allocation and static or dynamic scheduling algorithms for global system optimization.

The system in Figure 8.7 is a simplified model of a remote motor controller. It receives packeted messages from a central controller over a specialized simple field bus. The messages are encoded and any error condition must be signaled on the bus after a maximum time of $t_{IO1}$ after the error has been detected. Process $P_1$ is responsible for interface control. When complete, $P_2$ will process the message and decide whether the *motor control* must be adjusted. $P_1$ and $P_2$ are control tasks reacting to messages which arrive asynchronously at some unknown time. Thus, a bus message causes a chain of events. But, as an additional constraint, both must be fast enough not to miss any message on the

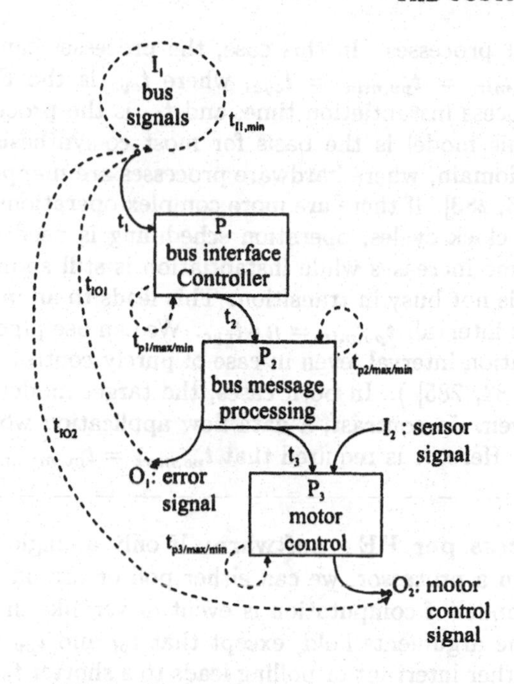


Figure 8.7 A remote motor controller.

bus. Here, we distinguish between two kinds of constraints. The *bus interface* controller is activated according to the signal rate of the *bus signals* (cycle time $t_{p1,max/min}$). Whenever a whole message word is received, it has to be decoded and passed to the *motor controller* by the *bus message controller* $P_2$. The $t_{p2,max/min}$ of this process corresponds to the bus signal rate divided by the message length. Process $P_3$ is activated independently of $P_1$ and $P_2$ with the rate $t_{p3,max/min}$.

The **inter process constraints** $t_{IO1}$ and $t_{IO2}$ influence the activation of a process chain. When a transmission error occurs, process $P_2$ must send an error message within $t_{IO1}$ after sending the first bit of the message. Also, the motor should react within $t_{IO2}$ after the the start of the message transfer.

In the new Cosyma system multi-way partitioning approach, **reactive** and **data flow oriented** processes are considered. Partitioning on the different processing elements (PE) and static scheduling are performed simultaneously by simulated annealing. Different configurations are considered:

- **single process per PE - hardware:** For control dominated systems without complex data operations, **RT-level FSMs** are often sufficient to implement processes. In this case, the processes can react instantaneously: $t_{pi,min} = t_{pv,min} = t_{cyc}$, where $t_{cyc}$ is the FSM cycle time, $t_{pi}$ is the process instantiation time, and $t_{pv}$ is the process instantiation interval. This model is the basis for most co-synthesis approaches in the control domain, where hardware processes are mapped to individual FSMs [10, 73, 283]. If there are more complex operations on data, which take several clock cycles, operation scheduling is needed and the FSM transition time increases while instantiation is still spontaneous as long as the FSM is not busy in transition. This leads to an increasing process instantiation interval: $t_{pv,min} = n * t_{cyc}$. We can use pipelining to reduce the instantiation interval, even in case of purely control dominated functions (e.g. [284, 285] ). In both cases, the target model of computation is event driven. In contrast, a data flow application would be executed periodically. Here, it is required that $t_{pe,max} = t_{pv,min,max}$, where $t_{pe}$ is the period.

- **single process per PE - software:** If only a single reactive process is running on a processor, we can either poll or use an interrupt. Here, the target model of computation is event driven like in hardware above and the same arguments hold, except that $t_{pi}$ and $t_{pv}$ will typically be larger. Whether interrupt or polling leads to a shorter $t_{pi}$ depends on the application and is architecture dependent.

- **multiple processes per PE:** If we assume that a PE has a single control flow, local scheduling is required for both hardware and software. Different strategies may be used for the various PEs depending on the processes mapped to the PE. Cosyma supports a **static scheduling approach** described in [286].

The optimization goal in scheduling and partitioning is to implement the system with minimal application specific hardware cost, regarding all time constraints. The target architecture template is used as a constraint during co-synthesis and can be adapted by the user for design space exploration. The application specific hardware cost is estimated with an approach that is based on earlier work of Vahid et al. [287].

In Table 8.3 we compare different implementation alternatives of a part of an *Ethernet bridge* benchmark which differ in the use of process pipelining (P) and the target architecture. Column 2 specifies whether processes are pipelined. The columns 3-5 specify the amount of hardware resources. The type and number of processor cores and their communication is provided by the user, while the algorithm minimizes the application specific hardware (16 bit coprocessor). The execution time of the partitioning tool on an UltraSparc

| *Benchmark* | *P* | *8051* | *SPARC* | *Gates (Coprocessor)* | *t(s)* |
|---|---|---|---|---|---|
| Bridge A | n | 1 | | 74195 | 42 |
| Bridge A | y | 4 | | | 133 |
| Bridge A | y | | 1 | | 40 |
| Bridge A | y | 1 | | 65658 | 123 |
| Bridge B | y | | 1 | 20626 | 47 |
| Bridge B | n | 1 | | 76450 | 91 |
| Bridge B | n | 1 | 1 | 19861 | 44 |
| Bridge B | y | 1 | 1 | 9204 | 52 |

**Table 8.3** The bridge examples

Workstation is given in the last column. The bridge benchmark consists of seven processes. In Bridge A the fastest process is executed with a cycle time of $50\mu s$, in Bridge B its cycle time is $100\mu s$. The A version can be mapped to a single *SPARC processor* as well as to four *8051 cores* without additional hardware. For all mixed HW/SW solutions with an 8051 holds that most of the design is mapped to the coprocessor, whereas the coprocessor replaces up to three 8051 cores. The last variant is mapped to a SPARC, an 8051, and a *coprocessor.*

### *8.8.2 SYMTA— Symbolic Timing Analysis*

The current simulation based profiling approach only calculates timing data for a set of input patterns. So, the user is responsible for selection of the worst case input data set. The new timing analysis approach called **SYMTA (symbolic hybrid timing analysis)**, which is a hybrid approach combining **simulation** and **formal analysis** using **symbolic simulation** provides tight upper and lower timing bounds and is adaptable to different architecture and program properties [279]. SYMTA analysis is based on program and architecture classification. SYMTA program path analysis uses symbolic simulation [288] to generate path information.

## 8.9 CONCLUSIONS

Cosyma is a platform for exploration of co-synthesis techniques. Beginning with rather small target architectures and single input programs it has developed into a design system for fairly complex time constrained multi process systems and larger heterogeneous target architectures. So far, the system has mainly been used for design-space exploration where it gives fast response times which are not available in a purely manual design process.

# 9 HARDWARE/SOFTWARE PARTITIONING USING THE LYCOS SYSTEM

Jan Madsen, Jesper Grode, and Peter V. Knudsen

Department of Information Technology
Technical University of Denmark
Lyngby, Denmark

## 9.1 INTRODUCTION

**Hardware/software partitioning** is often viewed as the synthesis of a target architecture consisting of a single processor and a single dedicated hardware component (full custom, FPGA, etc.) from an initial system specification, e.g. as in [289]. Even though the **single processor**, single **dedicated hardware** architecture is a special and limited example of a distributed system, the architecture is relevant in many areas such as **DSP** design, construction of embedded systems, software execution acceleration and hardware emulation and prototyping [290], and it is the most commonly used target architecture for automatic hardware/software partitioning.



*J. Staunstrup and W. Wolf (eds.), Hardware/Software Co-Design: Principles and Practice*, 283-305.

In this chapter we present the LYCOS (LYngby CO-Synthesis) system which is an experimental hardware/software co-synthesis system. In its current version, LYCOS may be used for hardware/software partitioning using a target architecture as described above. In this chapter we will focus on how LYCOS is used to do design space exploration in codesign. Details about the algorithms used within LYCOS can be found elsewhere [291, 292, 293, 294].

The chapter is organized as follows. Section 9.2 presents the problem of hardware/software partitioning and its relation to design space exploration. In Section 9.3 we give an overview of the LYCOSsystem. Section 9.4 gives a step by step walk-through of a partitioning session in LYCOS. Section 9.5 describes how LYCOSmay be used for design space exploration. Finally, Section 9.6 provides an evaluation of the system and some conclusions on the current version of LYCOS.

## 9.2 PARTITIONING AND DESIGN SPACE EXPLORATION

Consider the hardware/software **partitioning** and a set of requirements to be fulfilled by the partitioned system. In order to obtain a feasible partition, that is, a partition which fulfills the requirements, we need to know the target architecture, i.e. the processor on which to run the software, the technology of the dedicated hardware, and the interface used for communication between hardware and software. However, the choice of target architecture will greatly influence the outcome of the partition as each target architecture will have different "best" partitions. For instance, selecting a *fast* but expensive processor may lead to little (or no) hardware needed, while selecting a *slow* but cheap processor may require a large amount of dedicated hardware. Thus, in order to solve the problem efficiently we need a way to explore the design space.

Typically we will have an idea about possible suitable target architectures. These may be selected based on the designer's experience, the desire to reuse predesigned components or the use of third party components. One way to explore the design space is to find the best partition for each possible target architecture and select the best among these.

To find the best partition for a given target architecture, we need a model to represent computation and ways to estimate metrics of software, hardware, and communication.

It is important that the model of computation is independent of any particular implementation strategy, that being software or hardware. This will allow for an unbiased design space exploration as well as for translations from various specification languages. One model targeted towards partitioning is based on extracting chunks of computation, called basic scheduling blocks or just blocks. A partition in this model is an enumeration of each block indicating whether

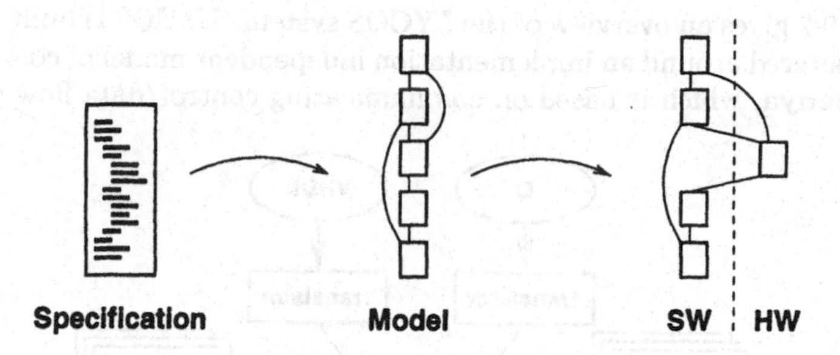


**Figure 9.1** The transformations involved in obtaining a hardware/software partition.

it is placed in software or hardware. Figure 9.1 illustrates the transformations involved in obtaining a partition from the initial specification.

The decision of whether to put a particular block in software or hardware has to be based on an evaluation of the metrics of interest for the entire system. This evaluation can be done in the ***physical domain*** by actual implementation, e.g. by synthesizing the hardware to a gate netlist on which accurate metrics for area and performance may be obtained, or it can be done in the ***model domain*** which is less accurate but much faster. For design space exploration we have to do the partitioning and evaluation over and over again, thus, the speed of the evaluation process is a critical issue. In practise, this means that the evaluation has to be done within the model domain, requiring efficient **estimation** techniques.

## 9.3 OVERVIEW OF THE LYCOS SYSTEM

In this section we describe the main ideas and motivations for building the LYCOS system and give an outline of how the system has been implemented, as well as an outline of the design trajectory for obtaining a hardware/software partitioning.

One of the main ideas of the LYCOS system is to have a system which supports an easy inclusion of new design tools and algorithms, as well as new design methods, e.g. the sequence in which tools have to be applied in order to obtain a solution. Thus, a key issue for LYCOS has been to be able to test new ideas and algorithms not as separate entities but as part of a complete design trajectory. This has to a large extend required the development of our own tools rather than trying to integrate existing tools, that being commercial or university tools. However, from the register transfer level and down to the final layout we are relying on existing commercial design tools.

Figure 9.2 gives an overview of the LYCOS system. LYCOS is built as a suite of tools centered around an implementation independent model of computation, called **Quenya**, which is based on communicating control/data flow graphs.

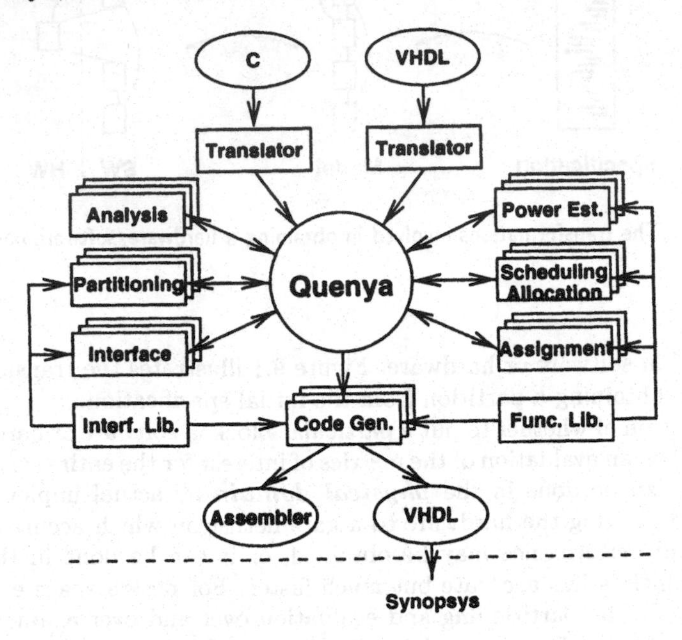


**Figure 9.2** Overview of the LYCOS system.

The information which is communicated between different tools in LYCOS basically consists of two parts: The central functionality of a design (i.e. the behavior derived from the input specification) and design hints obtained during synthesis (e.g. partitioning or scheduling information, profiling, and performance). In order to achieve the necessary flexibility to support the requirements of different kinds of synthesis tools, while preserving the semantic integrity of the description, this dichotomy has been reflected in the design of Quenya.

The functional behavior of a design is represented through a hierarchical network with strictly defined semantics. The network consists of a number of functional units communicating by asynchronous protocols, with the communication channels represented as shared variables. Thus, Quenya directly allows design partitioning to be represented, and supports representation of the system's environment in a uniform manner. The behavior of an individual functional unit is represented by a **control/data flow graph (CDFG)**. The basic CDFG provides a uniform representation of both data and control flow, utilizing lossless and self-timed communication of tokens, i.e. points of execution with associated data. This basic model for functional behavior has been extended to support inter-domain communication, allowing the communication between functional units to be modeled.

Different design problems require different combinations of tools and parameter settings in order to obtain a suitable solution, and as the right order for a given design problem is not evident, exploration is needed. Quenya has been designed with this in mind, i.e. Quenya is independent of the particular synthesis trajectory taken for a given design problem. Unlike the functional behavior, information added by analysis and synthesis tools has deliberately been kept outside the formally defined semantics of Quenya, allowing new information to be added without disturbing the semantics of the representation. This information is kept as annotations to the network and graphs. For more details on the Quenya CDFG representation, see [295, 293].

Figure 9.3 shows the synthesis trajectory for obtaining a hardware/software partitioning from a **behavioral specification** in **VHDL**. It also lists the names of the various tools involved in obtaining a partition.

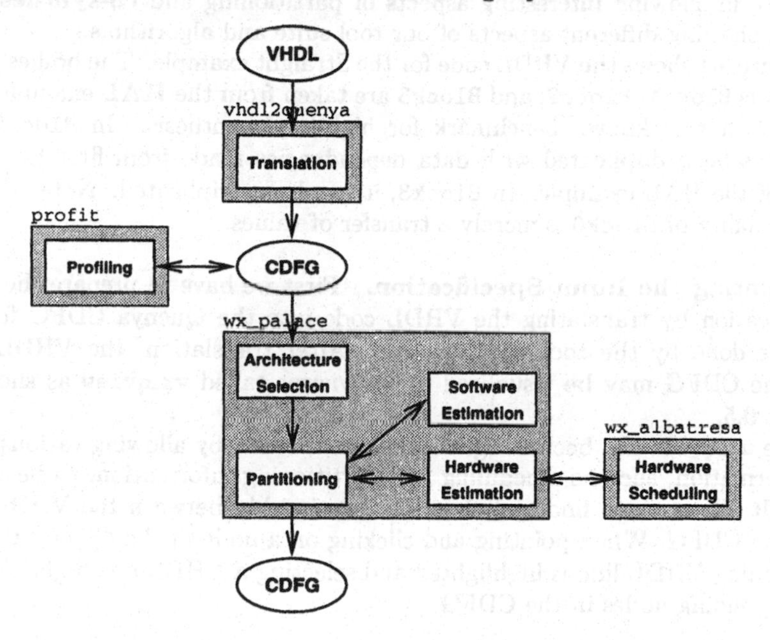


**Figure 9.3** Synthesis trajectory for a hardware/software partitioning session in LYCOS.

In the next section we will describe each step in detail. Here we will just outline the trajectory. The VHDL code is translated into a Quenya CDFG on which profiling is performed in order to gather execution statistics. The profiling information is used in the partitioner to identify time critical parts of the application. Before partitioning can be performed, the user has to select a target architecture which consists of a processor on which to execute the software and dedicated hardware, including the selection of hardware technology, size, and allocation of datapath resources. In the current version of LYCOS, the communication is fixed to **memory mapped I/O**, but we are currently working on making the selection of the communication scheme part of the target architecture selection. The result of partitioning is a Quenya CDFG annotated with the partitioning information.

## 9.4 A PARTITIONING SESSION IN LYCOS

This section describes a partitioning session in LYCOS. It is a walk-through of the steps and tools involved in obtaining a hardware/software partitioning from an initial input specification in VHDL. The input specification is a relatively simple application called Straight. This application has proven to be a good example in showing interesting aspects of partitioning and **co-synthesis**, as well as showing different aspects of our tool suite and algorithms.

Figure 9.4 shows the VHDL code for the Straight example. The bodies of the functions `Block1`, `Block2`, and `Block3` are taken from the **HAL** example [296] which is a well known benchmark for high-level synthesis. In `Block2`, the code has been duplicated with data dependencies made from first to second copy of the HAL example. In `Block3`, it has been triplicated. Note that the functionality of `Block0` is merely a transfer of values.

**Translating the Input Specification.** First we have to prepare the input specification by translating the VHDL code into the Quenya CDFG format. This is done by the tool **`vhdl2quenya`**. After translation, the VHDL code and the CDFG may be visualized in the viewer called `wx_qview` as shown in Figure 9.5.

The `wx_qview` has been designed to assist the user by allowing various types of information, such as scheduling and partitioning information, to be visualized. It allows us to find and show the relationship between the VHDL code and the CDFG. When pointing and clicking on a node in the CDFG, the corresponding VHDL line is highlighted and selecting a VHDL line highlights the corresponding nodes in the CDFG.

**Profiling the Input Specification.** From a partitioning point of view it is important to have some information about time critical parts of the application, since these may have to be moved into hardware in order to meet timing constraints. This information may be obtained by **profiling** which collects ex-

```
entity straight is
 port (go : in BOOLEAN;
   a_i, dz_i, z_i, u_i, y_i : in INTEGER;
   u_o, y_o, z_o : out INTEGER);
end straight;

architecture behavioral of straight is
begin
 process
  variable a, dz, z, u, y : INTEGER;
  variable z1, u1, y1, z2, u2, y2 : INTEGER;
  variable z3, u3, y3, z4, u4, y4 : INTEGER;
  variable z5, u5, y5, z6, u6, y6 : INTEGER;
  variable z7, u7, y7, z8, u8, y8 : INTEGER;
  variable z9, u9, y9 : INTEGER;

procedure Block0(a, dz, z, u, y : in INTEGER;
        u_o, y_o, z_o : out INTEGER) is
begin
 z_o := z; u_o := u; y_o := y;
end;

procedure Block1(a, dz, z, u, y : in INTEGER;
        u_o, y_o, z_o : out INTEGER) is
begin
 z_o := z + dz;
 u_o := u - (3*z*u*dz) - (3*y*dz);
 y_o := y + (u*dz);
end;

procedure Block2(a, dz, z, u, y : in INTEGER;
        u_o, y_o, z_o : out INTEGER) is
variable
 z1, u1, y1 : INTEGER;
begin
 z1 := z + dz;
 u1 := u - (3*z*u*dz) - (3*y*dz);
 y1 := y + (u*dz);
 z_o := z1 + dz;
 u_o := u1 - (3*z1*u1*dz) - (3*y1*dz);
 y_o := y1 + (u1*dz);
end;

procedure Block3(a, dz, z, u, y : in INTEGER;
        u_o, y_o, z_o : out INTEGER) is
 variable z1, u1, y1 : INTEGER;
 variable z2, u2, y2 : INTEGER;
begin
 z1 := z + dz;
 u1 := u - (3*z*u*dz) - (3*y*dz);
 y1 := y + (u*dz);
 z2 := z1 + dz;
 u2 := u1 - (3*z1*u1*dz) - (3*y1*dz);
 y2 := y1 + (u1*dz);
 z_o := z2 + dz;
 u_o := u2 - (3*z2*u2*dz) - (3*y2*dz);
 y_o := y2 + (u2*dz);
end;

 begin
  -- wait for the go signal
  if not go then
   wait until go;
  end if;

  -- Load the input values
  a := a_i;
  dz := dz_i;
  z := z_i;
  y := y_i;
  u := u_i;

  Block1(a, dz, z, u, y, u1, y1, z1);

  Block0(a, dz, z1, u1, y1, u2, y2, z2);

  Block1(a, dz, z2, u2, y2, u3, y3, z3);

  Block2(a, dz, u3, y3, z3, u4, y4, z4);

  Block2(a, dz, u4, y4, z4, u5, y5, z5);

  Block3(a, dz, u4, y4, z4, u6, y6, z6);

  -- Store the output values
  u_o <= u6;
  y_o <= y6;
  z_o <= z6;

  if go then
   wait until not go;
  end if;
 end process;
end behavioral;
```

**Figure 9.4** VHDL specification for the Straight example

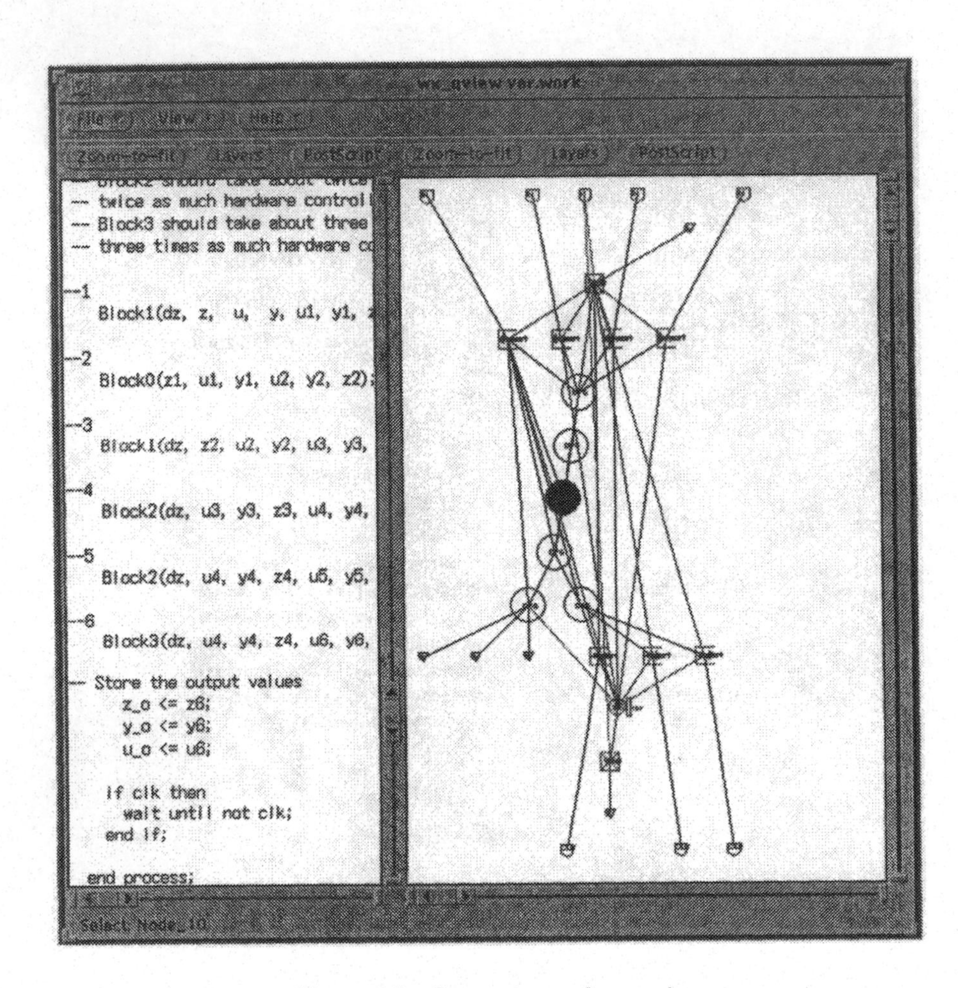


**Figure 9.5** Screen dump of `wx_qview`.

ecution statistics given a set of input data for the application. If the input data used for profiling is carefully selected, the profile will expose potentially time critical parts of the application. Since we allow applications with data dependent loops, conditionals, and synchronizations, it is in general not possible to determine a static execution profile for a given application.

It should be emphasized that the profile is a *hint* to the partitioning tool. It is not a worst case analysis and cannot be used when considering applications with hard timing constraints. Also, the profile (and hence the partitioning) could differ widely for different sets of input data. Profiling is performed on the CDFG and is therefore independent of the final implementation. The profile is obtained by translating the CDFG to C++ (using the tool called `quenya2c`). During the translation, code is added to collect the execution counts during execution. It is necessary to supply a test environment. This may either be written directly in C++ or supplied as a CDFG (translated from either C or VHDL). In the latter case the test environment should be translated to C++. Writing the test environment requires a minimal amount of work since it should merely implement the communication protocol. The profiling tool in LYCOS is called `profit`.

The Straight example used in this chapter contains no loops or branches so the individual counts will be equal for all parts of the application. However, since `Block1` and `Block2` are called twice, each will have two counts, therefore the *total* count for these blocks will be twice the counts for the other blocks.

**Loading the CDFG into the Partitioning Tool.** Hardware/software partitioning is performed by the tool `wx_palace` [297]. Figure 9.6 shows a screen dump of `wx_palace` after a complete partitioning session. In the following we will refer to this figure when describing each step in completing a partitioning session.

First, we have to load the Quenya file containing the CDFG representation of our input specification. The CDFG must contain profiling information. When the file has been loaded using the `File` menu, `wx_palace` creates the corresponding hierarchy of **basic scheduling blocks**, BSBs, which is the model on which the partitioning is performed. For details on BSBs, see [292, 293]. In Figure 9.6 the BSBs are shown in the lower left window. **BSB** indents indicate BSBs at different hierarchical levels. The top panel shows what have been loaded and, thus, indicates what we still have to do in order to complete the partitioning.

From the `CDFG ctrl.` menu we may change the partitioning **granularity**, i.e. we may choose whether to have a large number of rather small BSBs or to have a smaller number of larger BSBs. The number of BSBs has a high impact on the execution time of the partitioning algorithm. In this partitioning

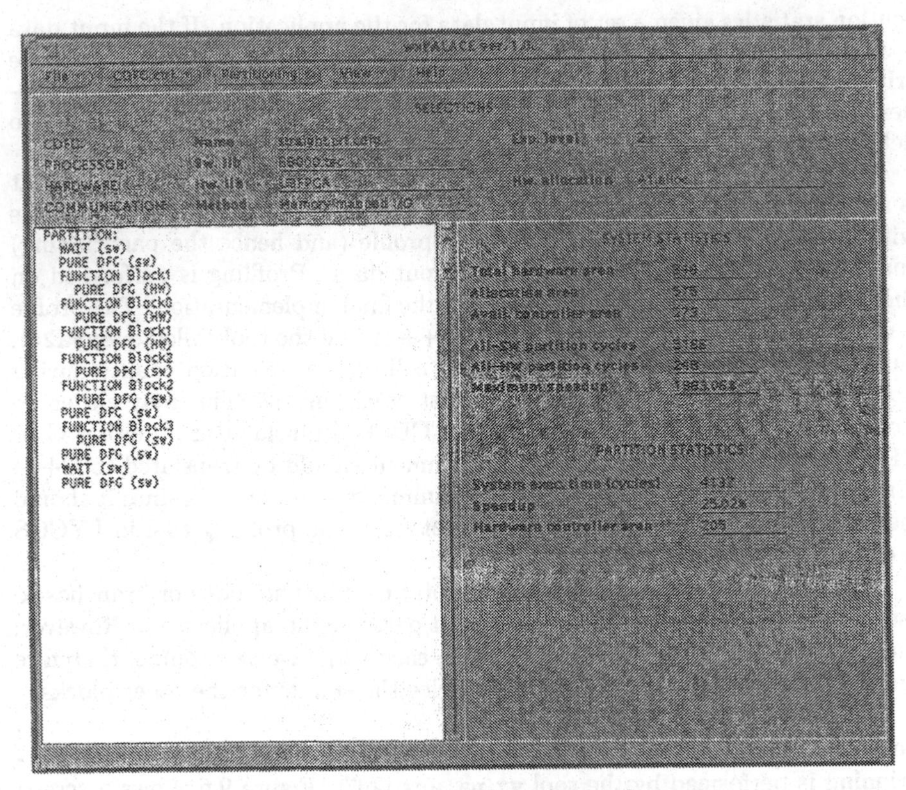


**Figure 9.6** Screen dump of `wx_palace`.

session we expand the BSB hierarchy fully, i.e. the smallest possible granularity is chosen. This is mainly due to the fact that our example application is indeed very small. It should be noted that the granularity may be changed at any point before executing the partitioning algorithm.

**Selecting a Processor.** Having loaded the CDFG, we have to decide upon the target architecture. First, we select the **processor** on which to run the software part. This is selected from the `File` menu by loading a processor technology file from the software library. In this example we have loaded the technology file for the **Motorola 68000** processor into `wx_palace`. The technology file contains a set of generic assembler instructions, each of which corresponds to a particular operation using a particular addressing mode. Each **generic instruction** is associated with its execution time and size in bytes.

This technology description is identical to the one used in [298]. The following is the first few lines of the 68000 technology file:

```
; OP    DESTINATION SOURCE1    SOURCE2    time(clock cycles) size(bytes)

(ALU    Register    Constant   Constant          12             2    )
(ALU    Register    Constant   Register          18             4    )
(ALU    Register    Register   Constant          10             6    )
(ALU    Register    Register   Register          10             4    )
(ALU    Register    DirectMem  Constant          18             6    )
(ALU    Register    Constant   DirectMem         20             4    )
(ALU    Register    DirectMem  Register          18             4    )
(ALU    Register    Register   DirectMem         12             4    )
```

We currently have technology files for the Motorola 68000, **Motorola 68020**, **Intel 8086**, and **Intel 80286** processors which enables us to perform design space exploration experiments where processor price/performance tradeoffs can be made.

**Selecting a Hardware Technology.** Having selected the processor, we have to decide how the **dedicated hardware** is to be implemented. In our example we have chosen an **Actel FPGA** by loading the hardware library file for the Actel ACT3 FPGA. The hardware library contains for each technology a set of functional components. For each component the library contains information on delay, latency, area, provided operations, storage capabilities, etc., as well as the number and types of ports. The following is the architecture and (functional) module specification for a serial multiplier component:

```
(FIXED-ARCHITECTURE mul-ser
  (TYPES
    (INTEGER-TYPE int 0x0000 0xFFFF)
  )
  (PORTS
    (IN int in1)
    (IN int in2)
    (OUT int out1)
  )
  (AREA-ESTIMATE 103 103 103)
  (MINIMUM-CYCLE-TIME 14.8)
)
```

```
(FUNCTIONAL-MODULE mul-ser
  (ARCHITECTURE mul-ser)
  (OPERATIONS
    (OPERATION multiply ((IN int a) (IN int b)
                         (RETURN int c))
      (LATENCY 15)
      (PFG
        (TRANSFER (mul-ser (in a) (in b)
                           (out c)))
      )
    )
  )
  (STORAGE)
  (INSTRUCTIONS
    (INSTRUCTION mul-ser ((IN int a) (IN int b)
                          (OUT int c))
      (DELAY 15)
      (TRANSFER
        (CYCLE 0
          (INPUT  a in1  (DELAY 0))
          (INPUT  b in2  (DELAY 0))
        )
        (CYCLE 15
          (OUTPUT c out1 (DELAY 14.8))
        )
      )
    )
  )
)
```

The `Architecture` defines the type `int`, which is the type of the three ports of the multiplier (two inputs, one output). Also, the estimated area (minimum, typical, and maximum) is listed together with the minimum cycle time for correct operation of the multiplier. In the `Functional Module`, the operations of the component are listed, in this case only one, and for each operation, the `PFG` (**protocol flow graph**) dictates how to interface to the operation. Each `Functional Module` is internally composed of a set of instructions to which the operations refer (one operation can refer to several instructions). Each instruction is defined by a list of transfers that describes the instruction in a ***cycle true*** manner.

In cycle 0, the values on input-ports `in1` and `in2` of the architecture are transferred (by the `PFG` of the operation `multiply`) to the instruction `mul-ser` through the parameters `a` and `b` of this instruction. After a delay of 15 cycles, the result is returned through the output parameter `c` to the output-port `out1`.

The component would have been pipelined if the latency was less than the delay (the operation could be re-invoked before the current computation(s) has produced its output). If the delay is zero, which implies the latency to be zero too, the component is combinatorial.

A detailed description of the hardware library and the PFG can be found in [291] and [294], respectively.

**Customizing the Hardware.** Having selected the hardware technology, we have to customize it, i.e. we have to decide how much area should be used for computation and how much for control. This is a very important step in the partitioning process.

In LYCOS, the total chip area is divided into a ***datapath area*** and a ***controller area***. Each BSB moved to hardware may be viewed as occupying a part of the datapath and a part of the controller. Figure 9.7a shows this model when one BSB has been moved to hardware.

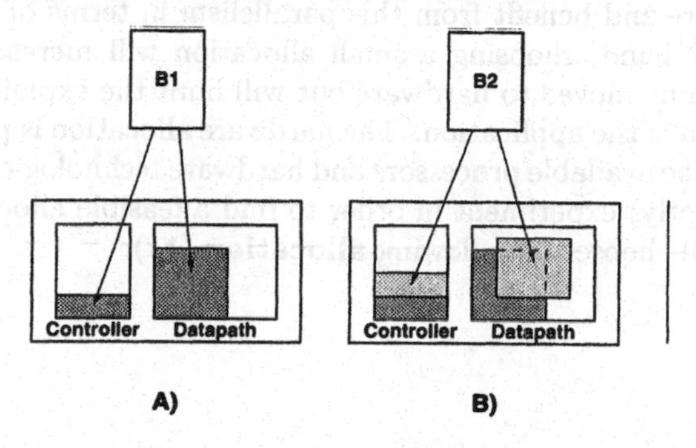


Figure 9.7 BSB area estimation which accounts for hardware sharing: a) Controller- and datapath area for a single BSB, b) When sharing hardware, the total area for multiple BSBs is less than the summation of the individual areas, c) Variable controller area and fixed datapath area for multiple BSBs with hardware sharing.

When more than one BSB are moved to hardware they may share hardware resources as they do not execute in parallel. Hence, estimating the area as the summation of datapath and control areas for all hardware BSBs will probably overestimate the total area. This problem is depicted in Figure 9.7b where the area of the datapath is ***not*** equal to the sum of the individual BSB datapaths.

In LYCOS, the datapath area, $a_{dp}$, is the area of a set of ***preallocated*** hardware resources in the datapath as illustrated in Figure 9.7c, i.e.

$$a_{dp} = \sum_{r \in \mathcal{R}} a_{dp,r}$$

where $\mathcal{R}$ is the set of preallocated resources and $a_{dp,r}$ is the area estimate of the individual resources (components) as obtained from the hardware library.

If the total chip area is denoted $a_{total}$, we can write the area left for the BSB controllers, $a_c$, as:

$$a_c = a_{total} - a_{dp} = a_{total} - \sum_{r \in \mathcal{R}} a_{dp,r}$$

The BSBs share the preallocated resources, and therefore we are only concerned with the area cost of implementing the BSB controllers (in the current version of LYCOS we do not consider the area of interconnects and registers). Thus, the hardware area of a BSB is estimated as the hardware area of the corresponding controller and will depend on the number of time-steps required for executing the BSB [297].

This approach, of course, leaves us with the problem of finding the *optimal* **hardware allocation** for the problem. Choosing a large allocation enables us to explore eventual parallelism in the application, but on the other hand leaves little space for controllers and therefore puts a limit on how much functionality can be moved to hardware and benefit from this parallelism in terms of faster execution. On the other hand, choosing a small allocation will increase the number of BSBs that can be moved to hardware but will limit the exploitation of the inherent parallelism of the application. The hardware allocation is part of the design space, just as the available processors and hardware technologies, and the designer must (currently) experiment in order to find a feasible allocation.

In our example we will choose the following **allocation** (A1):

```
(AvailableArea  848)

(add-sub-comb    1)
(constgen        1)
(div-comb        0)
(equal-comb      0)
(less-comb       0)
(mul-comb        1)
(mul-ser         1)
(simple-logic    0)
(importer        1)
(exporter        1)
```

This is actually the contents of the allocation file which is loaded through the `File` menu of `wx_palace`. The chosen FPGA has a total area of 848, where an area unit equals the area of a logic/sequential module in the FPGA. We have chosen 1 **ALU** and 2 **multipliers**, a fast combinatorial multiplier (`mul-comb`) and a slow but small serial multiplier (`mul-ser`). Further, we have chosen to have only 2 **interface ports** to the FPGA, an input port (`importer`) and an output port (`exporter`). In Figure 9.6, we can see that the allocation uses a datapath area of 575, leaving 273 area units for the controllers.

**Estimating Hardware/Software Execution Times.** When the target architecture is fully characterized, the **all-software** and **all-hardware** execution times are estimated. Besides the overall timing, each BSB is annotated with its execution time corresponding to placing it in hardware and software, respectively.

Software execution time is estimated within **wx_palace** using the CDFG and the selected processor technology file. The estimate is obtained by first linearizing the CDFG and then allocating the registers of the processor. After this, each operation in the CDFG can be mapped to a generic instruction (corresponding to the instructions in the processor technology file) using the appropriate addressing mode. Finally, the execution time is calculated based on the clock cycle times in the processor technology file.

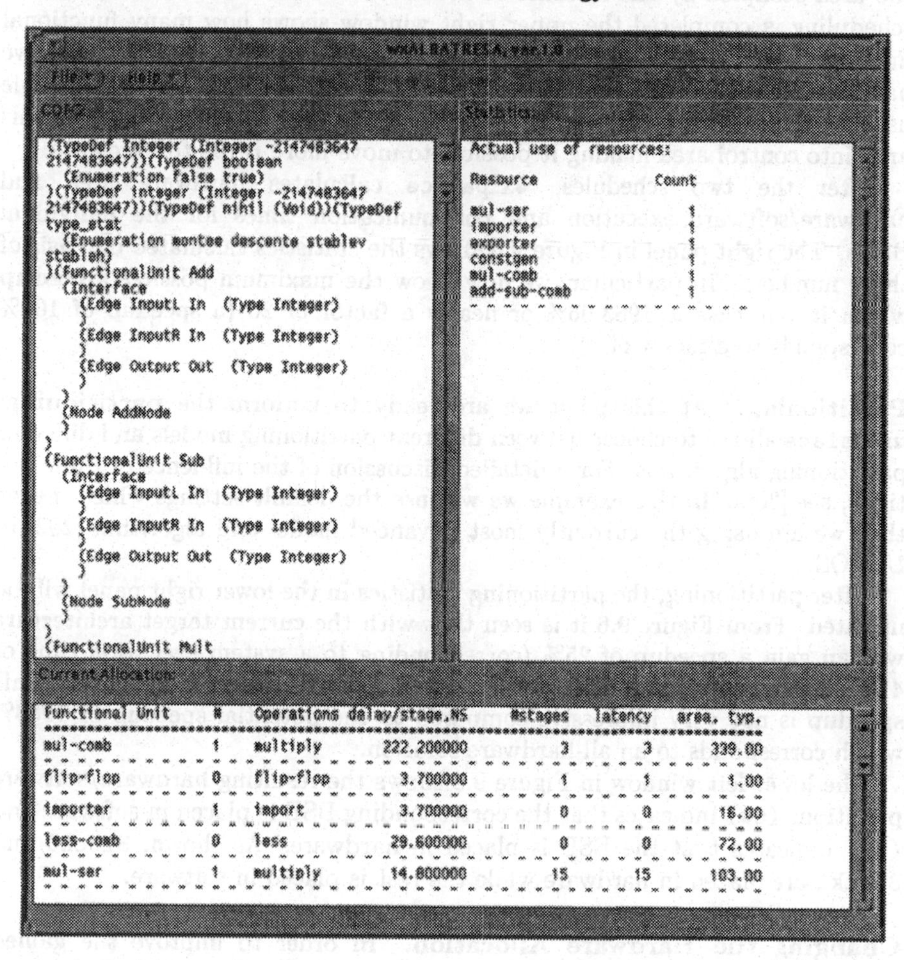


**Figure 9.8** Screen dump of **wx_albatresa**.

Hardware execution time is estimated by the tool **wx_albatresa** [299] which is a dynamic **list based scheduling** tool. **wx_albatresa** has been integrated with the partitioning tool, and thus pops up, whenever the hardware allocation is changed. Figure 9.8 shows a screen dump of **wx_albatresa**. The upper left window lists the Quenya code of the CDFG while the bottom window list the metrics of the selected hardware library and allocation, i.e. it shows for each functional unit the number of allocated units, the delay per. stage, the number of stages needed to complete the calculation, the latency in terms of stages, and the area occupied by one instance of the functional unit. When the hardware scheduling is completed the upper right window shows how many functional units were actually used. By comparing this number with the allocation, we have a first indication of the suitability of the allocation, i.e. if the schedule uses fewer functional units than allocated, we know that we should turn this area into control area making it possible to move more BSBs to hardware.

After the two schedules, **wx_palace** calculates hardware area and hardware/software execution and communication times for the individual BSBs. The right panel in Figure 9.6 shows the statistics calculated on basis of these numbers. In particular, we now know the maximum possible **speedup** which in our case is 1983.06% or nearly a factor of 20 (a speedup of 100% corresponds to a factor of 2).

**Partitioning.** At this point we are ready to perform the **partitioning**. **wx_palace** allows to choose between different partitioning models and different partitioning algorithms. For a detailed discussion of the influence of these options, see [300]. In this example we will use the default settings which means that we are using the currently most advanced model and algorithm [292] in LYCOS.

After partitioning, the partitioning statistics in the lower right panel will be updated. From Figure 9.6 it is seen that with the current target architecture we can gain a speedup of 25% (corresponding to a system execution time of 4132 clock cycles) using most of the available control area, i.e. 205 units. This speedup is not very impressive compared to the potential speedup of 1983% which corresponds to an all-hardware solution.

The lower left window in Figure 9.6 shows the resulting hardware/software partition. `(sw)` indicates that the corresponding BSB is placed in software and `(HW)` indicates that the BSB is placed in hardware. As shown, `Block0` and `Block1` are placed in hardware while the rest is placed in software.

**Changing the Hardware Allocation.** In order to improve the gained speedup, we will now try some other allocations as listed in table 9.1.

Figure 9.9 shows the partitioning results using the allocations from table 9.1. The vertical dotted line indicates the total available area, $a_{total}$, of the FPGA.

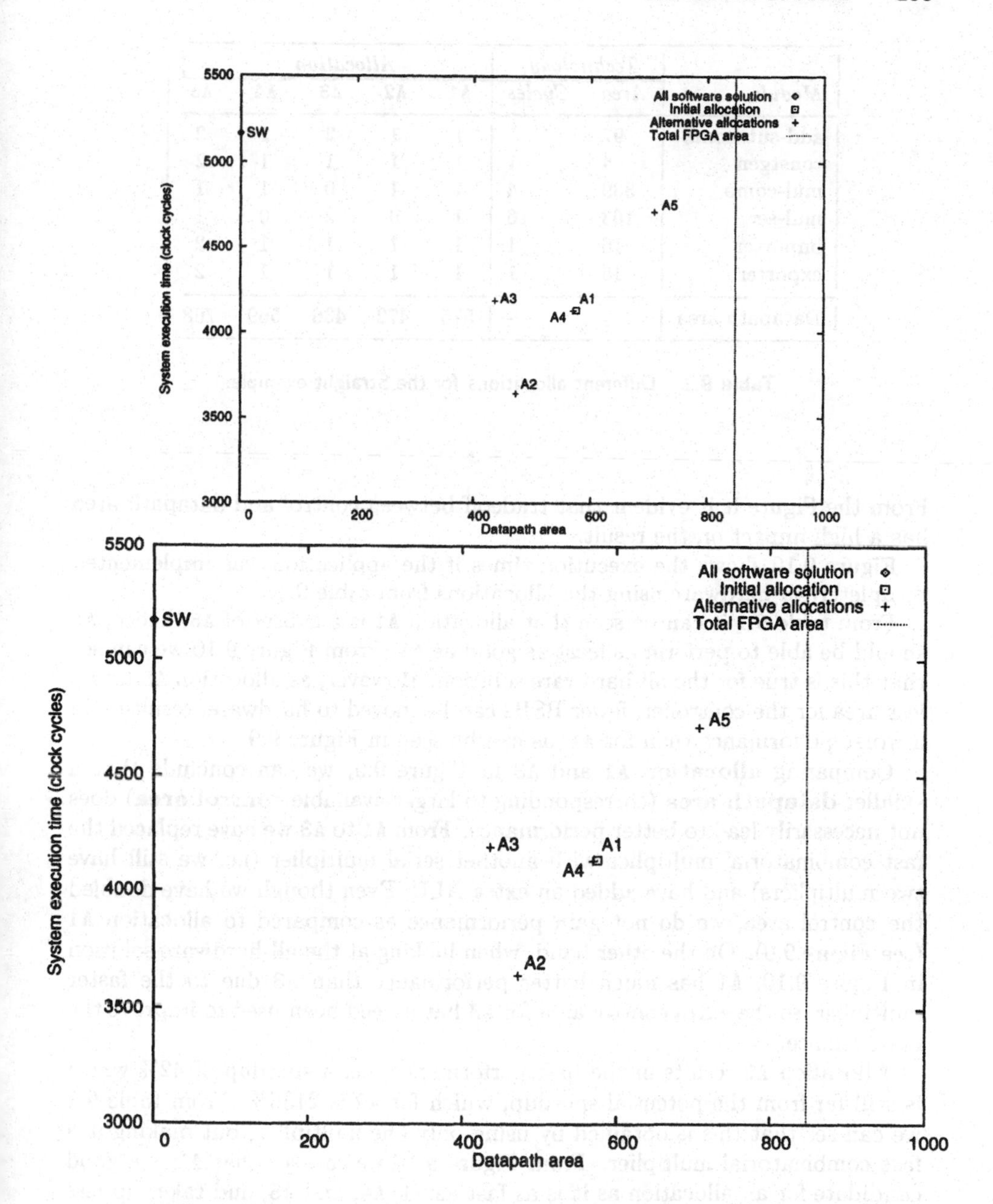


**Figure 9.9** Partitioning results using the allocations from table 9.1.

| | *Technology* | | *Allocation* | | | | |
|---|---|---|---|---|---|---|---|
| *Module* | *Area* | *Cycles* | A1 | A2 | A3 | A4 | A5 |
| add-sub-comb | 97 | 1 | 1 | 1 | 2 | 2 | 2 |
| constgen | 4 | 1 | 1 | 1 | 1 | 1 | 2 |
| mul-comb | 339 | 4 | 1 | 1 | 0 | 1 | 1 |
| mul-ser | 103 | 16 | 1 | 0 | 2 | 0 | 1 |
| importer | 16 | 1 | 1 | 1 | 1 | 1 | 2 |
| exporter | 16 | 1 | 1 | 1 | 1 | 1 | 2 |
| Datapath area | - | - | 575 | 472 | 436 | 569 | 708 |

**Table 9.1** Different allocations for the Straight example.

From the Figure it is evident that tradeoff between control and datapath area has a high impact on the result.

Figure 9.10 shows the execution times if the application was implemented completely in hardware using the allocations from table 9.1.

From table 9.1 it can be seen that allocation A1 is a subset of A5, hence, A5 should be able to perform as least as good as A1. From Figure 9.10 we can see that this is true for the all-hardware solution. However, as allocation A5 leaves less area for the controller, fewer BSBs can be moved to hardware, resulting in a worse performance than for A1, as can be seen in Figure 9.9.

Comparing **allocation** A1 and A3 in Figure 9.9, we can conclude that a smaller **datapath area** (corresponding to larger available **control area**) does not necessarily lead to better performance. From A1 to A3 we have replaced the fast combinatorial multiplier with another serial multiplier (i.e. we still have two multipliers) and have added an extra ALU. Even though we have doubled the control area, we do not gain performance as compared to allocation A1 (see Figure 9.9). On the other hand, when looking at the all-hardware solution in Figure 9.10, A1 has much better performance than A3 due to the faster multiplier, so the extra control area for A3 has indeed been used to improve the performance.

Allocation A2 results in the best performance, i.e. a speedup of 42% which is still far from the potential speedup, which for A2 is 2136%. From table 9.1 we can see that this is obtained by using only one multiplier, but making it a fast combinatorial multiplier. From Figure 9.10 we can see that A2 is a good candidate for an allocation as it is as fast as A1, A4, and A5, and takes up less datapath area. In general, however, it may not be an easy task to choose the right allocation.

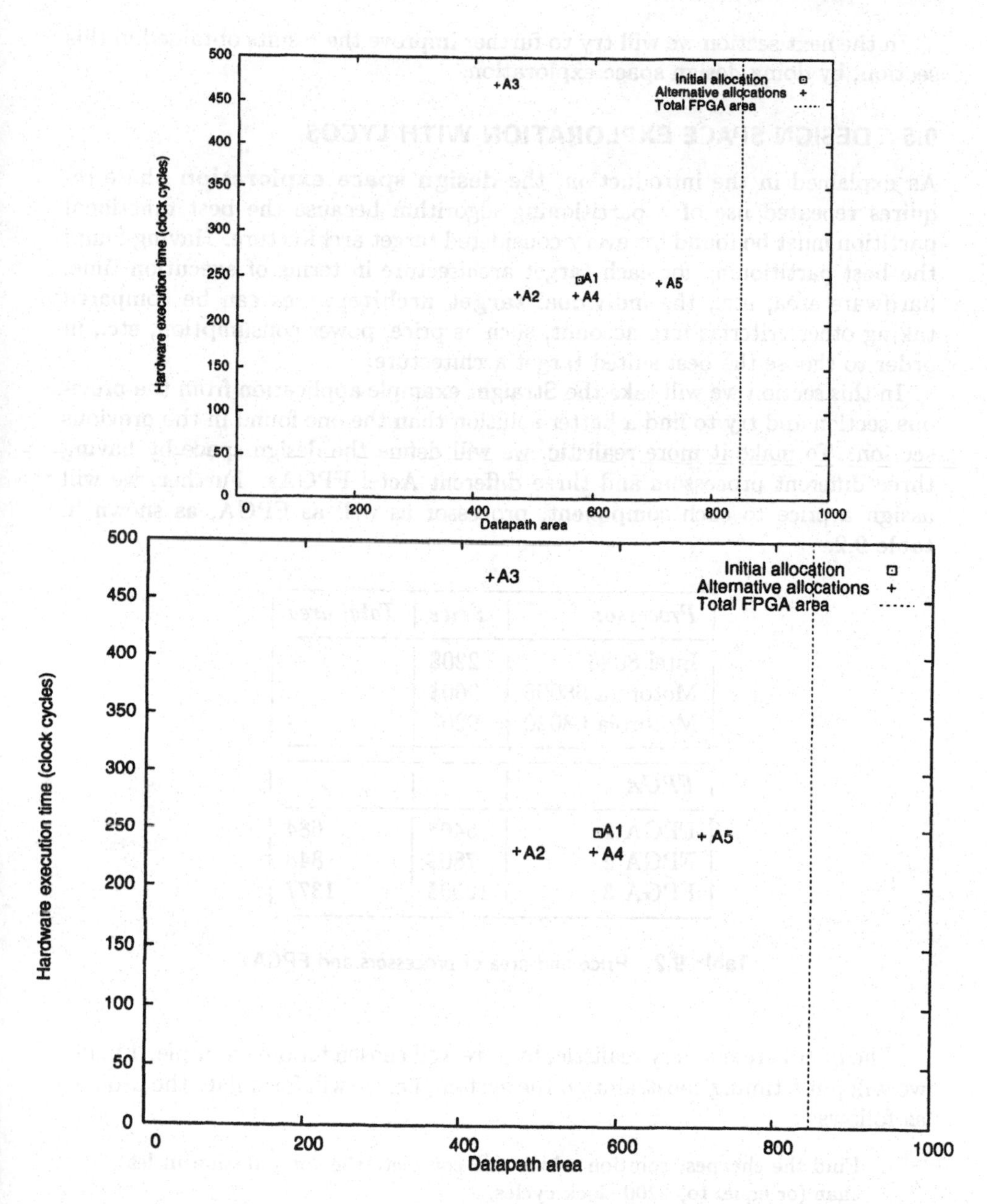


**Figure 9.10** Execution time for an all-hardware solution using the allocations from table 9.1.

In the next section we will try to further improve the results obtained in this section, by doing design space exploration.

## 9.5 DESIGN SPACE EXPLORATION WITH LYCOS

As explained in the introduction, the **design space exploration** phase requires repeated use of a partitioning algorithm because the best functional partition must be found for every considered target architecture. Having found the best partitioning for each target architecture in terms of execution time, hardware area, etc., the individual **target architectures** can be compared taking other criterias into account, such as price, power consumption, etc., in order to choose the best suited target architecture.

In this section we will take the Straight example application from the previous section and try to find a better solution than the one found in the previous section. To make it more realistic, we will define the design space by having three different processors and three different Actel FPGAs. Further, we will assign a price to each component, processor as well as FPGA, as shown in table 9.2.

| *Processor* | *Price* | *Total area* |
|---|---|---|
| Intel 8086 | 220$ | - |
| Motorola 68000 | 360$ | - |
| Motorola 68020 | 620$ | - |
| *FPGA* | | |
| FPGA 1 | 540$ | 684 |
| FPGA 2 | 780$ | 848 |
| FPGA 3 | 1200$ | 1377 |

**Table 9.2** Price and area of processors and FPGAs.

The prices are not very realistic, but are well chosen for our example. Finally, we will put a timing constraint on the system, i.e. we will formulate the problem as follows:

> **Find the cheapest solution which can complete the computation in less than (or equal to) 2200 clock cycles.**

Even though the example is very simple, the design space may be very large as there are many different allocation alternatives for each target architecture.

**Planning the Design Space Exploration.** The first thing we have to do is to plan how the design space is going to be searched. Even for a rather simple application, as in our case, a **complete search** through the design space may be infeasible or impossible.

| | *Software* | | |
|---|---|---|---|
| *Hardware* | 8086 | 68000 | 68020 |
| *None* | 220$ | 360$ | 620$ |
| FPGA 1 | 760$ | 900$ | 1160$ |
| FPGA 2 | 1000$ | 1140$ | 1400$ |
| FPGA 3 | 1420$ | 1560$ | 1820$ |

**Table 9.3** Price of the 12 possible target architectures.

Table 9.3 lists the price for each of the 12 possible target architectures. Rather than using a brute force method trying all possible configurations we will try to find the bounds on our design space, i.e.

- is it possible to find a solution requiring *no* additional hardware? Such a solution would in our case be the cheapest solution as seen from table 9.3.

- is it at all possible to solve the problem with the given architectural choices?

As the last question is the most important, we will try to find an answer to this. From the previous section we know that allocation **A2** is a good candidate for an efficient tradeoff between datapath and control area. Further the 68020 is the fastest processor and FPGA 3 is the largest FPGA, hence, we use these settings for a partitioning session in LYCOS as explained in the previous section. The resulting partition is able to perform the computation within 638 clock cycles, which means that it is indeed possible to meet the timing requirement with the given architectural choices. `wx_palace` also gives us an answer to the first question as an all-software solution using the fastest processor requires 2988 clock cycles, i.e. we do need dedicated hardware to solve the problem.

**Searching the Design Space.** Having established the bounds on our design space, how should we proceed? From the result of 638 clock cycles it is clear that we should be able to trade clock cycles for price, i.e. being able to find a cheaper solution which fullfils the timing requirement. Looking at table 9.3 we will select the cheaper processor 68000 and the cheaper FPGA 2. As all the FPGAs used in this example are Actel FPGAs only varying in their size, we may use the same hardware library. Hence, in `wx_palace` we do not need to reload a new hardware library and allocation when we chage the FPGA size. In the `Partitioning` menu of `wx_palace` we select the `Set options` field which pops up a window which in addition to selecting partitioning algorithm and partitioning model, allows us to change the size of the FPGA to 848 (see table 9.2). The 68000 processor is selected from the `File` menu. We can now perform a partitioning which results in a solution requiring 3644 clock cycles.

Changing FPGA 2 to FPGA 3 costs an extra 420$ while changing 68000 to 68020 only costs an extra 360$, thus, if we can fullfil the timing requirements by just changing the processor, this will be the cheapest solution, i.e. 1400$. A partitioning using this target architecture results in a solution requiring 2133 clock cycles which indeed meets the required time of 2200 clock cycles.

Before we can conclude that this is our solution, let us take a look at table 9.3. From our search so far we can conclude that using FPGA 2 requires the 68020 processor (the 68000 processor did not meet the timing requirements and neither will the even slower 8086 processor). So, if we are to find a cheaper solution we have to use a cheaper FPGA, i.e. FPGA 1. Performing a partitioning with this target architecture results in a solution which requires 2408 clock cycles and, thus, is not a feasible solution.

For the sake of completeness table 9.4 lists the partitioning results in clock cycles for all possible target architectures using allocation `A2`.

| | *Software* | | |
|---|---|---|---|
| *Hardware* | 8086 | 68000 | 68020 |
| *None* | 8647 | 5166 | 2988 |
| FPGA 1 | 6825 | 4128 | 2408 |
| FPGA 2 | 5975 | 3644 | 2133 |
| FPGA 3 | 1402 | 981 | 638 |

**Table 9.4** Partitioning results for the 12 possible target architectures using allocation A2 from table 9.1.

**Comments to the Design Space Exploration.** It should be noted that a well defined strategy for searching the design space does not exist. In this section we have tried to illustrate how the LYCOS system may be used to explore the design space, but the outcome of such an exploration is very much dependent on the user's knowledge, experience and intuition.

Also, it should be noted that the very simple example used in this chapter made it relatively easy to find a single allocation which was good for all the target architectures. However, in general such situations are very rare, as different target architectures may have different optimal allocations, as well as the best allocation may depend on the hardware size, i.e. the number of BSBs which can be moved to hardware as shown in [300, 292].

## 9.6 SUMMARY

In this chapter we have presented the LYCOS hardware/software co-synthesis system. We have focused on the use of LYCOS for design space exploration.

We have used a very simple example to demonstrate different aspects of the LYCOS system. The various tasks were performed within a few seconds. However, several real-life applications have been used to evaluate LYCOS. The most complex is an example taken from an image processing application in optical flow. The application calculates eigenvectors which are used to obtain local orientation estimates for the cloud movements in a sequence of Meteosat thermal images. It consists of 448 lines of behavioral VHDL. The corresponding CDFG contains 1511 nodes and 1520 edges. A partitioning session using this example fully expanded takes around 8 minutes; 40 sec. to load the CDFG file (19259 lines and 0.54 Mbytes), 3 min. 45 sec. to do the hardware scheduling, and 2 min. 30 sec. to execute the partitioning algorithm.

As demonstrated in this chapter, one of the important issues in obtaining a good partition is the selection of a good allocation. We are currently working on a tool which given a profiled application, a hardware library, and the ASAP/ALAP scheduling, is able to find a good allocation.

Further, we are working on supporting a more general target architecture than the single processor, single dedicated hardware architecture, a description of this can be found in [301]. Also, we are currently working on including the communication scheme as part of the target architecture, i.e. allowing the user to select among different **communication channels**.

## 9.7 ACKNOWLEDGMENTS

The work presented in this chapter has been supported by the Danish Technical Research Council under the Co-design program. We would like to thank the other members of the LYCOS team: Morten E. Petersen, Anna Haxthausen, and Peter Bjørn-Jørgensen for valuable contributions to the LYCOS system.

# 10 COSMOS: A TRANSFORMATIONAL CO-DESIGN TOOL FOR MULTIPROCESSOR ARCHITECTURES

C. A. Valderrama, M. Romdhani, J.M. Daveau, G.Marchioro, A. Changuel, and A. A. Jerraya

System Level Synthesis Group
TIMA Laboratory
Instut National Polytechnique de Grenoble (INPG)
Grenoble, France

## 10.1 INTRODUCTION

Co-design is becoming a bottleneck in the process of designing complex electronic systems under short time-to-market and low cost constraints. In this chapter, the word system means a multiprocessor distributed real time system composed of programmable processors executing software and dedicated hardware processors communicating through a complex networks. Such a system



*J. Staunstrup and W. Wolf (eds.), Hardware/Software Co-Design: Principles and Practice, 307-357.*

may be implemented as a single chip, a board or a geographically distributed system.

In a traditional design methodology, designers make the hardware/software partitioning at an early stage during the development cycle. The different parts of the system are designed by different groups. The integration of the different parts of the system leads generally to a late detection of errors meaning higher cost and longer delay needed for the integration step. Too-early early partitioning also restrains the ability to investigate a better partitioning trade-off, causing designers to use larger-than-necessary processing elements to ensure that performance requirements are met..

### *10.1.1 Requirements for Co-Design of Multiprocessor Systems*

A new generation of methods is emerging and maturing which can handle mixed hardware/software systems at the system-level. They are called co-design tools [302] [303] [304] [226] [305] [306] [307]. Co-design may provide a drastic increase in productivity by making easier concurrent design of the different parts of a distributed system and by the automation of the partitioning and the integration steps. However, co-design issues several challenges:

- The development of modern co-design methods has created an exaggerated hope for having general-purpose automatic partitioning tools that would start from a functional specification and produce an optimal solution reducing design time and cost. Indeed, several successful automatic partitioning approaches have been reported in the literature [308] [309] [310] [9] [311] [312]. However, most of these works restrict the problem to a single application domain or make use of simple estimation methods. These restrictions limit the applicability of these partitioning methods to complex systems including several hardware and software processors.

- The evaluation of a distributed architecture is a complex process depending on a large number of criteria such as: efficiency, reliability, manageability, portability, usability. Some of these criteria can entail a long list of sub-criteria. For example, efficiency may involve speed, cost, power consumption, volume or area. For each criterion a metric value has to be associated. In addition, the weights of these criteria may be different according to the application domain as well as the technology used. For instance, reliability will be the major criterion when dealing with systems concerning human life security. For other systems, such as portable multimedia systems, cost and power consumption will be the major criteria. It is then clear that it is quite hard to define a realistic evaluation procedure even for a specific application domain. We believe that a realistic goal for hardware/software partitioning is a semi-automatic method allowing to mix manual and automatic design.

- The designer needs to understand the results of automatic partitioning in order to be able to analyse it. This means that co-design tools should provide facilities that show correspondence between the initial specification and the resulting architecture.
- It is often the case when the designer has a good solution in mind before to start the co-design process. This may be a partial solution like fixing a communication model or fixing the number of processors of the resulting architecture. This means that co-design tools should take into account this partial solutions and allow the designer to control the co-design process.
- The design of complex system is generally an iterative process where several solutions need to be explored before finding the right one. This means that co-design should allow easy design space exploration.

As long as these mentioned challenges are not solved, co-design will remain restricted to specialists for specific applications.

### *10.1.2 Previous Work*

Most current researches in co-design fall in one of three categories:

- **ASIP (Application Specific Integrated Processor) co-design** In this case, the designer starts with an application, builds a specific programmable processor and translates the application into software code executable by the specific processor [313] [146] [314]. In this scheme the hardware/software partitioning includes instruction set design [315]. In this case the cost function is generally related to area, execution speed and/or power.
- **hardware/software synchronous system co-design** In this case the target architecture of co-design is a software processor acting as a master controller, and a set of hardware accelerators acting as co-processors. Within this scheme two kinds of partitioning have been developed: software oriented partitioning [304] [316] and hardware oriented partitioning [317]. Most of the published work in co-design fall in this scheme. They generally use a simple cost function related to area, to processor cost for software and to speed for hardware. Vulcan [317], Codes [318], Tosca [319], and Cosyma [304] are typical co-design tools for synchronous systems.

- **hardware/software for distributed systems** In this case, co-design is the mapping of a set of communicating processes (task graph) onto a set of interconnected processors (processor graphs). This co-design scheme includes behavioral decomposition, processor allocation and communication synthesis [48] [226]. Most of the existing partitioning methods restrict the cost function to parameters such as real time constraints [320] [9] or cost [321]. Coware [15] handles multiprocessor co-design during the latest design phases. However, it starts from a C/VHDL where partitioning is already done. Specsyn [322] is a precursor for the co-design of multiprocessors. It allows automatic partitioning and design space exploration. However, Specsyn doesn't help the designer to understand the produced architecture and it doesn't allow a co-design with partial solution. Siera [305] provides a powerful scheme for co-design of multiprocessor based on the re-use of components. It allows co-design with partial solutions. However it doesn't provide automatic partitioning. Ptolemy provides a powerful environment for co-design of multiprocessors [323]. However, its partitioning is restricted to DSP applications and doesn't allow partial solutions [309]. Wolf's group studies system-level analysis/performance and co-synthesis of distributed multiprocessor architectures [9]. They have developed an efficient algorithm to bound the computation time of a set of processes on a multiprocessor system, a synthesis method which takes into account communication and computation cost, and scheduling and allocation algorithms for multiprocessors.

In this paper we will restrict our discussion to co-design tools of the third category.

### *10.1.3 Contributions*

The main contribution of this chapter is to present a user-guided transformational approach for co-design. We present the underlying design methodology and show the efficiency of our refinement based approach for hardware/software co-design. The intention of this approach is to solve the challenges mentioned in the section 10.1.1. It covers the design process through a set of user guided transformations that allows semi-automatic partitioning synthesis with predictable results. The lack of realistic estimation method is compensated by the expertise of the designer. A large design space can be explored through multiple trials and a fast feedback as an implementation of the initial specification can be quickly obtained.

The approach described in this chapter is implemented in the Cosmos hardware/software co-design environment. Cosmos is a methodology and a tool intended to fill the gap between system-level tools and existing synthesis tools.

Several aspects of the Cosmos approach have already been presented in the literature. This chapter focuses on the methodological aspects from the designer's point of view and not on the algorithms and techniques used in Cosmos. Most of the details of the intermediate model Solar are given in [324], the behavioral transformation primitives are detailed in [303], the communication synthesis methods are explained in [309] and the code generation techniques can be found in [325]. However, for a better clarity of the chapter we will introduce the models and techniques used when needed.

The first section will present the main principles of Cosmos. Section 10.3 will focus on design models for co-design. The different models obtained after each refinement step will be detailed. Section 10.4 details the main steps of the co-design flow. This models and refinement steps will be clarified by an example, in section 10.5. Finally, we conclude shown the results and the strength of our methodology.

## 10.2 COSMOS: A GLOBAL VIEW

Within Cosmos, co-design is decomposed into four major steps (Figure 10.1). Functional decomposition is aimed to split large behaviors that need to be executed on several processors. This step may create additional communication. The virtual processor allocation fixes the number of processors and assign an execution processor to each function. Each abstract processor may be implemented in hardware or in software. Each of the different partitions will contain one or several functions coming from the initial specification. Several functions may be assigned to the same partition allowing them to share functional units or communication resources.

After the functional decomposition and virtual processor allocation follows the communication transformation where processors communicating through high-level communication scheme are transformed into processors communicating through buses and sharing communication control. The final step is prototyping. This includes the generation of C-VHDL models of the architecture and the mapping of this model onto the target architecture. In Figure 10.1, the processes P1 and P3 will be translated to a behavioral VHDL description for hardware synthesis and the process P2 will be translated to the C language in order to produce an executable code. In order to obtain the prototype, the C code have to be compiled onto software processors and VHDL needs to be synthesised onto ASICs or FPGAs.

The user guided transformational methodology assumes that the designer starts with an initial specification and an architectural solution in mind. System design from specification to implementation is performed through a set of primitives allowing the designer to transform the system, following an incremen-

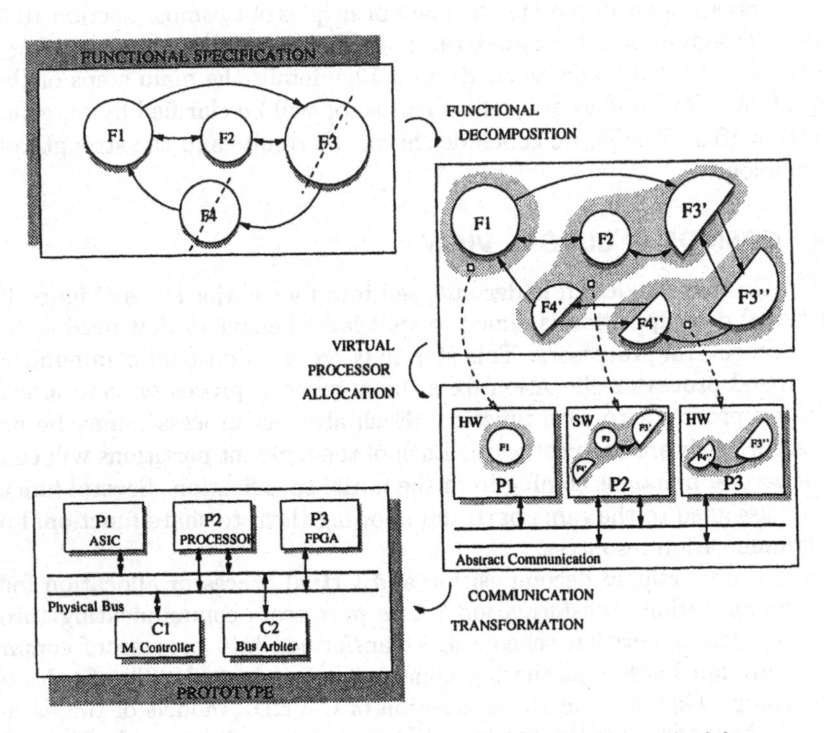


**Figure 10.1** Transformational partitioning in Cosmos methodology.

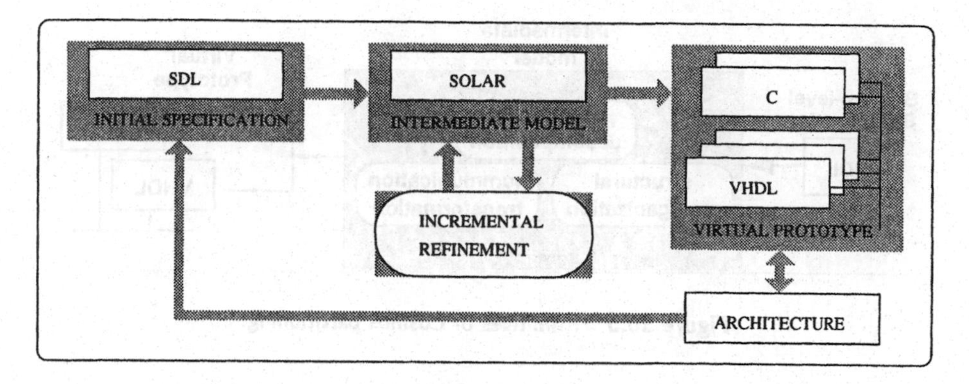


**Figure 10.2** Specification models used during co-design.

tal refinement scheme, in a distributed model that matches to the architectural solution. All the refinement transformations are performed automatically. The decisions are made by the designer who uses his knowledge and experience to achieve the desired solution.

Each step reduces the gap between specification and realization by fixing some implementation details (communication protocol, generating software or hardware code) or by preparing future implementation steps (merging and scheduling several processes to execute them in a single processor). The transformations must satisfy the designer's imposed constraints without changing the functionality of the system.

In the case of the Cosmos, the user guided transformational approach makes use of three specification formats, as shown in Figure 10.2. The initial specification is given in the system-level specification language SDL. The user control the refinement process through a set of transformation primitives. All the refinement process is based on an intermediate form called Solar. The output is a virtual prototype of the architecture given in a distributed C-VHDL model.

Within Cosmos, the partitioning steps are implemented through a set of primitives that performs basic transformations such as split, merge, move, flat and map. The system provides three set of primitives working on the system structure, behavior and communication. The designer guides the interaction process and chooses the transformations needed in order to obtain the desired solution. A new implementation can be obtained by changing the primitives sequence activation. The partitioning flow is shown in Figure 10.3.

The organization of partitioning into several small steps reduces the complexity of the problem. The designer controls the partitioning history within an interactive environment, through a fine grain control of the synthesis pro-

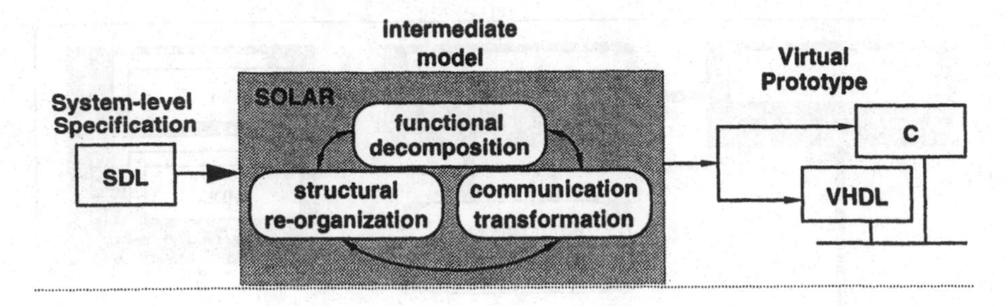


Figure 10.3 Primitives of Cosmos partitioning.

cess. This methodology can be seen as a human guided compilation where the designer spends additional effort to produce an efficient implementation [326].

To facilitate the user interaction for incremental transformations, a graphical interface has been developed. The interface provides a good control of the design process and a graphical view of Solar objects. This will be explained in the following sections.

## 10.3 DESIGN MODELS USED BY COSMOS

This section introduces the five design models used within the Cosmos co-design environment: SDL, Solar, the communication model, the mixed C/VHDL, and the target architecture. A special attention will be given to communication models and its transformations.

### *10.3.1 Target Architecture*

We use a modular and flexible architectural model. The general model, shown in Figure 10.4(a), is composed of three kind of components: software components (aimed to execute C programs), hardware components (implements the VHDL descriptions) and communication components. This model serves as a platform onto which a mixed hardware/software system is mapped. Communication modules come from a library, they correspond to existing communication models that may be as simple as a handshake or as complex as a layered network.

The proposed architectural model is general enough to represent a large class of existing hardware software platforms. It allows different implementation of mixed hardware/software systems, distributed architectures and several communication models. As shown in Figure 10.4(b), a typical architecture will be composed of several hardware modules, several software modules and communication modules linking hardware and/or software modules.

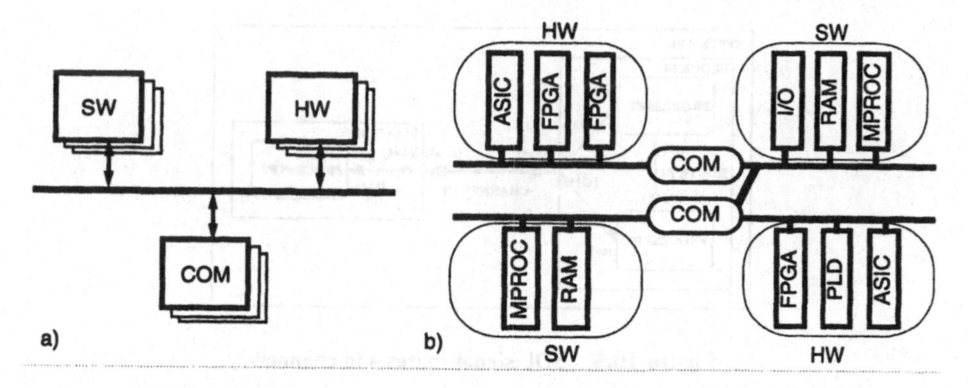


**Figure 10.4** Architectural model: a) architectural model. b) hardware/software platform.

### *10.3.2 System Specification With SDL*

SDL (Specification and Description Language) is intended for the modeling and simulation of real time, distributed and telecommunication systems and is standardized by the ITU [327]. A system described in SDL is regarded as a set of concurrent processes that communicate with each others using signals. SDL support different concepts for describing systems such as structure, behavior and communication. SDL is intended for describing large designs at the system level. There are two SDL formats, a textual and a graphical one.

**Structure.** The static structure of a system is described by a hierarchy of blocks. A block can contain other blocks, resulting in a tree structure or a set of processes to describe the behavior of a terminal block. Processes are connected with each other and to the boundary of the block by signalroutes. Blocks are connected together by channels. Channels and signalroutes are a way of conveying signals that are used by the processes to communicate. Signals exchanged by the processes follow a communication path made up of signalroutes and channels from the sending to the receiving process (Figure 10.5). SDL also support dynamic feature that are software oriented like dynamic process creation and dynamic addressing.

**Behavior.** The behavior of a system is described by a set of autonomous and concurrent processes. A process, described by a finite state machine, communicates asynchronously with other processes by signals. Each process has an input queue where signals are buffered on arrival. Signals are extracted from the input queue by the process in the order in which they arrived. In other words, signals are buffered in a first-in-first-out order. The arrival of an

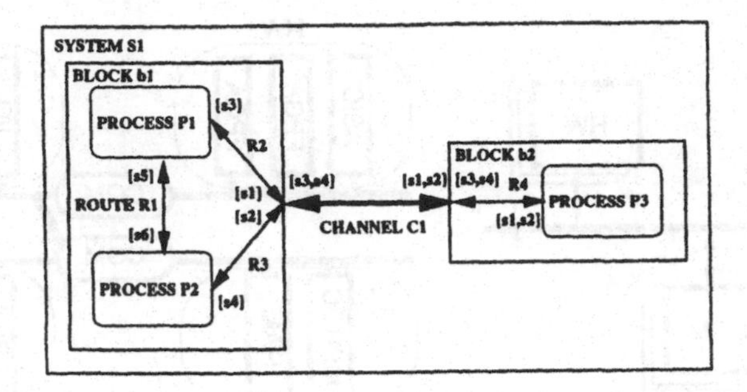


Figure 10.5 SDL signal-routes and channels.

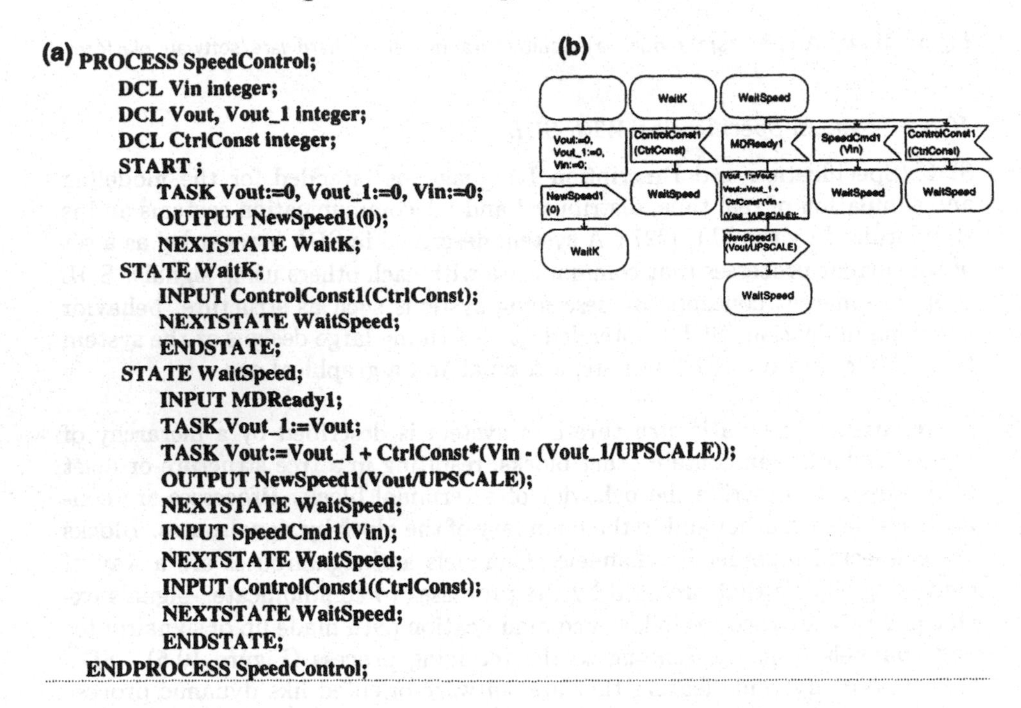

(a)
PROCESS SpeedControl;
DCL Vin integer;
DCL Vout, Vout_1 integer;
DCL CtrlConst integer;
START ;
TASK Vout:=0, Vout_1:=0, Vin:=0;
OUTPUT NewSpeed1(0);
NEXTSTATE WaitK;
STATE WaitK;
INPUT ControlConst1(CtrlConst);
NEXTSTATE WaitSpeed;
ENDSTATE;
STATE WaitSpeed;
INPUT MDReady1;
TASK Vout_1:=Vout;
TASK Vout:=Vout_1 + CtrlConst*(Vin - (Vout_1/UPSCALE));
OUTPUT NewSpeed1(Vout/UPSCALE);
NEXTSTATE WaitSpeed;
INPUT SpeedCmd1(Vin);
NEXTSTATE WaitSpeed;
INPUT ControlConst1(CtrlConst);
NEXTSTATE WaitSpeed;
ENDSTATE;
ENDPROCESS SpeedControl;
(b)
WaitK
WaitSpeed
Vout:=0,
Vout_1:=0,
Vin:=0;
ControlConst1
(CtrlConst)
MDReady1
SpeedCmd1
(Vin)
ControlConst1
(CtrlConst)
NewSpeed1
(0)
WaitSpeed
WaitSpeed
WaitSpeed
WaitK
NewSpeed1
(Vout/UPSCALE)
WaitSpeed


Figure 10.6 SDL process specification: a) textual form b) graphical form.

expected signal in the input queue validate a transition and the process can then execute a set of actions such as manipulating variables, procedures call and emission of signals. The received signal determines the transition to be executed. When a signal has initiated a transition it is removed from the input queue.

Figure 10.6 represents an SDL process specification. State Start represents the default state. Input represents the guard of a transition. This transition will be triggered when the specified signal is extracted from the input queue. Task represent an action to perform when the transition is executed and Output emit a signal with its possible parameters.

In SDL, variables are owned by a specific process and cannot be modified by others processes. The synchronisation between processes is achieved using the exchange of signals only. SDL includes communication through revealed and exported shared variables. However they are single-writer-multiple-readers. Shared variables are not recommended in the SDL92 standard [327]. Each process has an unique address (Pid) which identify it. A signal always carries the address of the sending and the receiving processes in addition to possible values. The destination address may be used if the destination process cannot be determined statically and the address of the sending process may be used to reply to a signal.

Signals are transferred between processes using signalroutes and channels. If the processes are contained in different blocks, signals must traverse channels. There are two major differences between channels and signalroutes. First, channels may perform a routing operation on signals; a channel routes a signal to different channels (signalroutes) connected to it at the frontier of a block depending on the receiving process. Communication through signalroutes is timeless while a communication through channel is delayed non-deterministically. No assumption can be made on the delay and no ordering can be presumed for signals using different delaying paths. Second, signal-routes only deliver signal to the destination process. A signalroute must not be connected to more than one channel. Figure 10.5 represents the structure and communication specification of an SDL system.

Channels and signalroutes may be both uni- and bidirectional. If many signals are transferred on the same channel or signalroute their ordering is preserved. When going through a channel or signal route, signals are not allowed to overtake each other. However there is no specific ordering of signals arriving in the input queue of a process that have followed different channels and signalroutes.

### *10.3.3 Solar: A System-Level Model for Co-Design*

Solar [263] is, the intermediate format used within Cosmos. The different refinement steps use Solar as an intermediate representation. Each step makes transformations on a Solar model. The use of an intermediate format for synthesis brings the advantages to make the system independent from description languages. For example, in the case of Cosmos, several input languages can be

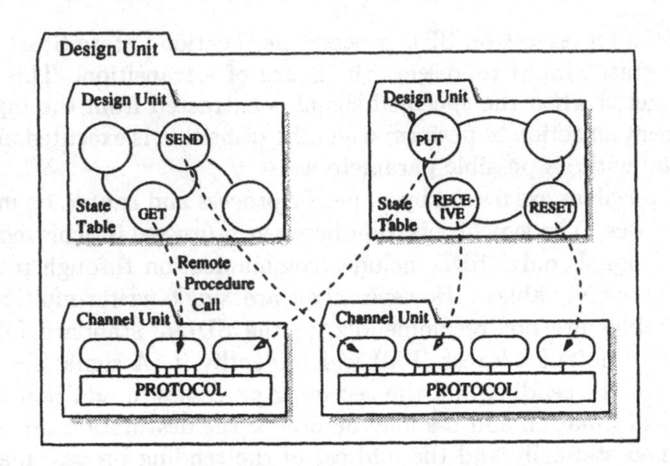


**Figure 10.7** EFSMs using RPC communication: the Solar model.

used as input specifications. In [263] Solar is used for co-design strategy from a multi-language specification combining SDL [328] and Statecharts [257]. Additionally, an intermediate format can be defined in such a way to be more suitable for synthesis algorithms. Finally, in this case, Solar has a graphical representation which makes easier to show the intermediate results to the designer.

Solar supports high-level communication concepts including channels and global variables shared over concurrent processes. It is possible to model most system-level communication schemes such as message passing, shared resources and other more complex protocols. Solar models system-level constructions in a synthesis-oriented manner. The underlying model combines two powerful system-level concepts: extended finite-state machines (EFSM) for behavioral description and remote procedure call (RPC) for the specification of high-level communication (Figure 10.7).

The basic concepts are: State Tables, Design Units and Channel Units.

- The State Table is the basic constructor for behavior description. A state table is an EFSM composed of an unlimited combination of states and state tables (behavioral hierarchy). All of these states can be executed sequentially, concurrently, or both. A state table have attributes to handle exceptions, global variables, reset and default states. Transitions between states are not level restricted. In other words, transitions may traverse hierarchical boundaries (global transitions).

- The Design Unit construct allows the structuring of a system description into a set of interacting subsystems (processes). These subsystems interact with the environment using a well defined boundary. A design unit can be specified as a set of communicating design units (structural hierarchy) or as a set of interacting state tables. The communication between design units can be performed in two different ways, first by means of classic port concept where single wires send data in one or two directions or by means of communication channels with a well defined protocol.

- The Channel Unit performs the communication between any number of design units. The model mixes the principles of monitors and message passing, also known as remote procedure call [329]. The use of a RPC to invoke channel services allows a flexible representation, with a clear semantic. These communication schemes can be described separately from the rest of the system, allowing modular design and specification. A channel unit consists of many individual connections and not only it acts as a transport mechanism for the communicated data, it also provides a handshaking interface to ensure both the synchronization and the avoidance of access conflicts. The channel is composed of a controller, a set of methods and a set of interconnected signals. The controller stores the resource current state. The access to the channel is governed by a fixed set of methods (services) that are the visible part of the channel. The utilization of an extensible library of protocol permits the reuse of existing components. If a communication unit that implements the protocol, services and average rate required is not found in the library, the designer can adapt an existing communication unit (increase the bus width or buffer size for example) rather than building from scratch.

Figure 10.8 shows a graphical and a textual view of a Solar description. In Figure 10.8(a) we can see two processes named DU_SERVER and DU_CLIENT represented by their respectively instances SERVER and CLIENT. These processes communicate with each other by means of signal nets (wr_req, rewr_rdy, read_req and data_io) connected to a channel controller named CHANNEL that implements a FIFO protocol. The CLIENT AbstractChannel and SERVER AbstractChannel correspond to abstract communication channels while CHANNEL represent an implementation of a communication protocol. Figure 10.8(b) shows a state table (FIFO_StateTable) and two states (Init_St and Rec_Send_St) belonging to the channel controller. A Solar textual view of each graphical element and their hidden structure can also be seen (Figure 10.8(c)).

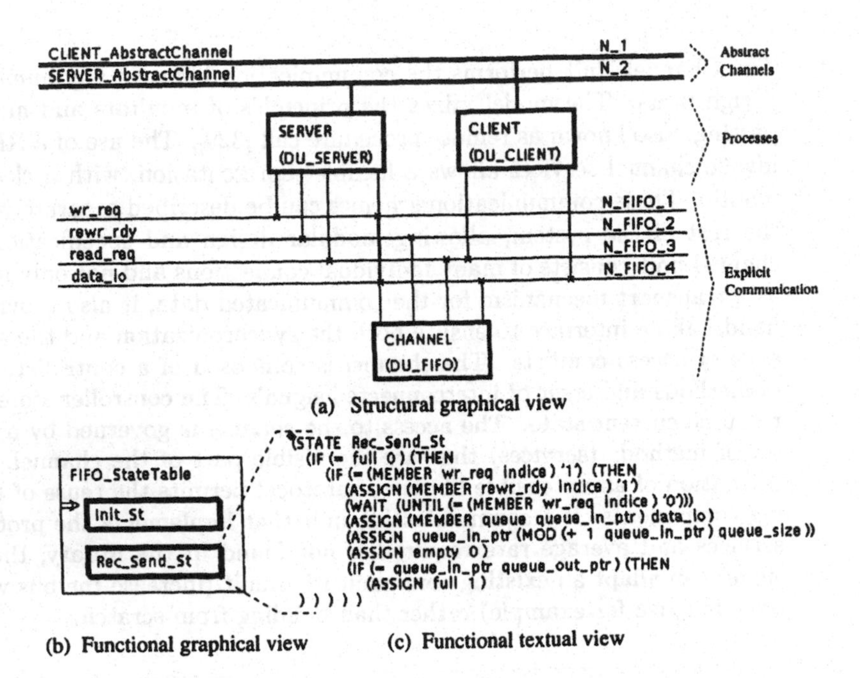


**Figure 10.8** Graphical and textual view of a Solar description.

### *10.3.4 Communication Modeling and Refinement Concepts*

In this section we describe our communication modeling scheme for system level synthesis. This model aims at representing and implementing the communication scheme used in specification languages. Our model is general enough to accommodate different communication schemes such as message passing or shared memory and allow an efficient implementation of a system level communication specifications.

At the system level, a design is represented by a set of processes communicating through abstract channels (Figure 10.9). An abstract channel is an entity able to execute a communication scheme invoked through a procedure call mechanism. It offers high level communication primitives that are used by the processes to communicate. Access to the channel is controlled by this fixed set of communication primitives and relies on remote procedures call [329] of these primitives. A process that is willing to communicate through a channel makes a remote procedure call to a communication primitive of that channel. Once the remote procedure call is done the communication is executed independently of the calling process by the channel unit. The communication primitives are transparent to the calling processes. This allows processes to communicate by means of high level communication schemes while making no assumption on the implementation of the communication.

There is no pre-defined set of communication primitives, they are defined as standard procedures and are attached to the abstract channel. Each application may have a different set of communication primitives (send_int, send_short, send_atm, etc.). The communication primitives are the only visible part of an abstract channel.

The use of remote procedure call allows to separate communication specification from the rest of the system. These communication schemes can be described separately. In our approach the detailed I/O structure and protocols are hidden in a library of communication components. Figure 10.9 shows a conceptual communication over an abstract communication network. The processes communicate through three abstract channels c1, c2 and c3. C1 and c2 offers services svc1, svc2 and c3 offers services svc3, svc4 (services svc1 and svc2 offered by abstract channel c2 are not represented).

We define a communication unit as an abstraction of a physical component. Communication units are selected from the library and instantiated during the communication synthesis step. Conceptually, the communication unit is an object that can execute one or several communication primitives with a specific protocol. A communication unit is composed of a set of primitives, a controller and an interface. The complexity of the controller may range from a simple handshake to a complex layered protocol. This modular scheme hides

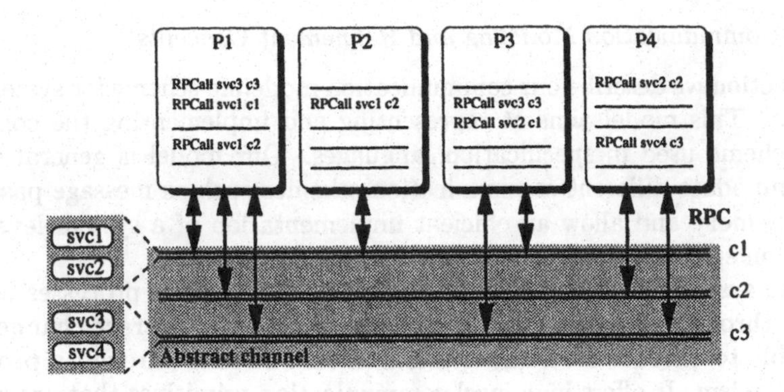


**Figure 10.9** Specification of communication with abstract channels.

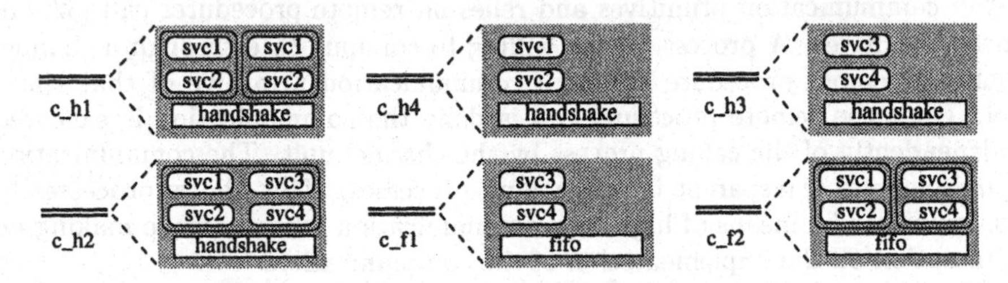


**Figure 10.10** Library of communication units.

the details of the realisation in a library where a communication unit may have different implementations depending on the target architecture. Communication abstraction in this manner enables a modular specification, allowing communication to be treated independently from the rest of the design.

### *10.3.5 Communication Refinement*

Communication synthesis aims to transform a system composed of processes that communicate via high level primitives through abstract channels into a set of interconnected processors that communicate via signals and share communication control. Starting from such a specification two steps are needed. The first is aimed to fix the communication network structure and protocols used for data exchange. This step is called protocol selection or communication unit allocation. The second step, called **interface synthesis**, adapts the interface of the different processes to the selected communication network.

Allocation of communication units starts with a set of processes communicating through abstract channels (Figure 10.10) and a library of communication units (Figure 10.11).

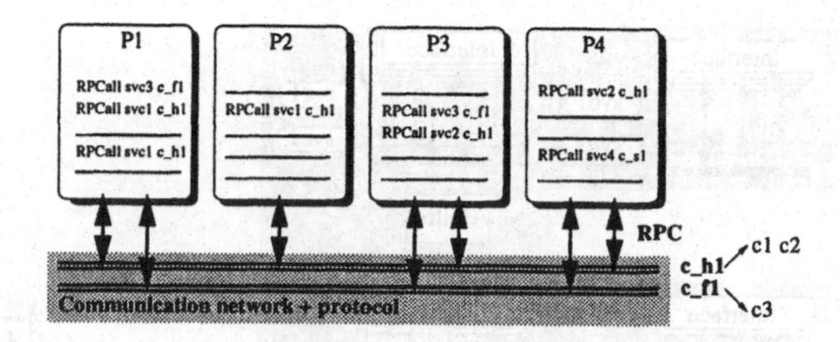


**Figure 10.11** System after allocation of communication units.

Each communication unit is an abstraction of some physical components. This step chooses the appropriate set of communication units from the library in order to provide the services required by the communicating processes. The communication between the processes may be executed by one of the schemes described in the library. The choice of a given communication unit will not only depend on the communication to be executed but also on the performances required and the implementation technology of the communicating processes. This step fixes the protocol used by each communication primitive by choosing a communication unit with a specific protocol for each abstract channel. It also determines the interconnection topology of the processes by fixing the number of communication units and the abstract channels executed on it.

An example of communication unit allocation for the system of Figure 10.9 is given in Figure 10.11. Starting with the library of communication units of Figure 10.10, the communication unit c_h1 has been allocated for handling the communication offered by the two abstract channels c1 and c2. The communication unit c_h1 is able to execute two independent communication requiring services svc1 and svc2. Communication unit c_f1 has been allocated for abstract channel c3.

Interface synthesis selects an implementation for each of the communication units from the implementation library (Figure 10.12) and generates the required interfaces and interconnections for all the processes using the communication units (Figure 10.13).

The library may contain several implementations of the same communication unit. Each communication unit is realized by a specific implementation selected from the library with regard to data transfer rates, memory buffering capacity, and the data bus width. The interface of the different processes are adapted according to the implementation selected and interconnected. The result of interface synthesis is a set of interconnected processors communicating through signals, buses and possible additional dedicated components selected from the implementation library such as bus arbiter, fifo, etc. With this approach it is

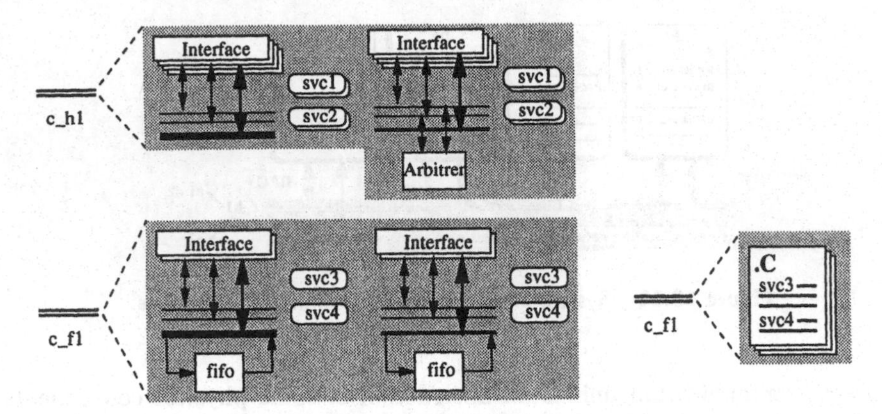


Figure 10.12 Implementation library.

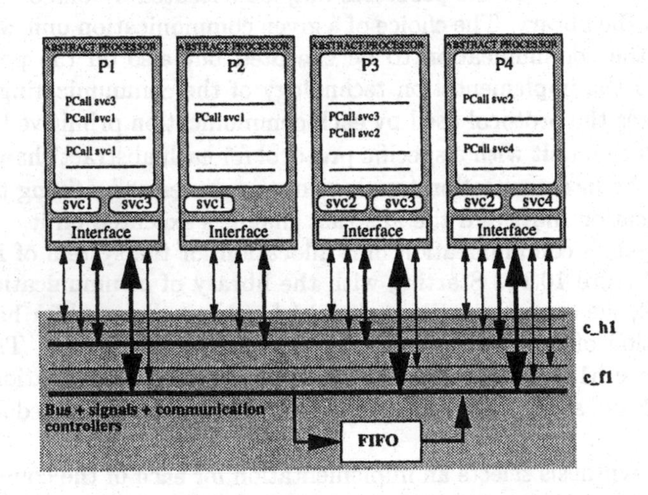


Figure 10.13 System after interface synthesis.

possible to map communication specification into any protocol, from a simple handshake to a complex protocol.

Starting from the system of Figure 10.11, the result of interface synthesis task is detailed in Figure 10.13. The communication unit c_h1 has two possible implementations, one with an external bus arbiter for scheduling the two communication, the other with the arbiter distributed in the interfaces. Any of the two implementation may be selected.

### *10.3.6 Virtual Prototyping Using C-VHDL Models*

The generated architecture, also called a **virtual prototype**, is a heterogeneous architecture composed of a set of distributed modules, represented in VHDL for hardware elements and in C for software elements, communicating through communication modules (located into a library of components). The virtual prototype is a simulatable model of the system. This model will be used for the production of the final prototype.

The prototyping (also called architecture mapping) produces an architecture that implements or emulates the initial specification. This step concentrates on the use of virtual prototype for both co-synthesis (mapping hardware and software modules onto an architectural platform) and co-simulation (that is the joint simulation of hardware and software components) into an unified environment [325].

The joint environment for co-synthesis and co-simulation provides support for multiple platforms aimed at co-simulation and co-synthesis, communication between C and VHDL modules and coherence between the results of co-simulation and co-synthesis. The model used by the environment allows to separate module behavior and communication. The interaction between the modules is abstracted using communication primitives that hide implementation details of the communication unit.

### *10.3.7 C-VHDL Communication Model*

The key issue in virtual prototyping is the modeling of hardware/software communication. In our case, we use the modular scheme provided by Solar, to produce a modular C-VHDL specification suitable for both, co-simulation and co-synthesis.

The use of modular description scheme, allows for separate evolution of the different modules. During the design process, several representations of the same object may be used at different design steps. During virtual prototyping, the modules are described at the behavioral level and the communication units are modelled as hardware or software processes. Later, after implementation, the communication units correspond to existing units.

In order to allow the use of a communication unit at different design steps we need to describe its communication procedures into different views. Additionally, in order to be able to connect a communication unit to both, hardware and software modules, we have hardware and software views of a communication procedure. The different views of communication procedures may be seen at different libraries required to link the design with different applications.

To support different applications, the number and type of views for each procedure will depend also on the co-simulation and co-synthesis environ-

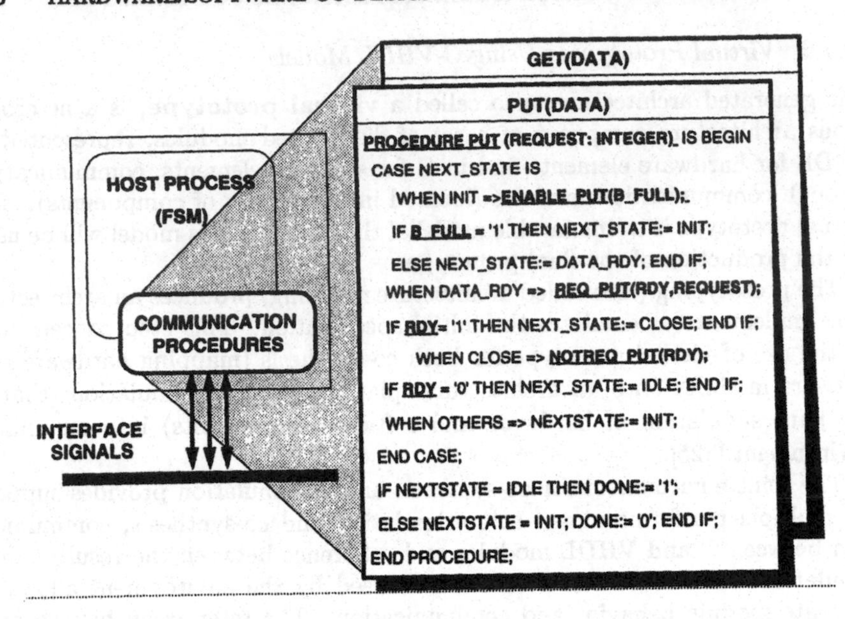


**Figure 10.14** Hardware view (VHDL) of a communication procedure.

ments used. The hardware view (given in VHDL) may be common to both co-simulation and co-synthesis. In the case where we use different synthesis systems supporting different abstraction levels (e.g. a behavioral synthesis and an RTL synthesis), we may need different views for the communication procedures. Figure 10.14 gives a hardware view for the procedure put for message passing based protocol. The procedure describes the interaction of the communication primitive with the controller by using internal signals and handshakes. This VHDL procedure, at the behavioral level, uses a FSM model to describe its interaction through the interface signals. Within the put procedure, specific commands (like ENABLE_PUT, REQ_PUT, or NOTREQ_PUT) or single signal assignment operations performs this interaction.

Figure 10.15 shows two different software views of the communication procedure put. The two software views are needed for co-simulation and co-synthesis respectively.

The software simulation view (used for simulation) hides the simulation environment. In the present version, we use the Synopsys C-language Interface (CLI) as the target architecture for simulation, then the procedure is expanded into CLI routines [330]. Of course, other co- simulation models can be used. For example, if we use the interprocesses communication (IPC) model of UNIX,

SW SIMULATION VIEW (CLI)
• HIDES THE SIMULATOR COMMUNICATION DETAILS

```
INT PUT(REQUEST)
INTEGER REQUEST;
{ SWITCH( NEXT_STATE )
  { CASE INIT : { /* INIT STATE STATEMENTS */ } BREAK;
   CASE DATA_RDY :
   {
      CLIOUTPUT(DATAIN,REQUEST);
      CLIOUTPUT(REQ,BIT_1);
    IF( CLIGETPORTVALUE(RDY) == BIT_1)
      NEXT_STATE = CLOSE; } BREAK;
  CASE CLOSE :{ /* CLOSE STATE STATEMENTS */ } BREAK;
  DEFAULT : NEXTSTATE := INIT; }
 IF( NEXTSTATE == IDLE ) RETURN 1;
 ELSE { NEXTSTATE == INIT; RETURN 0; } }
```

SW SYNTHESIS VIEW (IBM/PC)
• HIDES THE ARCHITECTURE COMMUNICATION DETAILS

```
INT PUT(REQUEST)
INTEGER REQUEST;
{ SWITCH( NEXT_STATE )
  { CASE INIT : { /* INIT STATE STATEMENTS */ } BREAK;
   CASE DATA_RDY :
   {
      OUTPORT(DATAIN,REQUEST);
       OUTPORT(REQ,BIT_1);
    IF(  INPORT(RDY) == BIT_1)
      NEXT_STATE = CLOSE; } BREAK;
  CASE CLOSE :{ /* CLOSE STATE STATEMENTS */ } BREAK;
  DEFAULT : NEXTSTATE := INIT; }
 IF( NEXTSTATE == IDLE ) RETURN 1;
 ELSE { NEXTSTATE == INIT; RETURN 0; } }
```

**Figure 10.15** Different software views of a communication procedure.

this communication procedure call will be expanded to system routines using the IPC mechanism [325].

The software synthesis view (used for synthesis) hides the compilation environment. The view will depend on target architecture selected. If the communication is entirely a software executing on a given operating system, communication procedure calls are expanded into system calls, making use of existing communication mechanisms available within the system. For example, if the communication is to be executed on a standard processor, the call becomes an access to a bus routine written as an assembler code. In the example, the software synthesis view makes use of existing mechanisms available within the IBM PC system. The communication can also be executed as an embedded software on a hardware data path controlled by a micro-coded controller, in which case, this communication procedure call will become a call to a standard micro-code routine.

In this case, we need one hardware view given in VHDL, one software simulation view given in C, and a software synthesis view specific to each target architecture. The difficulty found with the use of multiple views of a communication primitive is the management of the different views to support multiples applications and platforms within the environment. We assume the existence of trusted librarian able to produce coherent views. This is a general problem that exist when libraries are used. The librarian may be an automatic generation tool or a designer.

## 10.4 DESIGN STEPS

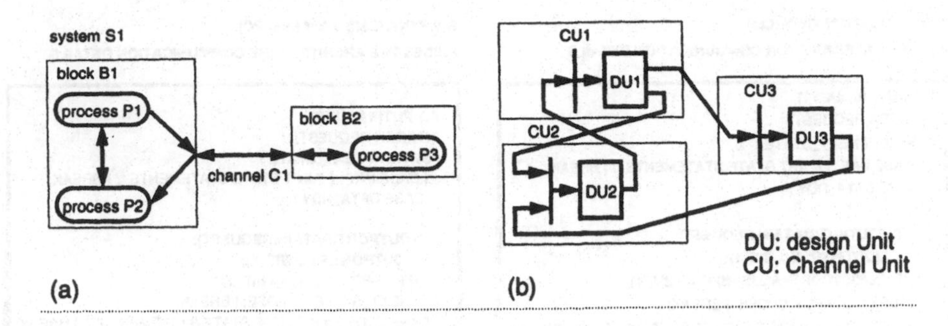


**Figure 10.16** Corresponding models: (a) SDL, (b) Solar.

Cosmos starts with a multi-thread description and produces a multiprocessor architecture. The main steps are system specification, compilation, partitioning, communication synthesis and architecture generation.

### *10.4.1 SDL Compilation*

The design flow begins with an SDL description, but all the synthesis steps use Solar. SDL allows to handle concepts as system, block, process and channel. Translating most SDL concepts (except communication concepts) into Solar is fairly straightforward.

However, translating of SDL communication concepts (channel and signal route) requires reorganizing the description. In SDL, communication is based on message passing. Processes communicate through signal routes and channels group all communication between blocks. Solar communication methodology lets systems containing processes and channels be modelled. Figure 10.16 summarizes this scheme. Figure 10.16(a) shows a system example with two communicating blocks. Block B1 contains two processes (P1, P2) that communicate with process P3, a process that belong to block B2. Translating this system onto Solar produces the structure shown in Figure 10.16(b). Each SDL process is translated into a Solar Design Unit (DU) containing an extended FSM and a Channel Unit (CU). The SDL-to-Solar translation is performed automatically.

### *10.4.2 Restrictions for Hardware Synthesis*

SDL supports a general and abstract communication model that is not well suited for hardware synthesis. This is mainly due to the fact that signals can be routed through channels. In other words the destination of a signal can be determined dynamically by the address of the receiver. In SDL, the dynamic routing scheme is mainly intended for use with the dynamic process creation feature. This feature is very software oriented and is difficult to map in hardware. Nevertheless we can restrict the SDL communication model for hardware synthesis without losing too much of its generality and abstraction. The restriction imposed on SDL will concern its dynamical aspects such as process creation and message routing. The SDL subet supported by Cosmos includes:

A wide subset of SDL is supported including:

- system, process, multiple instances;
- state, state*, state*(), save, save*, continuous signals, enabling condition;
- input, output, signals with parameters, task, label, join, nextstate, stop, decision, procedure call, imported and exported variables.

Feature not supported includes: dynamic creation of processes, Pid (supported for multiple instances processes), channel substructure, non determinism (input any, input none, decision any).

**Structure.** As stated in Section 10.4.2, dynamic aspects of SDL are not considered for hardware synthesis. To avoid any routing problem and obtain an efficient communication, we will restrict ourselves to the case where the destination process of a signal can be statically determined. Communication structure can then be flattened at compile time. A signal emitted by a process through a set of channels must have a single receiver among the processes connected to these channels. In such a case, channels only forward signals from one boundary of a block to another. No routing decision may be taken as there is only one path for a signal through a set of channels. Therefore channels and signalroutes won't be represented in the final system. A process that is emitting a signal will write it directly in the input queue of the destination process without going through several channels. Flattening the communication eliminates the communication overhead that occurs when traversing several channels.

Each SDL process will be translated into the corresponding finite state machine. During the partitioning step state machines may be split and merged to achieved the desired solution. This step may generate additional communication like shared variables. All the communication protocols and implementation details will be fixed by the communication synthesis step regardless from where it has been generated (initial specification or partitioning).

In SDL, each process has a single implicit queue used to store incoming messages. Therefore we will associate one abstract channel to each process (Figure 10.17). This abstract channel will stand for the input queue and will offer the required communication primitives. During the communication synthesis steps a communication unit able to execute the required communication

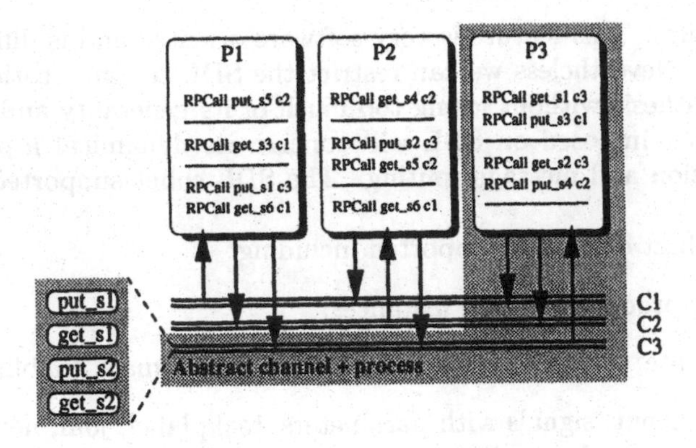


**Figure 10.17** Modeling SDL communication with abstract channels.

scheme will be selected from the library. This approach allows the designer to choose from the library a communication unit that provide an efficient implementation of the required communication. Despite the fact that SDL offers only one communication model, several different protocols may be allocated from the library for different abstract channels. Each signal will be translated as two communication primitives offered by the abstract channel to read and write the signal and its parameters in the channel.

Figure 10.17 represents the refined SDL model corresponding to the system of Figure 10.5 for synthesis. Each SDL process is mapped to a process containing the behavioral part of the specification and an abstract channel that offers communication primitives to send and receive signals. Each process will read from its own channel and write into other processes channel. An SDL specification is therefore represented by a set of processes and abstract channels. As stated before channels and signalroutes are not represented.

### *10.4.3 Hardware/Software Partitioning And Communication Refinement*

As stated before, the partitioning process may be composed of three types of transformations: functional decomposition, structural reorganization and communication transformation. Functional decomposition acts on the state tables allowing to refine behavioral descriptions. Structural reorganization acts on design units allowing to refine the structure of the system. Communication transformation acts on channel units allowing to refine the communication protocols. All these refinements make use of a set of five primitives called split, merge, move, flat and map allowing to decompose, compose and transform Solar objects.

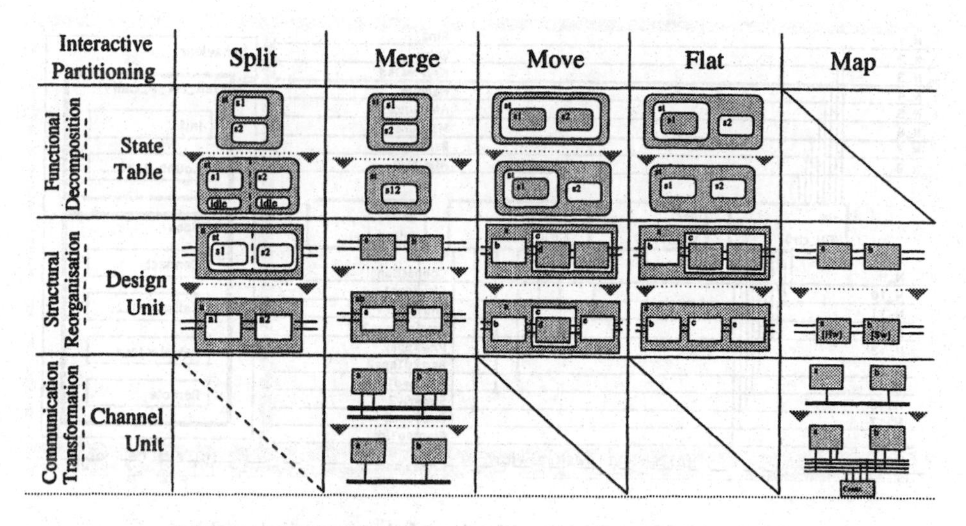


**Figure 10.18** Decomposition/composition and communication primitives.

For each transformation an applicability condition is associated. It defines constraints on the application of the transformations. Before the application of any transformation requested by the designer, the system checks the applicability conditions. The designer can apply the primitives in any order as long as the application conditions of these primitives are fulfilled. The effect of these primitives are explained in the following paragraphs using a co-design of an answering machine [8]. In this example we will perform a sequence of primitives calls in order to achieve a desired realization.

Figure 10.19(a) shows the initial system description coded in Solar. The answering machine is composed of four processes: a controller (ctrl), two decks (decl and dec2) one to read the announcement and another one to record the messages and, a time counter to measure delay. The controller is the main process which manages the resources utilization represented by the other processes. This description can be derived from SDL or another system-level language.

Figure 10.19 shows the four processes, represented by boxes, the external I/O ports (nets 1 to 8) and the internal signals (nets 9 to 18). The ctrl process has a behavior description (called behavior in Figure 10.19(b)), represented by a state table composed of two machines (Wait-For-A-Call and OffHook). The machines hierarchy are graphically represented by indented boxes in our representation. The boxes aligned vertically (horizontally) have a sequential (parallel) execution.

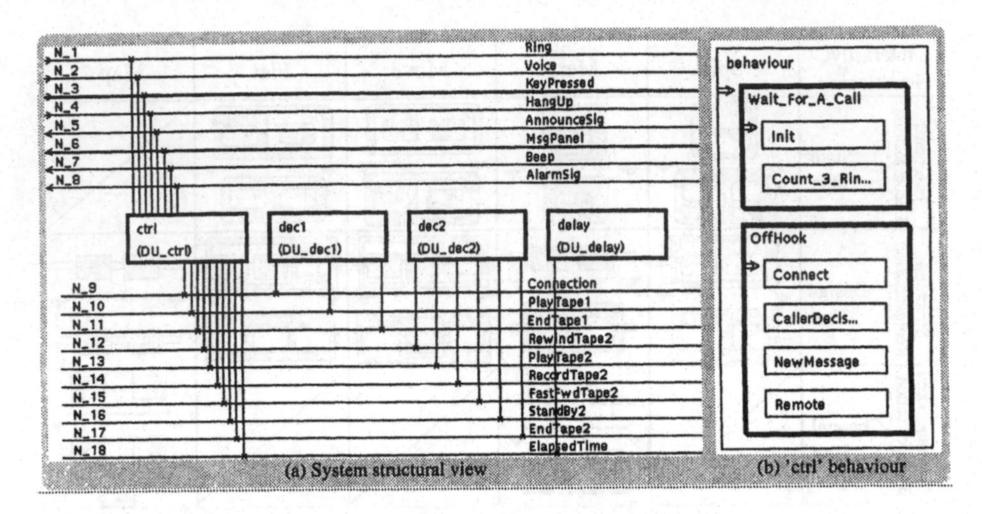


**Figure 10.19** Initial specification of the answering machine.

The desired implementation is guided to divide the ctrl behavior in two partitions. Functions belonging to the Remote machine will be realized in software and the other functions in hardware. The software will be converted into C code and executed by a standard processor. The hardware will be translated into behavioral VHDL and realized by an application specific or a programmable circuit. The appropriate set of functions belonging to each partition is a result of the designer's interactions.

**Functional Decomposition Primitives.** The functional decomposition primitives are used to transform behaviors. A detailed description of these primitives with algorithms can be found in [303]. A brief description follows:

- Split decomposes a machine into a set of subsystems (the term machine is used to represent a state or a state table).
- Merge groups a set of states or state tables into a unique machine.
- Move transforms the hierarchy of a given machine.
- Flat reduces the height of the machine hierarchy.

In the case of the answering machine, the partitioning begins with the functional decomposition primitives in the following sequence (Figure 10.20): a move of the Remote machine to a higher hierarchical level, a merge of the Wait-For-A-Call and OffHook machines and, a split of the behavior machine.

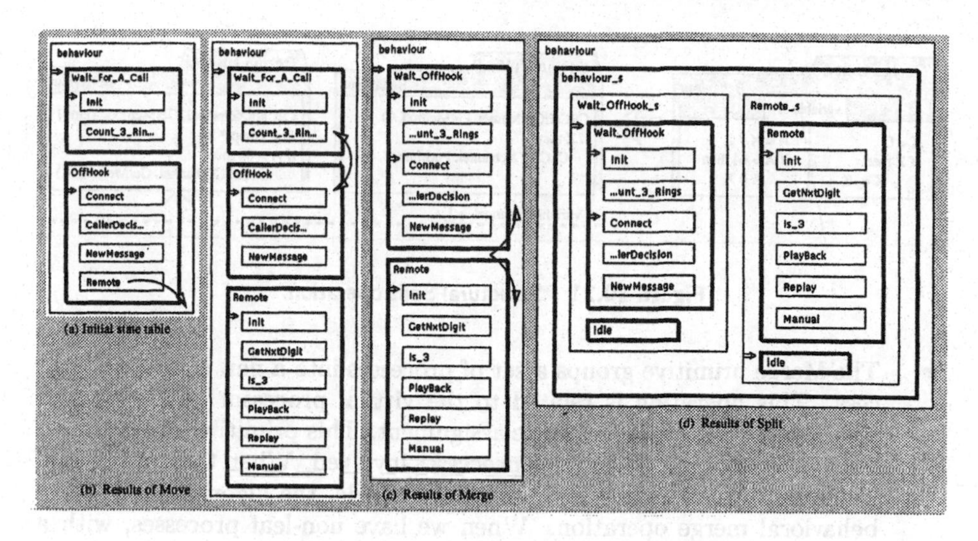


**Figure 10.20** Functional decomposition primitives.

We use the split primitive to transform the sequential machines into parallel machines in order to allow the separation in two communicating processes. To simulate a parallel execution, a control signal and an idle state are annexed to each machine.

The structural refinement goal is to distribute the behavior into a set of design units that correspond to the final processors. In this case, the application of the primitives will transform the structure hierarchy. Hereby follows a description of the structural reorganization primitives:

- The Split primitive works on the behavior of a process in order to cut it into a set of independent processes. Each process will communicate with the others by means of its I/O signals and channels. The data shared between the split machines (global variables), are converted to abstract channels in the new representation. These abstract channels are usually mapped to a shared variable communication protocol. In each generated process, global variable accesses are replaced to calls to remote services offered by the channel. The channel read and write services provide access to these variables. The concurrent accesses to variables in parallel processes is controlled by the channel controller. Figure 10.21 shows the result of splitting a system composed of two concurrent machines that share a common variable. The processes generated can only access the shared data through the abstract channel named X_variable_channel.

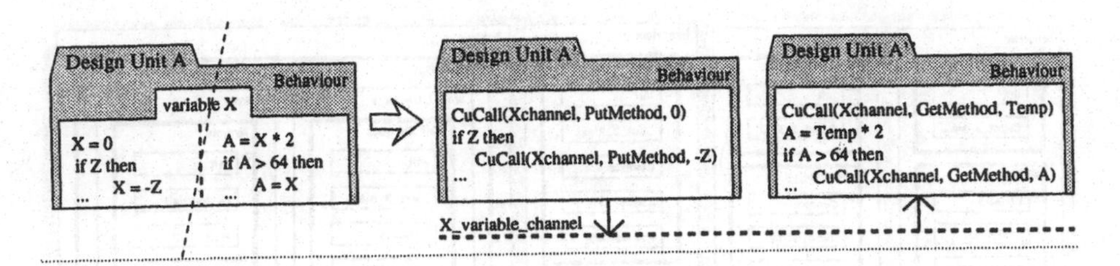


**Figure 10.21** Structural Split operation.

- The Merge primitive groups a set of processes into a new structural process. This operation is related to the virtual processor allocation and precedes the hardware/software assignment. This primitive works in two ways depending on the type of processes involved. When they are all leaf processes, with a behavioral description inside, the algorithm realizes a behavioral merge operation. When we have non-leaf processes, with a process hierarchy inside, the algorithm works on the structure. In the first case, the approach serves to create a behavioral process with a parallel execution of each process behavior involved in the merge operation. When possible, the communication channels between these processes are converted to internal global variables. In the second case, a structural process is created with instance calls to the processes involved on the merge operation. A new hierarchical level is created and the interfaces adjusted.

- he Move moves a design unit in the hierarchy. This primitive is usually activated before the merge primitive. The goal is to put together processes that will be mapped to the same processor in the target architecture. We adjust the signal and channel nets to this new structure.

- Flat performs a structural flattening operation on the hierarchy. An hierarchy level is removed and the signal and channel nets adjusted.

- Map permits the identification of hardware and software realization options for each process.

Figure 10.22 is a continuation of our example and it shows the structural reorganization primitives. In Figure 10.22a we have the machine behavior_s composed of two concurrent machines, Wait-OffHook-s and Remote-s. The use of the structural reorganization primitives begins with the utilization of a split to divide the behavior_s in two different processes. In this case, four extra channels are inserted, each one representing a global variable.

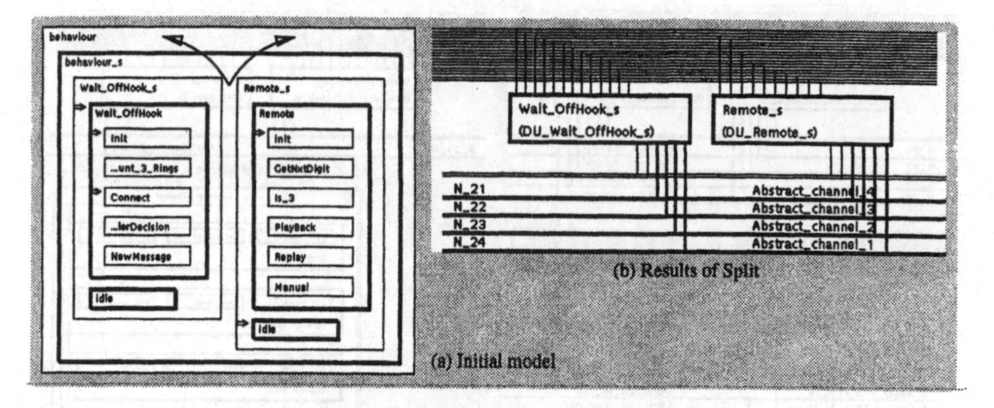


**Figure 10.22** Structural reorganization primitives.

This step transforms a system composed of processes that communicate via high-level primitives (through abstract channels) into interconnected processes that communicate via signals. This step was introduced in section 10.3.5.

Protocol selection is the first step in the communication transformation. The user interactively chooses the appropriate communication unit from a library to implement the protocol [302]. This choice must also match the performances required and realization costs. This operation replaces the abstract channel calls by procedure calls to remote services belonging to the communication unit selected.

The interface building step adds to each process the needed code to control the communication according to the selected protocol. A communication controller may also be inserted. At this moment the channel controller can be seen as a normal process (design unit).

Cosmos makes use of a channel library composed of basic protocol templates. Two primitives are available for the communication transformation:

- Map generates the interfaces (Figure 10.23(c)).
- Merge transformation allows to replace two abstract channels by another abstract channel. In several cases this leads to a cost reduction. The application of this transformation is restricted, the resulting channel needs to have an implementation in the library.

Figure 10.23 shows the use of the primitives merge and map on channels. In this example, we start with a system composed of two design units (Wait-OffHook-s and Remote-s) communicating through four abstract channels (figure 25(a)). Figure 10.23(b) shows the results of the application of the merge

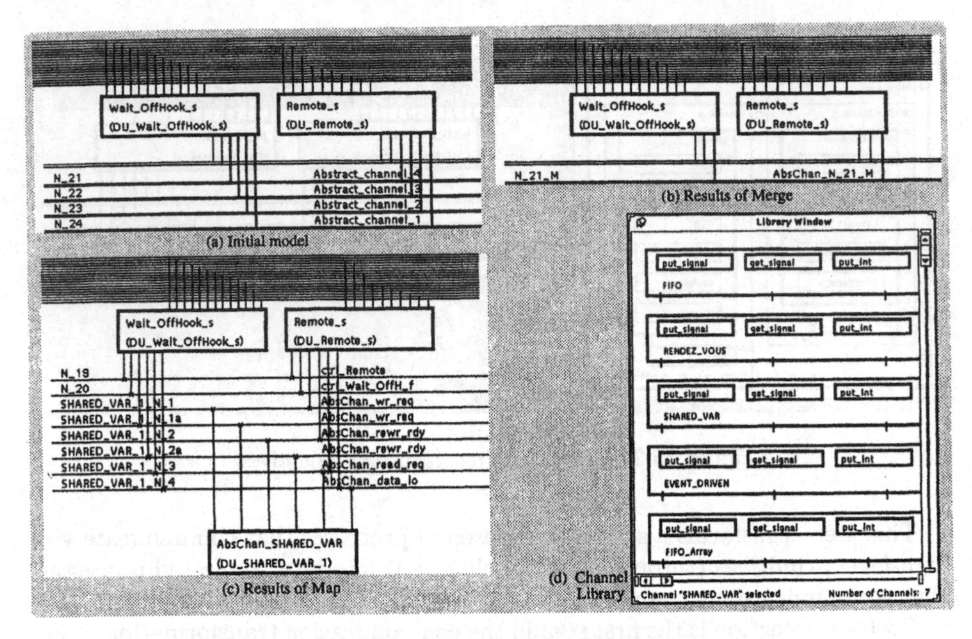


**Figure 10.23** Communication transformation primitives.

transformation. The four abstract channels are merged into a single new one. Figure 10.23(c) shows the results of the application of the map transformation. In this case the abstract channel is expanded using a specific implementation (shared variables) from the library (Figure 10.23(d)).

### *10.4.4 Architecture Generation*

Architecture generation translates Solar into executable code (C and VHDL) to allows the co-simulation and co-synthesis of the system. The virtual prototype generated is a heterogeneous architecture composed of a set of distributed modules, represented in VHDL for hardware elements and in C for software elements, communicating through communication modules from a library of components. The virtual prototype is a simulatable model of the system. The final step is prototyping. The prototyping or architecture mapping, produces an architecture that implements or emulates the initial specification. The general model, introduced in section 10.3.6, allows to separate the behavior of the modules (hardware and software) and the communication units. Inter-module interaction is abstracted using communication primitives that hide the implementation details of the communication units.

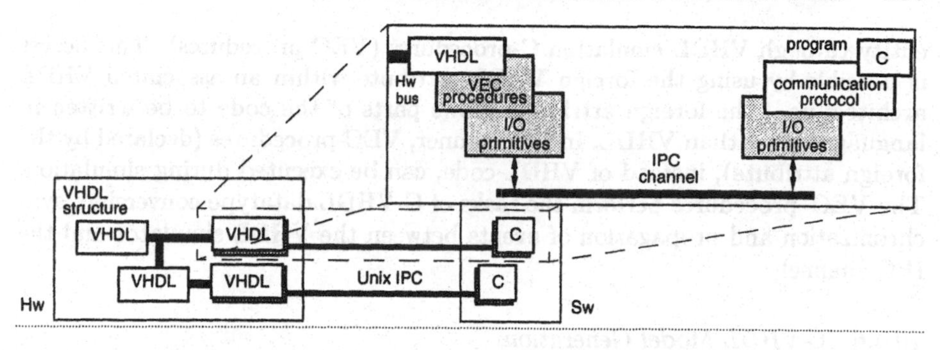


**Figure 10.24** Distributed C-VHDL co-simulation model.

### *10.4.5 VHDL/C Cosimulation Interface*

A distributed co-simulation environment is composed of VHDL simulators for hardware modules and C-debuggers for software modules (C programs). The link between hardware and software simulation environments and the elements used to interface VHDL and C modules during co-simulation are automatically created by a C-VHDL interface generation tool called VCI [331]. The tool takes as input a user defined configuration file, which specifies the desired configuration characteristics (I/O interface and synchronization between debugging tools) and produces a ready-to-use VHDL/C co-simulation interface.

The system configuration is described in VHDL as a set of interconnected blocks. Figure 10.24 details a system composed of four interconnected modules (VHDL structural view). Two of them are software modules. Each software module is encapsulated in a VHDL entity and its behavior is given by a C-program.

The top box of Figure 10.24 shows the detail of the C-VHDL communication structure for one of the software modules. All the dashed parts are generated automatically using VCI. The C-program is connected to the rest of the system through a communication protocol. The protocol allows the C program to exchange information with the external world. When the program interacts with a set of Hw modules, I/O primitives are used by the communication protocol to perform I/O operations over the ports of the hardware part (Hw bus). On the hardware side, the C-program is connected to the hardware bus through a VHDL entity that encapsulates the ports used by the I/O primitives.

The C-program and the set of hardware modules are treated as separated Unix processes communicating through the IPC mechanism. In other words, an IPC channel is the link between the I/O primitives used by the C-program and the I/O ports of the VHDL entity. The channel is accessed by the VHDL entity through VHDL emulation C-procedures (VEC-procedures). This access is possible by using the foreign VHDL attribute within an associated VHDL architecture. The foreign attribute allows parts of the code to be written in languages other than VHDL. In this manner, VEC-procedures (declared by the foreign attribute), instead of VHDL-code, can be executed during simulation. The VEC- procedures perform the tasks of C-VHDL datatype conversion, synchronization and propagation of events between the VHDL simulator and the IPC channel

### *10.4.6 C-VHDL Model Generation*

Figure 10.25 shows the overall C-VHDL generation flow. Starting from a refined Solar model, a first tool, called s2cv produces C and VHDL models, for software and hardware processes respectively. S2cv produces also an interface file that specifies the C-VHDL interface. In a second step, the C- VHDL interface is generated. The first step makes use of trivial transformations and will not be explained here. The rest of this section will detail the C-VHDL interface generation tool (VCI).

The elements used to interface VHDL and C modules during simulation are automatically created by the C-VHDL interface generation tool (VCI). According to the input file (VCI interface description), VCI generates the VHDL and C files needed to interconnect the C-program to the VHDL structure during co-simulation. It is important to note that the tool can alternatively be used to generate the interconnection between only software modules.

The input of VCI is a file that specifies the C-program interface and additional information for co-simulation. The resulting output consists of the VHDL/IPC interface on the hardware side, the C/IPC interface on the software side, and a make-file.

The VHDL/IPC interface is composed of three parts: the VHDL entity (that encapsulates the C program interface), a VHDL architecture (contains the declaration of the VEC-procedures introduced by the foreign VHDL attribute) and the VEC/IPC interface (C-procedures executed by the VHDL simulator that communicates with the software part through the IPC channel).

The C/IPC interface is composed of the I/O primitives (used by the communication protocol of the software module to communicate during simulation through the IPC channel). This primitives and the communication protocol are compiled and linked to the C program creating the executable code. Finally the make-file execution generates automatically the hardware side that will be part of the VHDL structure to be simulated.

Figure 10.26 shows a co-simulation screen corresponding to a distributed model composed of two C-modules and a VHDL module. The right side of the

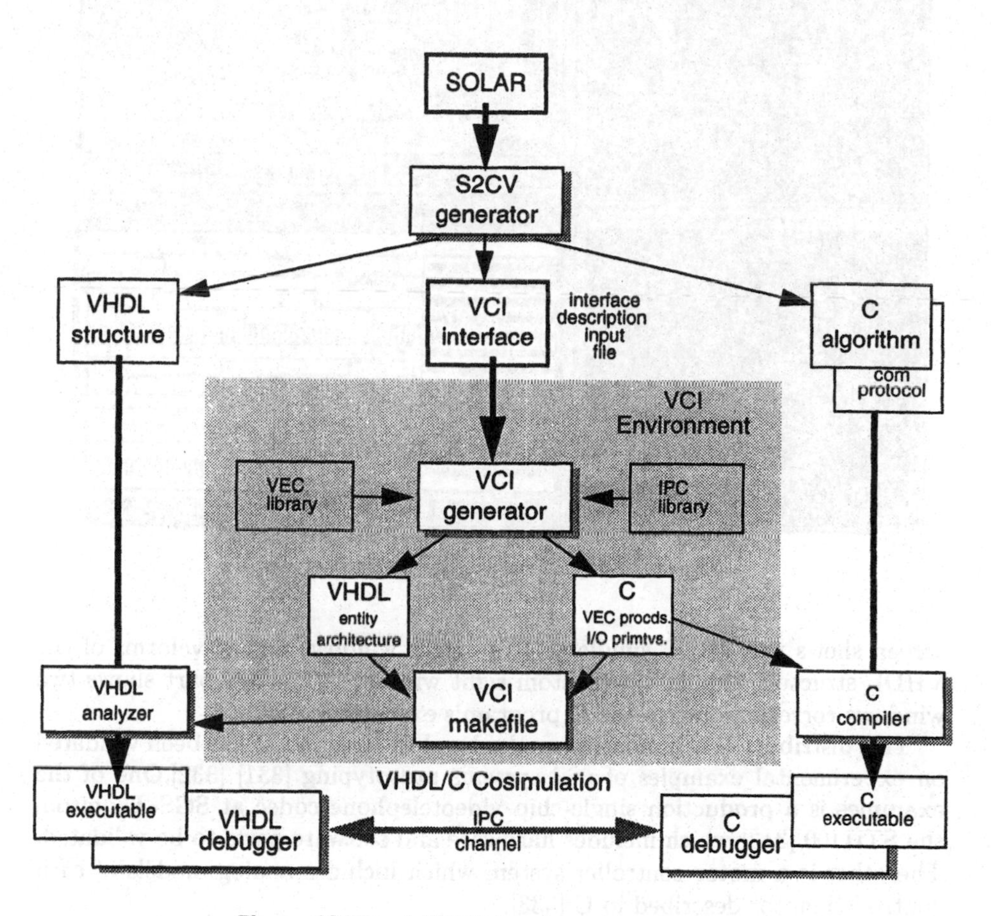


Figure 10.25 C-VHDL interface generation flow.

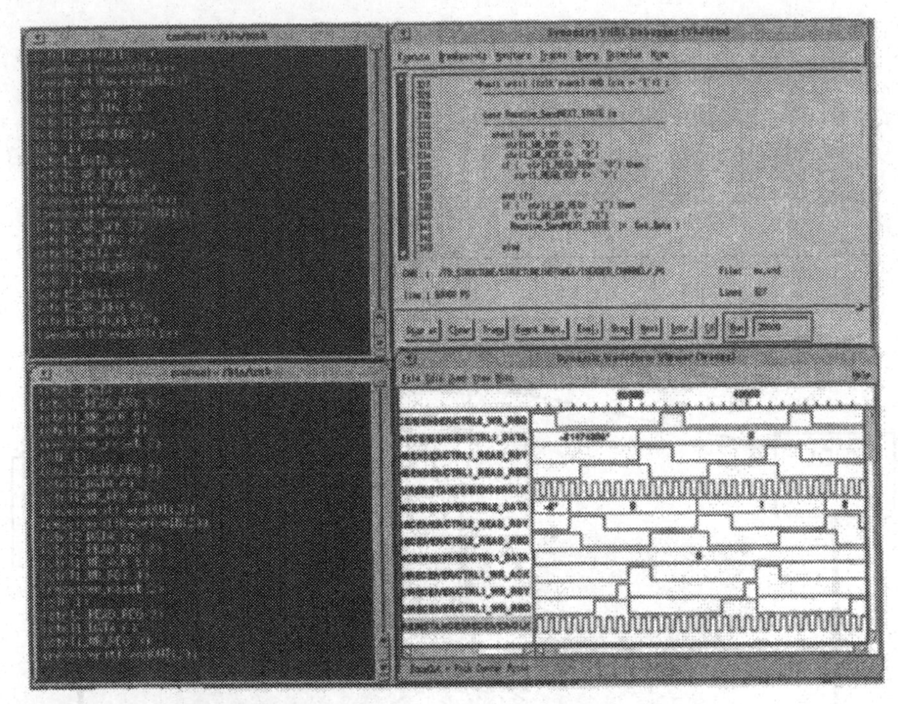

**Figure 10.26** A co-simulation screen dump.

screen shows the VHDL simulator (top- right window) and waveforms of the VHDL structure simulation (bottom-right window). The left part shows two windows corresponding to the C program's execution trace.

This distributed co-simulation method and the tool (VCI) has been validated on experimental examples of real systems prototyping [331] [332].One of the examples is a production single-chip videotelephone codec at SGS-Thomson, the STi1100 [313] which includes hardware and software parts to be validated. The other is a motor controller system which includes analog models of each controlled motor described in C [333].

### *10.4.7 Prototyping*

The starting point of prototyping is the virtual prototype, composed of a set of interconnected hardware modules (in VHDL), software modules (in C), and communication modules. The result of prototyping is a functional architecture integrating all the modules into processors, FPGAs and already existing circuits (Figure 10.27) [333].

The prototyping consists of the realization of an application prototype. The prototype can be a first version of the final machine or a temporary realization of the application. This step consists of integrating the software parts and the hardware parts on a physical support. The result can be a circuit, a circuit board or a complex system. In the first case, the Sw, hardware and the communication modules are integrated within a single chip. In other cases, one can use the processors and partially existing architectures that can be adapted to validate the application. This step is necessary to understand the real performance of the solution. It consists of assembling the different parts of the system.

The C-VHDL model used in Cosmos allows to combine co-simulation and co-synthesis within a single environment supporting a large variety of architectures and platforms. VHDL and C descriptions can be simulated together to validate the specification at different steps of the prototyping process. The simulated descriptions are also used for co-synthesis. The co-synthesis will produce code and/or hardware that will execute on a real architecture. In general, this step consists on mapping the system onto a Hw-Sw platform that includes a processor to execute the software and a set of ASIC's to realize the Hw.

As shown in Figure 10.27, the resulting prototype is generated by using the executable code generators for the software parts to be realized, the logic synthesis, placement and routing tools for the hardware parts to be realized. The compilers translate the C-programs into an assembler code to be executed on the processors, whereas, the synthesis tools transform the VHDL descriptions down to ASICs or FPGAs.

### *10.4.8 Hardware Design*

The design of complex hardware modules necessitates several iterations in the design flow. These iterations permit the designer in refining the selected hardware components as well as their codes of descriptions. The starting point of the hardware design is a behavioral description of the module in VHDL. The final result is a circuit netlist. The design flow consists of two synthesis steps: An architectural synthesis and a logic synthesis. The architectural synthesis permits us to analyse trade-offs between different architectures algorithmically and to pass from a behavioral level to register-transfer level (RTL). The RTL description obtained, describes the functional architecture of the circuit. The logic synthesis compiles the VHDL RTL description to optimize the design for area, performance and power and generates a netlist of the circuit. At each level of abstraction (behavior, RTL, netlist), one conducts a co-simulation of the hardware module with that of software (Figure 10.27). This permits us to

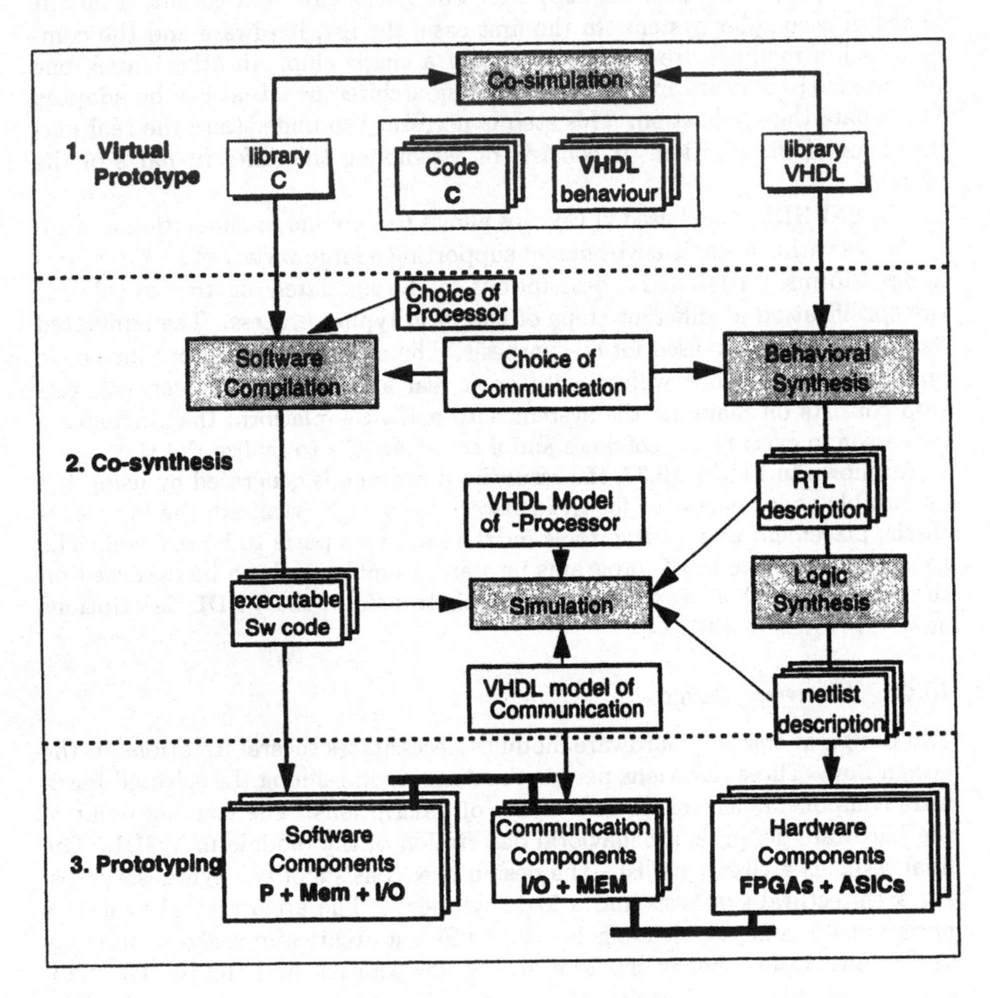


**Figure 10.27** Prototyping design flow.

validate the functionality of the circuit at each step of the synthesis process (namely, architectural synthesis and logic synthesis).

### 10.4.9 Software Design

A software module is described in the ANSI standard C language. This description is used at the beginning for co-simulation. The same description, accompanied by the procedures and functions described for the software synthesis, are compiled on a microprocessor based architecture. Only, the principal program code is the same for co-simulation and co-synthesis. On the other hand, for the procedures and functions of calculations and of communication, one is provided with the necessary libraries for co-simulation and co-synthesis. For establishing the library necessary for the compilation of the software code, one needs to know the type of the architecture upon which the software will execute and in particularly the type of the microprocessor.

The software module consists of a processor executing a machine code. The performance as well as the cost of this module is decided principally by the type of the processor. The C-program is generated independently of the processor. This allow to postpone the processor selection decision to a late stage in the design process. This choice is indispensable since for the compilation of the C-code, one needs to describe the procedures used by the software conforming to the microprocessor on which they will be executed. The processor can be a standard micro processor or an Application Specific Instruction Processor (ASIP).

The performance requirements for the execution of the software code will determine the type of processor. For a signal processing application, where the software code will consist of several functions of complex calculations (example: sum of products, operations on matrices, etc.), a DSP processor is more suitable. On the other hand, for simple calculations on data types, a standard microprocessor would be sufficient.

The software module described and co-simulated in C is compiled with the procedures of calculation and communication in order to obtain a code that can be executed by the microprocessor of the software module. This compilation is realized either directly on a software module or by a compiler specific to the microprocessor.

An interesting method for simulating this code consists of using a VHDL processor model that can execute the code. This description executes all the machine code supported by this processor and gives the same result as can be obtained from a real architecture. At this level, one can simulate the hardware modules described at the RTL or netlist level with the software modules composed of the compiled code and the processor model upon which they will be executed.

## 10.5 APPLICATION

This section details the application of the transformational approach to a co-design example. The example is a large application, a robot arm controller. In the example we will illustrate the overall design flow of Cosmos, from a system-level specification given in SDL to a distributed hardware/software architecture described in C/VHDL. The use of SDL allows for fivefold reduction of the size of system specification when compared to distributed C-VHDL models. All the transformations applied in this section are fast enough to look instantaneous during an interactive session. None of this primitives require more than 5 seconds CPU time on a Sparc 20 workstation. The robot arm controller implementation regarding to the design space exploration and high-level synthesis can be found in [334] and [333] respectively.

### *10.5.1 Robot Arm controller co-design example.*

The robot arm controller adjusts the position and speed parameters of eighteen motors belonging to a robot arm. This computation is intended to avoid discontinuous motors operation problems. The change in motors speed should follow a smooth curve for acceleration and deceleration for mechanical reasons.

The basic block diagram of this system, in SDL, is shown in the Figure 10.28(a. The AdaptativeSpeedControl block receives positions from the HostMachine block. The AdaptativeSpeedControl calculates the speed that each motor needs, to arrive at the same time at the desired position. The motor with the longest distance to go runs at the maximal speed and the speed of the other motors are adapted to this one. The output speed desired for each motor is send in the form of speed control pulses to the MotorSender block, which converts the pulses into control signals to each motor.

- The Host machine process is the interface between the user and the system, it takes care of storage segments and motors constants calculation.
- The Distance calculation stores up to date information about the distance to go for every motor according to the current position.
- The Distribution controller identifies the motor that must run at maximal speed and sends scaled commands to the speed controller.
- The Speed controller converts the input steps into a curve that respects the worst case maximum load acceleration and deceleration curves of the motors.

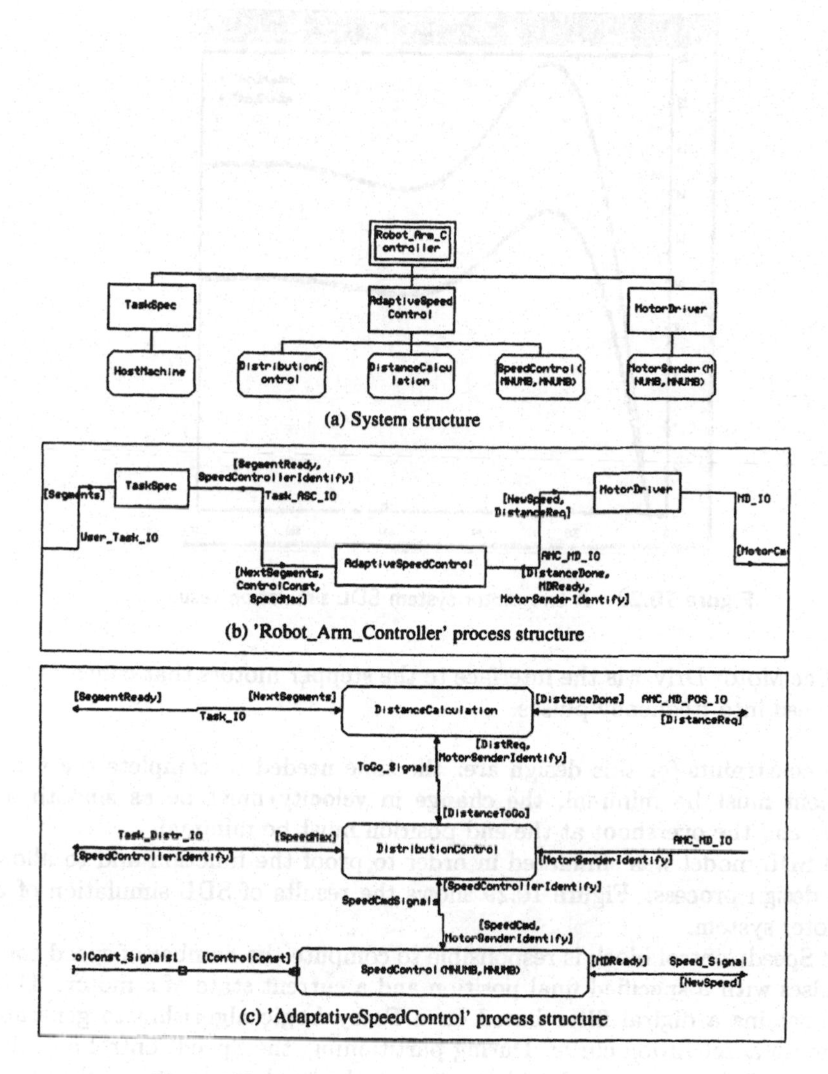


(a) System structure

(b) 'Robot_Arm_Controller' process structure

(c) 'AdaptativeSpeedControl' process structure

**Figure 10.28** SDL input graphical representation.

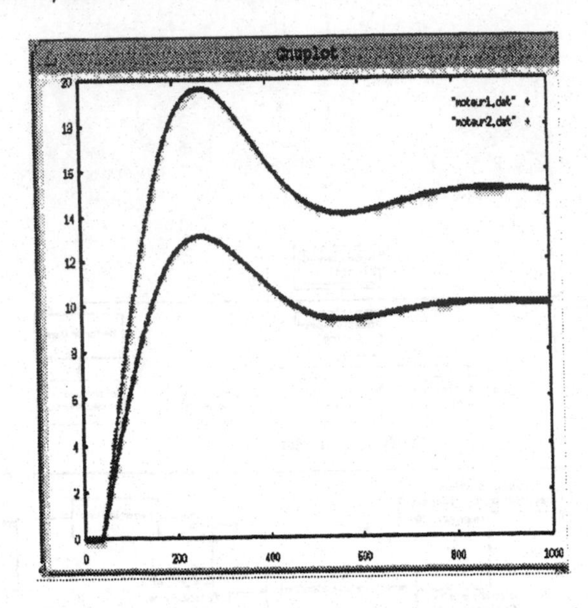


**Figure 10.29** A two-motor system SDL simulation results.

- The Motor Driver is the interface to the stepper motors that converts the speed into frequency pulses.

The constraints for this design are: the time needed to complete the total movement must be minimal, the change in velocity must be as smooth as possible and the overshoot at the end position must be minimal.

The SDL model was simulated in order to proof the function and to allow the co-design process. Figure 10.29 shows the results of SDL simulation of a two-motor system.

The SpeedControl block is responsible to compute the number of speed control pulses with a specified final position and a current state of a motor. This block contains a digital filter based on a Fuzzy Logic algorithm to generate the smooth acceleration curve. During partitioning, the SpeedControl and the MotorSender blocks were assigned to compose the hardware part and the other blocks the software part. In this architecture implementation, the processes with extensive computation were assigned to have a hardware implementation to respect the system time constraints.

Figure 10.30 shows the use of refinement primitives on the Solar representation. Figure 10.30(a) shows the top hierarchy translated from SDL to Solar. The three top modules are connected to 2 external communication units and to four internal ones. Figure 10.30(b) shows the structure of the adaptive speed

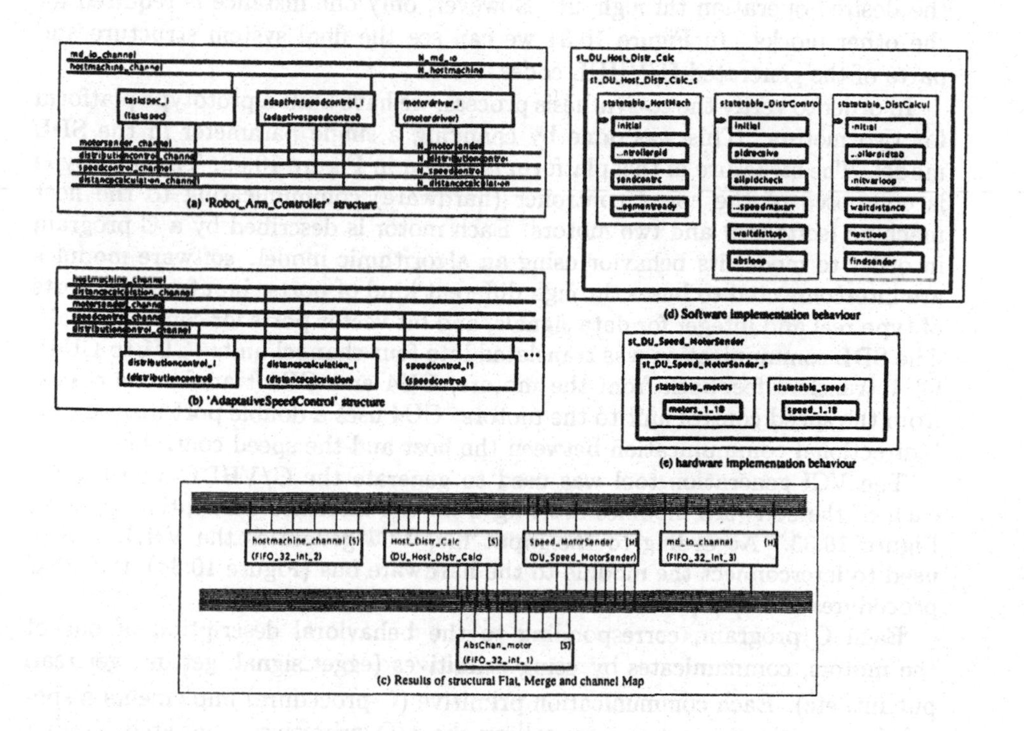


(a) 'Robot_Arm_Controller' structure

(b) 'AdaptativeSpeedControl' structure

(c) Results of structural Flat, Merge and channel Map

(d) Software implementation behaviour

(e) hardware implementation behaviour

**Figure 10.30** Refinement steps.

control block. Figure 10.30(c) shows the result of a set of transformations, including flat and merge operations on the initial structure and, merge and map on the channels. The abstract channels were mapped on protocols from the library. In Figure 10.30(d) and Figure 10.30(e) we can see a graphical representation of the finite state machines resulting after the application of the structural merge operations.

The initial design requires to duplicate the SpeedControl and MotorSender blocks eighteen times. An instance for each motor was necessary to arrive to the desired operation throughput. However, only one instance is required for the other blocks. In Figure 10.31 we can see the final system structure and parts of the generated C/VHDL codes.

In order to verify the co-synthesis process, we have built a prototype platform for two motors. This was done by changing a single parameter in the SDL model. The structure of this platform is shown in Figure 10.32. The prototype is composed of the speed controller (hardware) communicating to the host machine (software) and two motors. Each motor is described by a C program in order to model its behavior using an algorithmic model. software modules are interconnected to buses through different kind of ports: input-output ports of type real and integer for data signals, and bit vector ports for control signals. The SDL communication was translated into four channel units (CU1 to CU4). CU1 transfers feedback from the motors, CU4 and CU2 transfer commands from the speed control unit to the motors. CU4 uses a double port memory for bidirectional communication between the host and the speed controller.

The VCI generation tool was used to generate the C/VHDL interface for each of the software modules starting from the interface description given in Figure 10.33. According to the input file, VCI generates the VHDL entity used to interconnect the module to the hardware bus (Figure 10.34), the VEC procedures and I/O primitives.

Each C program, corresponding to the behavioral description of one of the motors, communicates by using primitives (eġget_signal, get_int, get_real, put_int, etc). Each communication primitive (C-procedure) implements a specific communication scheme by calling the I/O primitives generated by VCI (motor1itfSendOut, and motor1itfReceiveIn). This scheme is explained in Figure 10.34.

### *10.5.2 C-VHDL Co-Simulation*

The co-simulationof the example requires a VHDL simulator and three C programs, two C modules for modeling the motors and the third to be executed on the host machine. Each software module was debugged in its corresponding debugging tool and the speed of each motor was traced by using gnuplot

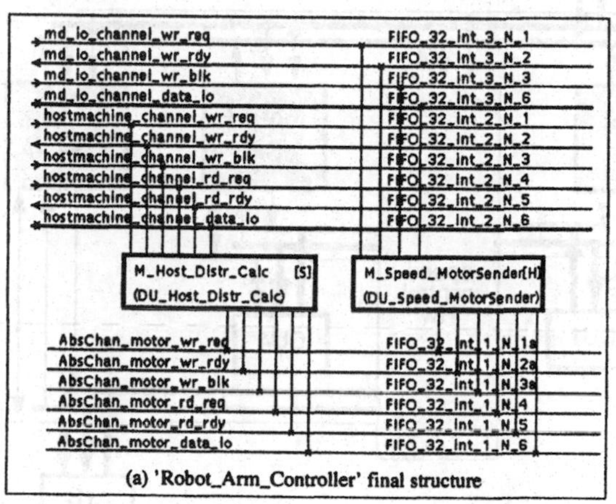


(a) 'Robot_Arm_Controller' final structure

```
/*
                C-file distancecalculation.c
*/
int distancecalculation()
{
  ........

  switch(statetable_distancecalculationNextState)
   {
     case initialState :
     {
       hostmachine_channel_putsdl_signalProc(&(segmentready));
       updated= FALSE;
       motornr=  0 ;
       statetable_distancecalculationNextState= initloopState;
     } break;
     case waitdistanceState :
     {
       distancecalculation_channel_getsdl_signalProc(&(received_signal));
       if ((received_signal)==( 1 ))
       {
         distributioncontrol_channel_putsdl_marrProc(&(distancetogo),&(togo));
         statetable_distancecalculationNextState= waitdistanceState;
         break;
       }
       else
       if ((received_signal)==( 2 ))
       {
         distancecalculation_channel_getsdl_marrProc(&(tmparr));
         updated= TRUE;
         motornr=  0 ;
         statetable_distancecalculationNextState= nextloopState;
         break;
       }
        .......
   }
}
/* end distancecalculation.c */
```

(b) Partial C-code of 'distancecalculation' process

```
/*
                          VHDL-file speedcontrol1.vhd
*/
library IEEE;
   use IEEE.STD_LOGIC_1164.all;
   use IEEE.STD_LOGIC_ARITH.all;
.....
entity speedcontrol1 is
   Generic ( IPCKEY: INTEGER:= 1);
   Port (
     CLK: IN BIT;
     RST: IN BIT;
     motorsender1_channel_wr_req: OUT BIT;
     motorsender1_channel_wr_rdy: IN BIT;
     motorsender1_channel_wr_blk: OUT BIT;
.....
   );
end speedcontrol1;

architecture speedcontrol1_behaviour of speedcontrol1 is
 begin
   process
.....
   begin
     StateTable_statetable_speedcontrol1: loop
       case statetable_speedcontrol1_NextState is
         when(initial) =>
             vin_1:= 0;
             motorsender1_channel_putsdl_integer( sdl_signal=> 3, param_1=> 0);
             statetable_speedcontrol1_NextState:= waitk;
             exit StateTable_statetable_speedcontrol1;
         when(waitk) =>
             speedcontrol1_channel_getsdl_signal( sdl_signal=> received_signal);
             if (received_signal = 1) then
                speedcontrol1_channel_getsdl_integer( param_1=> ctrlconst);
                statetable_speedcontrol1_NextState:= waitspeed;
                exit StateTable_statetable_speedcontrol1;
             end if;
.....
```

(b) Partial VHDL-code of 'speedcontrol' process

**Figure 10.31** Virtual prototype.

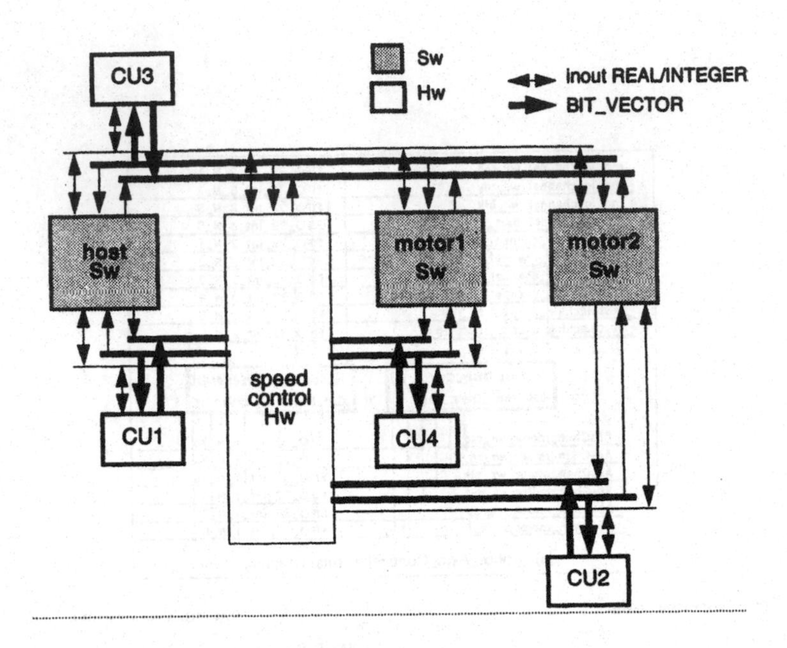


**Figure 10.32** Adaptative motor controller system.

motor1itf VCI

```
!
!PORT                        DIR    TYPE{[RANGE]}SENSITIVITY
!
CLK                          IN     BIT         R
RST                          IN     BIT         N
algo1_channel_wr_req         OUT    BIT         N
algo1_channel_rewr_rdy       IN     BIT         N
algo1_channel_read_req       OUT    BIT         N
algo1_channel_data_int       INOUT  INTEGER     N
algo1_channel_data_real      INOUT  REAL        N
control_channel_wr_req       OUT    BIT         N
control_channel_rewr_rdy     IN     BIT         N
control_channel_read_req     OUT    BIT         N
control_channel_data_int     INOUT  INTEGER     N
control_channel_data_real    INOUT  REAL        N
```

**Figure 10.33** VCI input file for the motor application.

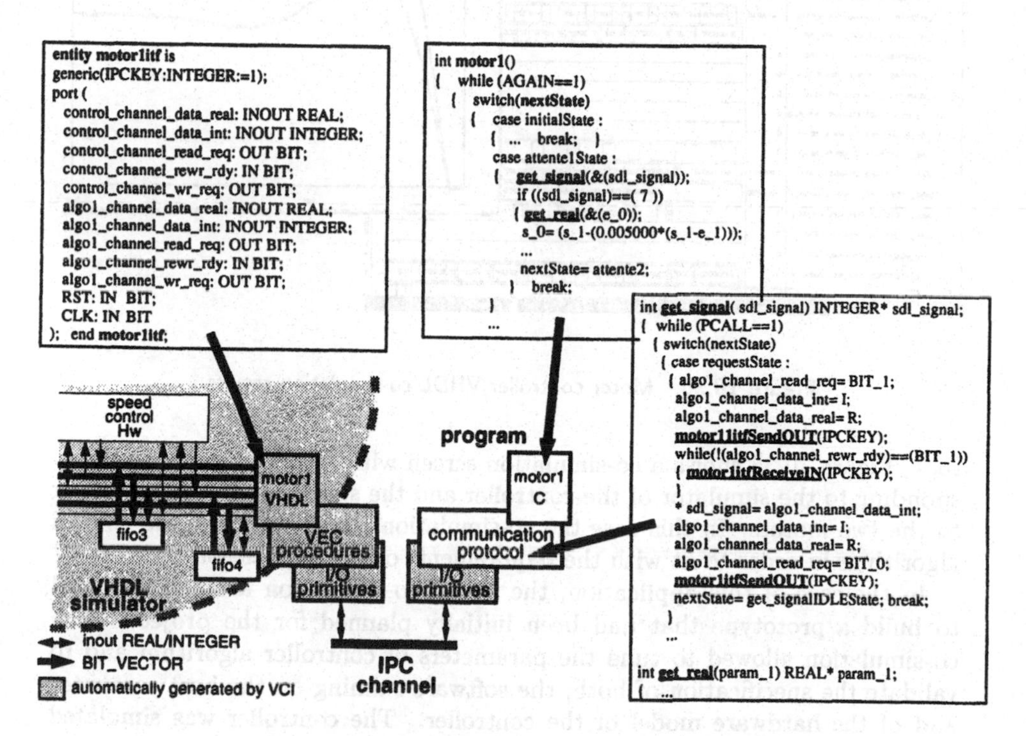
entity motor1itf is
generic(IPCKEY:INTEGER:=1);
port (
control_channel_data_real: INOUT REAL;
control_channel_data_int: INOUT INTEGER;
control_channel_read_req: OUT BIT;
control_channel_rewr_rdy: IN BIT;
control_channel_wr_req: OUT BIT;
algo1_channel_data_real: INOUT REAL;
algo1_channel_data_int: INOUT INTEGER;
algo1_channel_read_req: OUT BIT;
algo1_channel_rewr_rdy: IN BIT;
algo1_channel_wr_req: OUT BIT;
RST: IN BIT;
CLK: IN BIT
); end motor1itf;
int motor1()
{ while (AGAIN==1)
{ switch(nextState)
{ case initialState :
{ ... break; }
case attente1State :
{ get_signal(&(sdl_signal));
if ((sdl_signal)==( 7 ))
{ get_real(&(e_0));
s_0= (s_1-(0.005000*(s_1-e_1)));
...
nextState= attente2;
} break;
}
...
int get_signal( sdl_signal) INTEGER* sdl_signal;
{ while (PCALL==1)
{ switch(nextState)
{ case requestState :
{ algo1_channel_read_req= BIT_1;
algo1_channel_data_int= I;
algo1_channel_data_real= R;
motor11itfSendOUT(IPCKEY);
while(!(algo1_channel_rewr_rdy)==(BIT_1))
{ motor1itfReceiveIN(IPCKEY);
}
* sdl_signal= algo1_channel_data_int;
algo1_channel_data_int= I;
algo1_channel_data_real= R;
algo1_channel_read_req= BIT_0;
motor1itfSendOUT(IPCKEY);
nextState= get_signalIDLEState; break;
}
...
int get_real(param_1) REAL* param_1;
{ ...
speed control Hw
motor1 VHDL
VEC procedures
I/O primitives
fifo3
fifo4
VHDL simulator
program
motor1 C
communication protocol
I/O primitives
IPC channel
inout REAL/INTEGER
BIT_VECTOR
automatically generated by VCI

**Figure 10.34** C-VHDL motor1 co-simulation models.

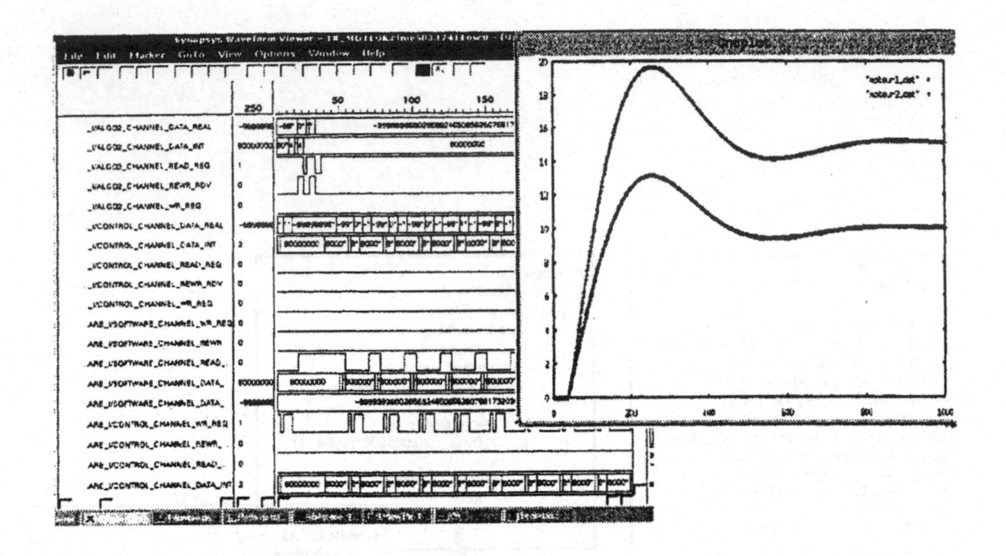

**Figure 10.35** Motor controller VHDL co-simulation results.

tool. Figure 10.35 shows a co-simulation screen with VHDL waveforms corresponding to the simulator of the controller and the speed curve corresponding to the two motors. In this case the co-simulation allowed to tune the control algorithms in order to fit with the requirements of the application.

In the case of this application, the use of co-simulation avoided the need to build a prototype that had been initially planned for the project. The co-simulation allowed to tune the parameters of controller algorithm and to validate the specification of both, the software running on the host computer and of the hardware model of the controller. The controller was simulated at three abstraction levels [333]. The initial model was given in behavioral VHDL. The application of a behavioral synthesis tool produced a VHDL-RTL model that was also co-simulated in the same environment. The application of RTL synthesis produced a gate model that was also simulated. The same hardware/software interface generated by VCI was used for the three co-simulations.

### *10.5.3 Architecture Generation*

We implemented the modules using standard design tools, such as C compilers for software modules tool Amical (developed at the INPG/TIMA laboratory) and the Synopsys logic synthesis tool [309] for hardware.

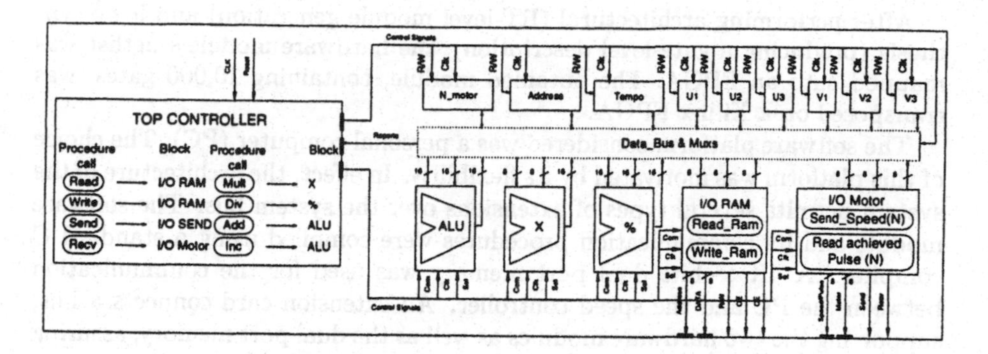


**Figure 10.36** Generated hardware architecture.

The speed calculation module is specified in behavioral VHDL. The high-level synthesis process is performed in three steps: scheduling, functional units allocation and architecture generation. Figure 10.36 represents the global architecture of the speed calculation hardware module, automatically generated. The initial description (in behavioral VHDL) contains about 150 lines of VHDL code. The generated architecture is described into 3740 lines of RT-level VHDL code. The architectural synthesis generates an architecture composed of a controller composed of 39 states and a FSM with 108 transitions. The controller commands the data path using 49 control lines. This controller sequences the operations executed by the functional units. It is generated automatically during the synthesis process. The data-path is constituted by 5 functional units (ALU, 8x8 multiplier, 16/6 divider, I/O RAM and I/O motor), 9 registers of 8 bits and muxes (3 Mux(2), 5 MUx(30 and 8 Mux(4), where Mux(i) means a multiplier with i entries).

The speed calculation unit is connected with the software distribution module and the motor interface module. The hardware/software communication is assured by a shared memory to store the I/O parameters. Thus, the speed control enclose two additional functional units (Figure 10.36): IO_RAM (between the speed control and the dual port RAM memory) and IO_Motor (between the speed control and the motors interface). Each unit executed a specific protocol synchronized by an external clock signal. The IO_RAM unit executes two operations: Read_Ram (read memory operation) and Write_RAM (writes the memory). The unit executes two operations: Send_Speed (send the speed parameters) and Recv_Pulse (receiving the consumed pulses). The motor interface module, allows to transform the speed value (8-bit value) issued by the speed controller into a width-modulated pulse to run the motor.

After performing architectural (RT-level module generation) and logic synthesis (producing a gate-level description), the hardware module's netlist was mapped onto an FPGA. The obtained module, containing 10,000 gates, was transposed on a Xilinx FPGA.

The software platform considered was a personal computer (PC). The choice of this platform was motivated by its flexibility. In effect, the architecture of the system permits several types of extensions over the system bus. The software module and its communication procedures were compiled using a standard C compiler. A 4K×8 bits dual port memory was used for the communication between the PC and the speed controller. An extension card connects a bus, supporting the two hardware modules as well as the dual port memory, assuring the hardware/software communication.

## 10.6 EVALUATION

The design methodology and prototyping facilities provided by Cosmos project permits to have an executable model much earlier within the design cycle. A system level specification serves as a basis for deriving an implementation. The use of system-level specification languages offers concepts and methods adapted to the description of complex systems, providing formal methods for verifying an validating the system behavior. The co-design process is an interactive process where several solutions can be explored at an early stage of the development cycle. In this context, the Cosmos project is fast enough to allow the exploration of several solutions in short time intervals.

Figure 10.37 shows the co-design flow and gives the synthesis time for different abstraction levels associated to the motor controller development example. While we can perform a fast simulation and verification at the system level (SDL), because the size of system-level descriptions is usually 5-10 times smaller than equivalent lower-level descriptions, the gap between the specification and the realization includes many design decisions and implementation details that need to be performed in short time intervals. The Cosmos methodology allows a fast transformation of the system specification given in SDL to a distributed hardware/software architecture.

The co-verification of the resulting hardware/software virtual prototype is less expensive than working directly with its prototype. This co-verification facilitates the understanding of the hardware/software communication and to detect errors present in the initial specification. In other words, the time necessary for the co-simulation of software modules (each processor executing less than 1000 lines of code per second) with high-level descriptions (10 to 100 thousand lines per second) is smaller as compared to the co-simulation using the low-level ones.

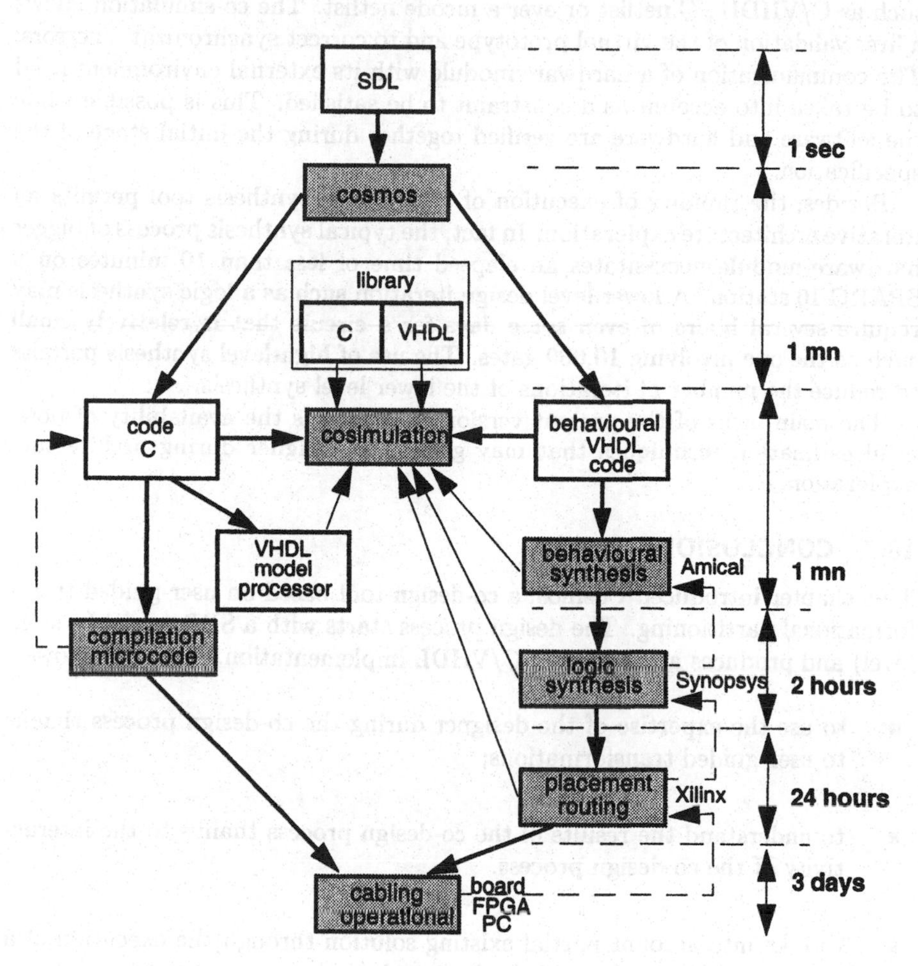


**Figure 10.37** Time for traversing the design process.

The link between co-simulation and co-synthesis within a unified environment permit to realize a multi-level co-simulation using the generated VHDL during the different intermediate phases of synthesis. In this manner, it becomes possible to cosimulate a system described at several levels of abstraction such as C/VHDL , C netlist or ever a mcode netlist. The co-simulation allows a first validation of the virtual prototype and to correct synchronization errors. The communication of a hardware module with its external environment need to be taken into account as a constraint to be satisfied. This is possible while the software and hardware are verified together during the initial stage of the specification.

Besides, the rapidity of execution of a high-level synthesis tool permits an iterative architecture exploration. In fact, the typical synthesis process of bigger hardware module necessitates an elapsed time of less than 10 minutes on a SPARC-10 station. A lower level design iteration such as a logic synthesis may require several hours of even some days for a circuit that is relatively small such as the one involving 10,000 gates. The use of high-level synthesis permits to reduce the number of iterations of the lower level synthesis.

The main lacks of the present version of cosmos is the availability of powerful estimation techniques that may guide the designer during architecture exploration.

## 10.7 CONCLUSIONS

This chapter introduced Cosmos, a co-design tool based on user-guided transformational partitioning. The design process starts with a SDL model (system level) and produces a distributed C/VHDL implementation. The tool allows:

- to use the expertise of the designer during the co-design process thanks to user-guided transformations;

- to understand the results of the co-design process thanks to the interactivity of the co-design process.

- to take into account partial existing solution through the execution of a specific set of transformations that produce the required target solution.;

- a large design space exploration thanks to a fast and interactive interface;

- to start with a very high level specification language, SDL, in order to produce an implementation of the system.

## 10.8 ACKNOWLEDGEMENTS

This work was supported by France-Telecom/CNET under Grant 941B113, SGS-Thomson, Aerospatiale, PSA, ESPRIT programme under project COMITY 23015 and MEDEA programme under Project SMT AT-403. (*) under grant supported by CAPES/COFECUB-144-94.

## References


[1] G. D. Micheli, *Synthesis and Optimization of Digital Circuits.* McGraw Hill, 1994.

[2] D. D. Gajski, *Principles of Digital Design.* Prentice Hall, 1997.

[3] E. A. Lee and D. G. Messerschmitt, "Statis scheduling of synchronous data flow programs for digital signal processing," *IEEE Transactions on Computers*, vol. C-36, pp. 24–35, January 1987.

[4] A. A. Jerraya, H. Ding, P. Kission, and M. Rahmouni, *Behavioral Synthesis and Component Reuse with VHDL.* Kluwer Academic Publishers, 1997.

[5] D. Drusinsky and D. Harel, "Using statecharts for hardware description and synthesis," *IEEE Transactions on Computer Aided Design*, 1989.

[6] D. Harel, "Statecharts: A visual formalism for complex systems," *Science of Computer Programming*, vol. 8, 1987.

[7] F. Balarin, P. Giusto, A. Jurecska, C. Passerone, E. Sentovich, B. Tabbara, M. C. H. Hsieh, L. Lavagno, A. Sangiovanni-Vincentelli, and K. Suzuki, *Hardware-Software Co-Design of Embedded Systems.* Kluwer Academic Publishers, April 1997.

[8] D. Gajski, F. Vahid, S. Narayan, and J. Gong, *Specification and Design of Embedded Systems.* Prentice Hall, 1994.

[9] W. Wolf, "Hardware-software co-design of embedded systems," *Proceedings of the IEEE*, vol. 82, pp. 967–989, July 1994.

[10] M. Chiodo, P. Giusto, H. Hsieh, A. Jurecska, L. Lavagno, and A. Sangiovanni-Vincentelli, "Hardware-software codesign of embedded systems," *IEEE Micro*, pp. 26–36, August 1994.

[11] J. A. Stankovic, M. Spuri, M. D. Natale, and G. C. Buttazzo, "Implications of classical scheduling results for real-time systems," *IEEE Computer*, vol. 28, pp. 16–25, June 1995.

[12] C. L. Liu and J. W. Layland, "Scheduling algorithms for multiprogramming in a hard-real-time environment," *Journal of the ACM*, vol. 20, pp. 46–61, January 1973.

[13] J. Lehoczky, L. Sha, and Y. Ding, "The rate monotonic scheduling algorithm: exact characterization and average case behavior," in *Proceedings, Real-Time Systems Symposium*, pp. 166–171, IEEE Computer Society Press, 1989.

[14] M. B. Srivastava, T. I. Blumenau, and R. W. Brodersen, "Design and implementation of a robot control system using a unified hardware-software rapid-prototyping framework," in *Proceedings, ICCD '92*, IEEE Computer Society Press, 1992.

[15] B. Lin, S. Vercauteren, and H. D. Man, "Embedded architecture co-synthesis and system integration," in *Proceedings, Fourth International Workshop on Hardware/Software Codesign*, pp. 2–9, IEEE Computer Society Press, 1996.

[16] S. Vercauteren and B. Lin, "Hardware/software communication and system integration for embedded architectures," *Design Automation for Embedded Systems*, vol. 2, pp. 359–382, May 1997.

[17] W. Hardt and W. Rosenstiel, "Speed-up estimation for HW/SW-systems," in *Proceedings, Fourth International Workshop on Hardware/Software Codesign*, pp. 36–43, IEEE Computer Society Press, 1996.

[18] J. Henkel and R. Ernst, "The interplay of run-time estimation and granularity in HW/SW partitioning," in *Proceedings, Fourth International Workshop on Hardware/Software Codesign*, pp. 52–58, IEEE Computer Society Press, 1996.

[19] D. Ku and G. de Micheli, *High-level Synthesis of ASICs under Timing and Synchronization Constraints*. Boston: Kluwer Academic Publishers, 1992.

[20] R. K. Gupta and G. D. Micheli, "Specification and analysis of timing constraints for embedded systems," *IEEE Transactions on Computer-Aided Design of Integrated Circuits and Systems*, vol. 16, pp. 240–256, March 1997.

[21] J. Henkel, R. Ernst, U. Holtmann, and T. Benner, "Adaptation of partitioning and high-level synthesis in hardware/software co-synthesis," in *Proceedings, ICCAD-94*, pp. 96–100, IEEE Computer Society Press, 1994.

[22] A. Kalavade and E. A. Lee, "A global criticality/local phase driven algorithm for the constrained hardware/software partitioning problem," in *Third International Workshop on Hardware-Software Codesign*, pp. 42–48, IEEE Computer Society Press, 1994.

[23] D. Kirovski and M. Potkonjak, "System-level synthesis of low-power hard real-time systems," in *Proceedings, 1997 Design Automation Conference*, pp. 697–702, ACM Press, 1997.

[24] S. Prakash and A. C. Parker, "SOS: Synthesis of application-specific heterogeneous multiprocessor systems," *Journal of Parallel and Distributed Computing*, vol. 16, pp. 338–351, 1992.

[25] J. G. D'Ambrosio and X. Hu, "Configuration-level hardware/software partitioning for real-time embedded systems," in *Third International Workshop on Hardware-Software Codesign*, pp. 34–41, IEEE Computer Society Press, 1994.

[26] X. S. Hu and J. G. D'Ambrosio, "Hardware/software communication and system integration for embedded architectures," *Design Automation for Embedded Systems*, vol. 2, pp. 339–358, May 1997.

[27] J. K. Adams and D. E. Thomas, "Multiple-process behavioral synthesis for mixed hardware-software systems," in *Proceedings, $8^{th}$ International Symposium on System Synthesis*, pp. 10–15, IEEE Computer Society Press, 1995.

[28] T.-Y. Yen and W. Wolf, "Performance estimation for real-time distributed embedded systems," in *Proceedings, ICCD '95*, pp. 64–69, IEEE Computer Society Press, 1995.

[29] F. Balarin and A. Sangiovanni-Vincentelli, "Schedule validation for embedded reactive real-time systems," in *Proceedings, 1997 Design Automation Conference*, pp. 52–57, ACM Press, 1997.

[30] Y. Li and W. Wolf, "A task-level hierarchical memory model for system synthesis of multiprocessors," in *Proceedings, 34th Design Automation Conference*, pp. 153–156, ACM Press, 1997.

[31] S. Agrawal and R. K. Gupta, "Data-flow assisted behavioral partitioning for embedded systems," in *Proceedings, 1997 Design Automation Conference*, pp. 709–712, ACM Press, 1997.

[32] W. Wolf, "An architectural co-synthesis algorithm for distributed, embedded computing systems," *IEEE Transactions on VLSI Systems*, vol. 5, pp. 218–229, June 1997.

[33] T.-Y. Yen and W. Wolf, "Sensitivity-driven co-synthesis of distributed embedded systems," in *Proceedings, 8th International Symposium on System Synthesis*, pp. 4–9, IEEE Computer Society Press, 1995.

[34] S. Antoniazzi, A. Balboni, W. Fornaciari, and D. Sciuto, "A methodology for control-dominated systems codesign," in *Third International Workshop on Hardware-Software Codesign*, pp. 2–9, IEEE Computer Society Press, 1994.

[35] A. Balboni, W. Fornaciari, and D. Sciuto, "Partitioning and exploration strategies in the TOSCA design flow," in *Proceedings, Fourth International Workshop on Hardware/Software Codesign*, pp. 62–69, IEEE Computer Society Press, 1996.

[36] J. Hou and W. Wolf, "Process partitioning for distributed embedded systems," in *Proceedings, Fourth International Workshop on Hardware/Software Codesign*, pp. 70–75, IEEE Computer Society Press, 1996.

[37] Y. Shin and K. Choi, "Software synthesis through task decomposition by dependency analysis," in *Digest of Papers, ICCAD '96*, pp. 98–102, IEEE Computer Society Press, 1996.

[38] J. A. Rowson and A. Sangiovanni-Vincentelli, "Interface-based design," in *Proceedings, 1997 Design Automation Conference*, pp. 178–183, ACM Press, 1997.

[39] L. Lavagno, J. Cortadella, and A. Sangiovanni-Vincentelli, "Embedded code optimization via common control structure detection." Presented at the Fifth International Workshop on Hardware/Software Codesign, March 1997.

[40] F. Balarin, M. Chiodo, A. Jurecska, L. Lavagno, B. Tabbara, and A. Sangiovanni-Vincentelli, "Automatic generation of a real-time operating system for embedded systems." Presented at the Fifth International Workshop on Hardware/Software Codesign, March 1997.

[41] H. Hsieh, L. Lavagno, C. Passerone, C. Sansoè, and A. Sangiovanni-Vincentelli, "Modeling micro-controller peripherals for high-level co-simulation and synthesis," in *Fifth International Workshop on Hardware/Software Codesign*, pp. 127–130, IEEE Computer Society Press, 1997.

[42] M. Chiodo, D. Engels, P. Giusto, H. Hsieh, A. Jurecska, L. Lavagno, K. Suzuki, and A. Sangiovanni-Vincentelli, "A case study in computer-aided co-design of embedded controllers," *Design Automation for Embedded Systems*, vol. 1, pp. 51–67, January 1996.

[43] P. Chou, E. A. Walkup, and G. Borriello, "Scheduling for reactive real-time systems," *IEEE Micro*, vol. 14, pp. 37–47, August 1994.

[44] R. B. Ortega and G. Borriello, "Communication synthesis for embedded systems with global considerations," in *Fifth International Workshop on Hardware/Software Codesign*, pp. 69–73, IEEE Computer Society Press, 1997.

[45] J.-M. Daveau, T. B. Ismail, and A. A. Jerraya, "Synthesis of system-level communication by an allocation-based approach," in *Proceedings, $8^{th}$ International Symposium on System Synthesis*, pp. 150–155, IEEE Computer Society Press, 1995.

[46] T. B. Ismail, M. Abid, and A. Jerraya, "COSMOS: a codesign approach for communicating systems," in *Third International Workshop on Hardware-Software Codesign*, pp. 17–24, IEEE Computer Society Press, 1994.

[47] D. Verkest, K. Van Rompaey, I. Bolshens, and H. De Man, "CoWare—a design environment for heterogeneous hardware/software systems," *Design Automation for Embedded Systems*, vol. 1, October 1996.

[48] T.-Y. Yen and W. Wolf, "Communication synthesis for distributed embedded systems," in *Proceedings, ICCAD-95*, pp. 64–69, IEEE Computer Society Press, 1995.

[49] S. Note, J. V. Ginderdeuren, P. V. Lierop, R. Lauwereins, M. Engels, B. Almond, and B. Kiani, "Paradigm rp: A system for the rapid pro-

totyping of real-time dsp applications," *DSP Applications*, vol. 3, no. 1, 1994.

[50] D. Bittruf and Y. Tanurhan, "A survey of hardware emulators," tech. rep., ESPRIT Basic Research Project No. 8135, 1994.

[51] H. Owen, U. Kahn, and J. Hughes, "FPGA based ASIC hardware emulator architectures," tech. rep., School of Electrical and Computer Engineering, Georgia Institute of Technology, 1993.

[52] J. Staunstrup, *A Formal Approach to Hardware Design*. Kluwer Academic Publishers, 1994.

[53] U. Weinmann, "FPGA partitioning under timing constraints," in *Field Programmable Logic Workshop*, 1993.

[54] R. Camposano and W. Rosenstiel, "Synthesizing circuits from behavioral descriptions," *IEEE Transactions on CAD*, vol. 8, no. 2, 1989.

[55] P. Gutberlet, J. Müller, H. Krämer, and W. Rosenstiel, "Automatic module allocation in high level synthesis," in *Proceedings, European Design Automation Conference*, IEEE Computer Society Press, 1992.

[56] P. Gutberlet and W. Rosenstiel, "Scheduling between basic blocks in the CADDY synthesis system," in *Proceedings, European Conference on Design Automation*, IEEE Computer Society Press, 1992.

[57] G. Koch, U. Kebschull, and W. Rosenstiel, "A prototyping architecture for hardware/software codesign in the COBRA project," in *Proceedings of 3$^{rd}$ International Workshop on Hardware/Software Codesign*, IEEE Computer Society Press, 1994.

[58] Hyperstone Electronics, *Hyperstone E1 32-Bit-Microprocessor User's Manual*. 1990.

[59] J. Stankovic and K. Ramamritham, *Tutorial on Hard Real-Time Systems*. IEEE, 1988.

[60] P. Lala, *Fault Tolerant & Fault Testable Hardware Design*. Prentice Hall, 1985.

[61] D. Pradhan, *Fault Tolerant Computing Theory and Techniques, Vol. I and II*. Prentice-Hall, 1986.

[62] H. Ott, *Noise Reduction Techniques in Electronic Systems*. Wiley & Sons, 1988.

[63] D. Gajski, N. Dutt, A. Wu, and Y. Lin, *High-Level Synthesis : Introduction to Chip and System Design.* Boston, Massachusetts: Kluwer Academic Publishers, 1992.

[64] G. D. Micheli, *Synthesis and Optimization of Digital Circuits.* Mc Graw Hill, 1994.

[65] J. Hennessy and D. Patterson, *Computer Architecture: A Quantitative Approach.* Morgan-Kaufmann, second ed., 1996.

[66] M. Flynn, "Some computer organizations and their effectiveness," *IEEE Transactions on Computers*, pp. 948–960, 1972.

[67] Digital Equipment, *DECchip 21030, Reference Manual.* 1994.

[68] Analog Devices, *ADSP-2106x SHARC User's Manual.* 1995.

[69] Texas Instruments, *TMS320C80 (MVP) Collection of User's Guides.* 1995.

[70] K. Vissers, "Architecture and programming of two generations video signal processors," *Microprocessing and Microprogramming*, pp. 373–390, 1995.

[71] P. Lippens, J. van Meerbergen, A. van der Werf, W. Verhaegh, B. McSweeney, J. Hiusken, and O. McArdle, "PHIDEO: A silicon compiler for high speed algorithms," in *Proceedings, EDAC 91*, pp. 436–441, IEEE Computer Society Press, 1991.

[72] M. Mirinda, F. Catthoor, M. Janssen, and H. D. Man, "ADOPT: Efficient hardware address generation in distributed memory architectures," in *Proceedings, Ninth ISSS*, pp. 20–25, IEEE Computer Society Press, 1996.

[73] A. Balboni, W. Fornaciari, and D. Sciuto, "Cosynthesis and co-simulation of control-dominated embedded systems," *Design Automation of Embedded Systems*, vol. 1, no. 3, 1996.

[74] L. Lavagno, A. Sangiovanni-Vincentelli, and H. Hsieh, "Embedded system co-design: Synthesis and verification," in *Hardware/Software Co-Design* (G. DeMicheli and M. Sami, eds.), pp. 213–242, Kluwer, 1996.

[75] C. Carreras, J. Lopez, M. Lopez, C. Delgado-Kloos, N. Martinez, and L. Sanchez, "A co-design methodology based on formal specification and high-level estimation," in *Proceedings, Fourth International Workshop on Hardware-Software Codesign*, IEEE Computer Society Press, 1996.

[76] M. Gasteier and M. Glesner, "Bus-based communication synthesis on system-level," in *Proceedings, Ninth ISSS*, p. 65, IEEE Computer Society Press, 1996.

[77] Motorola, *MC68332 Reference Manual*. 1991.

[78] T. Shanley and D. Anderson, *PCI System Architecture*. Addison-Wesley, 1995.

[79] Siemens, *SAB 80C515 Users Manual*. 1992.

[80] Philips, *80C552 Product Specification*. 1993.

[81] P. Chou, R. Ortega, and G. Borriello, "Interface co-synthesis techniques for embedded systems," in *Proceedings, ICCAD 95*, pp. 280–287, IEEE Computer Society Press, 1995.

[82] Motorola, *TPU Reference Manual*. 1991.

[83] Siemens, *SAB 80C166 Users Manual*. 1990.

[84] A. Oppenheim and R. Schafer, *Digital Signal Processing*. Prentice Hall, 1975.

[85] Motorola, *MC56000 Digital Signal Processor User's Manual*. 1990.

[86] Inmos, *Transputer Reference Manual*. Prentice Hall, 1988.

[87] Texas Instruments, "Parallel processing with TMS320C40." Application Note, `http://www.ti.com/`.

[88] IEEE, "VME bus, ANSI/IEEE standard 1014-1987," tech. rep., IEEE, 1987. Available from IEEE Customer Service, 445 Hoes Lane, PO Box 1331, Piscataway, NJ 08855-1331.

[89] M. Tremblay, J. M. O'Connor, V. Narayanan, and L. He, "VIS speeds new media processing," *IEEE Micro*, August 1996.

[90] K. Jack., *Video Demystified*. HighText Publications, 1996.

[91] J. M. Holtzmann and D. J. Goodman, *Wireless and Mobile Communications*. Kluwer, 1994.

[92] ARM, "Digital waveband platform for digital wireless telephone." Press release at http://www.arm.com, October 1996.

[93] Motorola, *RCPU Reference Manual*. 1996.

[94] Digital Equipment, *DECchip 21066, Reference Manual.* 1994.

[95] Advanced RISC Machines, "ARM CPU core." `http://www.arm.com/`.

[96] Motorola, "ColdFire: Variable-length RISC processors." `http://www.mot.com/`.

[97] D. P. Ryan, "Intel's 80960. an architecture optimized for embedded control," *IEEE Micro*, pp. 63–76, 1988.

[98] Intel, "i960 processor overview," tech. rep., 1996. http://www.intel.com/.

[99] J. Wilberg and R. Camposano, "VLIW processor codesign for video processing," *Design Automation of Embedded Systems*, vol. 2, pp. 79–119, 1996.

[100] P. Athanas and H. Silverman, "Processor reconfiguration through instruction set metamorphosis," *IEEE Transactions on Computers*, pp. 11–18, March 1993.

[101] M. Kozuch and A. Wolfe, "Compression of embedded system programs," in *Proceedings, ICCD 94*, pp. 270–277, IEEE Computer Society Press, 1994.

[102] H. Tomiyana and H. Yasuura, "Optimal code placement of embedded software for instruction caches," in *Proceedings, ED&TC 96*, pp. 96–101, IEEE Computer Society Press, 1996.

[103] H. Schmit and D. Thomas, "Address generation for memories containing multiple arrays," in *Proceedings, ICCAD 95*, pp. 510–514, IEEE Computer Society Press, 1995.

[104] K. V. Rompaey, D. Verkest, J. Bolsens, and H. D. Man, "CoWare: A design environment for heterogeneous hardware/software systems," in *Proceedings, EURODAC 96*, IEEE Computer Society Press, 1996.

[105] P. Paulin, C. Liem, M. Cornero, F. Naçabal, and G. Goossens, "Embedded software in real-time signal processing systems: Application and architecture trends," *Proceedings of the IEEE*, vol. 85, pp. 419–435, March 1997.

[106] B. Case, "First Trimedia chip boards PCI bus," *Microprocessor Report*, vol. 9, no. 15.

[107] M. Ikeda *et al.*, "A hardware/software concurrent design for a real-time SP@ML MPEG2 video-encoder chip set," in *Proceedings, European Design & Test Conference*, pp. 320–326, 1996.

[108] C. Liem *et al.*, "An embedded system case study: The firmware development environment for a multimedia audio processor," in *Proceedings, 1997 Design Automation Conference*, 1997.

[109] L. Gwennap, "Intel's MMX speeds multimedia," *Microprocessor Report*, pp. 1, 6–10, 1996. March 5.

[110] Y. Yao, "PC graphics reach new level: 3D," *Microprocessor Report*, pp. 14–19, 1996. January 22.

[111] P. L. et. al., "A chip set for 7 kHz handfree telephony," in *Proceedings, IEEE Custom Integrated Circuits Conference*, 1994.

[112] P. Vanoostende, "Retargetable code generation: Key issues for successful introduction." $1^{st}$ Workshop on Code Generation for Embedded Processors, Dagstuhl, August 1994.

[113] A. Picciriello, "Italtel activities in ASIP design." Digital Signal Processing in ASICs (Commett Course), Unterspremstatten, Austria, April 1995.

[114] P. Paulin, C. Liem, T. May, and S. Sutarwala, "Flexware: A flexible firmware development environment for embedded systems," in *Code Generation for Embedded Processors* (P. Marwedel and G. Goossens, eds.), Kluwer Academic Publishers, 1995.

[115] C. Liem, T. May, and P. Paulin, "Instruction-set matching and selection for DSP and ASIP code generation," in *Proceedings, European Design and Test Conference*, pp. 31–37, 1994.

[116] I. B. et. al., "Hardware/software co-design of digital telecommunication systems," *Proceedings of the IEEE*, vol. 85, pp. 391–418, March 1997.

[117] C. Liem, *Retargetable Compilers for Embedded Core Processors: Methods and Experiences in Industrial Applications.* Kluwer Academic Publishers, 1997.

[118] L. Bergher *et al.*, "Mpeg audio decoder for consumer applications," in *Proc. of the IEEE Custom Integrated Circuits Conference*, 1995.

[119] P. Paulin *et al.*, "High-level synthesis and codesign methods: An application to a videophone codec," in *Proceedings, EuroVHDL/EuroDAC*, 1995.

[120] C. Liem, F. Naçabal, C. Valderrama, P. Paulin, and A. Jerraya, "System-on-a-chip cosimulation and compilation," *IEEE Design & Test of Computers*, 1997.

[121] M. Harrand *et al.*, "A single chip videophone video encoder/decoder," in *Digest of Technical Papers, IEEE International Solid-State Circuits Conference*, pp. 292–293, 1995.

[122] M. S. Inc., "Motorola DSP product overview." http://www.mot.com/SPS/DSP/home/prd/prodover.html.

[123] C. Liem, P. Paulin, and A. Jerraya, "Address calculation for retargetable compilation and exploration of instruction-set architectures," in *Proceedings, 1996 Design Automation Conference*, pp. 597–600, 1996.

[124] R. Woudsma *et al.*, "EPICS, a flexible approach to embedded DSP cores," in *Proceedings*, $5^{th}$ *Int'l Conference on Signal Processing and Applications and Technology*, (Dallas TX), 1994.

[125] R. Goering, "Design worlds take integration route," *Electronic Engineering Times*. Jan 6, 1997, issue 935.

[126] J. Turley and P. Lapsley, "Motorola, TI extends 16-bit DSP families," *Microprocessor Report*, pp. 14–15, 1996. March 5.

[127] M. C. et. al., "Code generation requirements for industrial embedded systems." $2^{nd}$ International Workshop on Code Generation for Embedded Processors, Leuven, Belgium, March 1996.

[128] M. Gold, "TI's Tartan takeover," *Electronic Engineering Times*. July 29, 1996, issue 912.

[129] V. Zivojnovic *et al.*, "DSPstone: A DSP-oriented benchmarking methodology," 1994.

[130] V. Zivojnovic, J. M. Velarde, and C. Schlager, "DSPstone: A DSP-oriented benchmarking methodology," tech. rep., Aachen University of Technology, August 1994.

[131] B. Cole, "DSP embedded use grows," *Electronic Engineering Times*, pp. 80–82. April 17, 1995.

[132] P. Paulin, C. Liem, T. May, and S. Sutarwala, "DSP design tool requirements for embedded systems: A telecommunications industrial perspective," *Journal of VLSI Signal Processing*, vol. 9, pp. 23–47, March 1995.

[133] T. B. Ismail, K. O'Brien, and A. Jerraya, "Interactive system-level partitioning with partif," in *Proceedings, European Design & Test Conference*, 1994.

[134] F. Naçabal, O. Deygas, P. Paulin, and M. Harrand, "C-VHDL co-simulation: Industrial requirements for embedded control processors," in *Proceedings, EuroDAC/EuroVHDL Designer Sessions*, pp. 55–60, 1996.

[135] C. Valderrama, F. Naçabal, P. Paulin, and A. Jerraya, "Automatic generation of interfaces for distributed c-vhdl cosimulation of embedded systems: an industrial experience," in *Proceedings, International Workshop on Rapid Systems Prototyping*, pp. 72–77, 1996.

[136] E. B. et. al., "Combined control-flow dominated and data-flow dominated high-level synthesis," in *Proceedings, 1996 Design Automation Conference*, pp. 573–578, 1996.

[137] G. Goossens, J. van Praet, D. Lanneer, W. Geurts, P. Paulin, and C. Liem, "Embedded software in real-time signal processing systems: Design technology," *Proceedings of the IEEE*, vol. 85, pp. 436–454, March 1997.

[138] A. Aho, R. Sethi, and J. Ullman, *Compilers: Principles, Techniques and Tools*. Reading MA: Addison-Wesley, 1988.

[139] C. Fischer and R. LeBlanc, *Crafting a Compiler with C*. Redwood City CA: The Benjamin/Cummings Publishing Co., 1991.

[140] R. Stallman, "Using and porting GNU CC," tech. rep., Free Software Foundation, June 1994.

[141] M. Ganapathi, C. Fisher, and J. Hennessy, "Retargetable compiler code generation," *ACM Computing Surveys*, vol. 14, pp. 573–593, 1982.

[142] P. Marwedel, "Tree-based mapping of algorithms to pre-defined structures," in *Proceedings, ICCAD-93*, pp. 586–593, 1993.

[143] R. Leupers and P. Marwedel, "Retargetable generation of code selectors from hdl processor models," in *Proceedings, European Design & Test Conference*, pp. 140–144, 1997.

[144] A. Fauth and A. Knoll, "Automated generation of DSP program development tools using a machine description formalism," in *Proceedings, ICASSP '93*, 1993.

[145] A. Fauth, "Beyond tool specific machine descriptions," in *Code Generation for Embedded Processors* (P. Marwedel and G. Goossens, eds.), Kluwer Academic Publishers, 1995.

[146] D. L. et. al., "Chess: Retargetable code generation for embedded DSP processors," in *Code Generation for Embedded Processors* (P. Marwedel and G. Goossens, eds.), Kluwer Academic Publishers, 1995.

[147] B. Wess, "On the optimal code generation for signal flow graph computation," in *Proceedings, International Symoposium on Circuits and Systems*, pp. 444–447, 1990.

[148] S. Liao, S. Devadas, K. Keutzer, S. Tjiang, and A.Wang, "Code optimization techniques for embedded DSP microprocessors," in *Proceedings, 1997 Design Automation Conference*, 1997.

[149] R. Mueller, M. Duda, and S. O'Haire, "A survey of resource allocation methods in optimizing microcode compilers," in *Proceedings, 17th Annual Workshop on Microarchitecture*, pp. 285–295, 1984.

[150] G. Chaitin, "Register allocation & spilling via graph coloring," *SIGPLAN Notices*, vol. 17, pp. 98–105, June 1982.

[151] L. Hendren, G. Gao, E. Altman, and C. Mukerji, "A register allocation framework based on hierarchical cyclic interval graphs," in *Proceedings of the International Conference on Compiler Construction*, 1992.

[152] H. Feuerhahn, "Data-flow driven resource allocation in a retargetable microcode compiler," in *Proc. of the 21st Workshop on Microprogramming and Microarchitecture*, pp. 105–107, 1988.

[153] C. Liem, T. May, and P. Paulin, "Register assignment through resource classification for ASIP microcode generation," in *Proceedings, ICCAD-94*, pp. 397–402, 1994.

[154] K. Rimey and P. Hilfinger, "A compiler for application specific signal processors," in *VLSI Signal Processing III*, pp. 341–351, IEEE Press, 1988.

[155] R. Hartmann, "Combined scheduling and data routing for programmable asic systems," in *Proceedings, European Design Automation Conference*, pp. 486–490, 1992.

[156] G. Araujo and S. Malik, "Optimal code generation for embedded memory non-homogeneous register architectures," in *International Symposium on System Synthesis*, pp. 36–41, 1995.

[157] S. Novak, A. Nicolau, and N. Dutt, "A unified code generation approach using mutation sceduling," in *Code Generation for Embedded Processors* (P. Marwedel and G. Goossens, eds.), Kluwer Academic Publishers, 1995.

[158] T. Wilson, G. Grewal, S. Henshall, and D. Banerji, "An ILP-based approach to code generation," in *Code Generation for Embedded Processors* (P. Marwedel and G. Goossens, eds.), Kluwer Academic Publishers, 1995.

[159] A. Lioy and M. Mezzalama, "Automatic compaction of microcode," *Microprocessors and Microsystems*, vol. 14, pp. 21–29, January/February 1990.

[160] D. Bacon, S. Graham, and O. Sharp, "Compiler transformation for high-performance computing," *ACM Computing Surveys*, vol. 26, pp. 345–420, December 1994.

[161] U. Banerjee, *Loop Parallelization.* Kluwer Academic Publishers, 1994.

[162] M. Lam, "Software pipelining: An effective scheduling technique for vliw machines," in *ACM SIGPLAN Conference on Programming Language Design and Implementation*, vol. 23, pp. 318–328, July 1988.

[163] G. Goossens, J. Vandewalle, and H. DeMan, "Loop optimization in register-transfer scheduling for DSP-systems," in *Proceedings, Design Automation Conference*, pp. 826–831, 1989.

[164] H. Tomiyama and H. Yasuura, "Optimal code placement of embedded software for instruction caches," in *Proceedings, European Design & Test Conference*, pp. 96–101, 1996.

[165] P. Panda, N. Dutt, and A. Nicolau, "Memory organization for improved data cache performance in embedded processors," in *Proceedings, International Symposium on System Synthesis*, pp. 90–95, 1996.

[166] A. Sudarsanam and S. Malik, "Memory bank and register allocation in software synthesis for ASIPs," in *Proceedings, ICCAD-95*, pp. 388–392, 1995.

[167] P. Hilfinger and J. Rabaey, "DSP specification using the SILAGE language," in *Anatomy of a Silicon Compiler* (R. Brodersen, ed.), Kluwer Academic Publishers, 1992.

[168] N. Gehani, *C: An Advanced Introduction, ANSI C Edition.* New York: Computer Science Press, Inc., 1988.

[169] R. Deodhar, "Optimizing compiler technology streamlines complex systems," *Electronic Design*, pp. 153–160, 1997. May 1.

[170] N. Ramsey and D. Hanson, "A retargetable debugger," in *ACM SIGPLAN Conference on Programming Language Design and Implementation*, vol. 27, pp. 22–31, July 1992.

[171] O'Reilly & Associates, *Understanding and Using COFF.* O'Reilley & Associates. See http://www.ora.com.

[172] U. I. P. L. S. I. Group, "Dwarf debugging information format." Revision 2.0.0, July 1993.

[173] I. Huang and A. Despain, "Synthesis of application specific instruction sets," *IEEE Transactions on Computer-Aided Design of Integrated Circuits and Systems*, vol. 14, pp. 663–675, June 1995.

[174] B. Holmer and A. Despain, "Viewing instruction set design as an optimization problem," in *Proc. of the 24th International Symposium on Microarchitecture*, pp. 153–162, 1991.

[175] A. A. et. al., "An ASIP instruction set optimization algorithm with functional module sharing constraint," in *Proc. of the International Conference on CAD*, pp. 526–532, 1993.

[176] M. Imai, A. Alomary, J. Sato, and N. Hikichi, "An integer programming approach to instruction implementation method selection problem," in *Proceedings, European Design Automation Conference*, pp. 106–111, 1992.

[177] R. Rauscher and M. Koegst, "A system for microcode reduction," in *Proc. of the IFIP Int. Workshop on Logic and Architecture Synthesis*, (Grenoble, France), pp. 379–386, 1996.

[178] O.-J. Dahl, E. Dijkstra, and C. Hoare, *Structured Programming.* Academic Press, 1972.

[179] W. Wolf, A. Wolfe, S. Chinatti, R. Koshy, G. Slater, and S. Sun, "Lessons from the design of a PC-based private branch exchange," *Design Automation for Embedded Systems*, vol. 1, pp. 297–314, October 1996.

[180] K. M. Chandy and J. Misra, *Parallel Program Design: A Foundation.* Addison-Wesley, 1988.

[181] C. Hoare, "Communicating sequential processes," *Communications of the ACM*, vol. 21, pp. 666–677, August 1978.

[182] A. J. Martin, "Compiling communicating processes into delay-insensitive VLSI circuits," *Distributed Computing*, vol. 1, no. 4, pp. 226–234, 1986.

[183] E. W. Dijkstra, "Guarded commands, nondeterminacy, and formal derivation of programs," *Communications of the ACM*, vol. 18, no. 8, pp. 453–457, 1975.

[184] M. O. Rabin, "Probabilistic algorithms," in *Algorithms and Complexity: New Directions and Recent Results* (J. Traub, ed.), pp. 21–39, Academic Press, 1976.

[185] N. Wirth, "Pl360, a programming language for the 360 computer," *Journal of the ACM*, vol. 15, no. 1, 1968.

[186] P. B. Hansen, *The Architecture of Concurrent Programs.* Prentice-Hall Inc., 1977.

[187] J. Ichbiah *et al.*, *Reference Manual for the Ada Programming Language.* United States Department of Defense, 1980.

[188] K. Arnold and J. Gosling, *The Java Programming Language.* Addison-Wesley Publishing Company, 1996.

[189] IEEE, New York, *VHDL Language Reference Manual*, std 1076-1987 ed., 1988.

[190] D. Thomas and P. Moorby, *The Verilog Hardware Description Language.* Kluwer Academic Publishers, 1991.

[191] A. J. Martin *et al.*, "The design of an asynchronous microprocessor," in *Decennial Caltech Conference on Research in VLSI* (C. L. Seitz, ed.), pp. 351–357, MIT Press, 1989.

[192] C. H. van Berkel, C. Niessen, M. Rem, and R. J. J. Saeijs, "VLSI programming and silicon compilation: A novel approach from Philips research," in *Proceedings of the 1988 IEEE International Conference on Computer Design*, pp. 150–166, IEEE, 1988.

[193] A. Benveniste and G. Berry, "The synchronous approach to reactive and real-time systems," *Proceedings of the IEEE*, vol. 79, pp. 1270–1282, September 1991.

[194] J. L. van de Snepscheut, "A derivation of a distributed implementation of Warshall's algorithm," *Science of Computer Programming*, vol. 7, pp. 55–60, 1986.

[195] F. Andrè, D. Herman, and J.-P. Verjus, *Synchronization of Parallel Programs.* The MIT Press, 1985.

[196] E. Dijkstra, "Co-operating sequential processes," in *Programming Languages* (F. Genuys, ed.), Academic Press, 1965.

[197] E. Brinksma, *On the Design of Extended LOTOS.* Technical University Twente, Holland, 1988. Dissertation.

[198] P. B. Hansen, "The nucleus of a multiprogramming system," *Communications of the ACM*, vol. 13, pp. 238–50, April 1970.

[199] P. B. Hansen, *Operating System Principles.* Prentice-Hall Inc., 1973.

[200] C. Hoare, "Monitors: an operating system structuring concept," *Communications of the ACM*, vol. 17, pp. 549–57, Oct. 1974.

[201] R. Milner, *Communication and Concurrency.* Series in Computing Science, Prentice-Hall, 1989.

[202] T. J. Chaney and C. E. Molnar, "Anomalous behavior of synchronizer and arbiter circuits," *IEEE Transactions on Computers*, vol. 22, pp. 421–422, April 1973.

[203] P. B. Hansen, "Structured multiprogramming," *Communications of the ACM*, vol. 15, pp. 145–50, July 1972.

[204] J.P. Billon and J.C. Madre, "Original concepts of PRIAM, an industrial tool for efficient formal verification of combinational circuits," in *The Fusion of Hardware Design and Verification* (G.J. Milne, ed.), (Glasgow, Scotland), pp. 487–501, IFIP WG 10.2, North-Holland, 1988. IFIP Transactions.

[205] K.L. McMillan, *Symbolic Model Checking.* Norwell Massachusetts: Kluwer Academic Publishers, 1993.

[206] A. Gupta, "Formal hardware verification methods: A survey," *Formal Methods in Systems Design*, vol. 1, pp. 151–238, October 1992.

[207] R. Floyd, "Assigning meanings to programs," in *Proceedings of the Symposium in Applied Mathematics* (J. Schwartz, ed.), vol. 19, pp. 19–32, American Mathematical Society, 1967.

[208] N. Mellergaard, *Mechanized Implementation Verification.* PhD thesis, Department of Computer Science, Technical University of Denmark, 1995.

[209] L. Lamport and M. Abadi, "The existence of refinement mappings," *Theoretical Computer Science*, vol. 2, no. 82, pp. 253–284, 1991.

[210] T. Kropf, *Formal Hardware Verification - Methods and Systems in Comparison*, vol. 1287. Springer Lecture Notes, 1997.

[211] R. Bryant, "Graph-based algorithms for boolean function manipulation," *Transactions on Computers*, vol. 8, no. C-35, pp. 677–691, 1986.

[212] R. Bryant, "Symbolic Boolean manipulation with ordered binary-decision diagrams," *Computing Surveys*, vol. 24, pp. 293–318, Sept. 1992.

[213] R. S. Boyer and J. S. Moore, *A Computational Logic*. Academic Press, 1979.

[214] M. Gordon, "HOL A Proof Generating System for Higher-Order Logic," *VLSI Specificacion, Verification and Synthesis*, 1987.

[215] S. J. Garland and J. V. Guttag, "An overview of LP: the Larch Prover," in *Proceedings of the Third International Conference on Rewriting Techniques and Applications*, Springer-Verlag, 1989.

[216] J. Staunstrup and N. Mellergaard, "Localized verification of modular designs," *Formal Methods in System Design*, vol. 6, pp. 295–320, June 1995.

[217] P. Girodias and E. Cerny, "Interface timing verification with delay correlation using constraint logic programming," in *Proceedings of ED&TC'97*, (Paris), March 1997.

[218] V. M. T. Sakamoto and G. D. Micheli, "Run-time scheduler syntheis for hardwrae-software systems and application to robot control design," in *Proceedings of the CHDL'97* (IEEE, ed.), pp. 95–99, March 1997.

[219] J.-M. Daveau, G. F. Marchioro, and A. Jerraya, "VHDL generation from SDL specification," in *Proceedings of CHDL* (C. D. Kloos and E. Cerny, eds.), pp. 182–201, IFIP, Chapman-Hall, April 1997.

[220] C. D. Kloos, M. Lopez, *et al.*, "From Lotos to VHDL," *Current Issues in Electronic Modelling*, vol. 3, September 1995.

[221] D. Hermann, J. Henkel, and R. Ernst, "An approach to the adaptation of estimated cost parameters in the Cosyma system," in *Proc. Third Int'l Workshop on Hardware/Software Codesign*, pp. 100–107, IEEE CS Press, 1994.

[222] J. Henkel and R. Ernst, "A path-based estimation technique for estimating hardware runtime in HW/SW-cosynthesis," in *Proceedings, Eighth Int'l Symposium on System Level Synthesis*, pp. 116–121, IEEE Computer Society Press, 1995.

[223] D. Ku and G. DeMicheli, "HardwareC—a language for hardware design," Tech. Rep. CSL-TR-88-362, Computer Systems Laboratory, Stanford University, August 1988.

[224] G. Berry and L. Cosserat, "The esterel synchronous programming language and its mathematical semantics," language for synthesis, Ecole National Superieure de Mines de Paris, 1984.

[225] G. Berry, "Hardware implementation of pure esterel," in *Proceedings of the ACM Workshop on Formal Methods in VLSI Design*, January 1991.

[226] D. Gajski, F. Vahid, and S. Narayan, "A system-design methodology: Executable-specification refinement," in *the European Conference on Design Automation*, February 1994.

[227] S. Narayan, F. Vahid, and D. Gajski, "System specification and synthesis with the speccharts language," in *Proc. Int'l Conf. on Computer-Aided Design (ICCAD)*, pp. 226–269, IEEE CS Press, November 1991.

[228] F. Vahid and S. Narayan, "Speccharts: A language for system-level synthesis," in *Proceedings of CHDL*, pp. 145–154, Apr. 1991.

[229] F. Vahid, S. Narayan, and D. Gajski, "SpecCharts: A VHDL front-end for embedded systems," *IEEE Transactions on CAD of Integrated Circuits and Systems*, vol. 14(6), pp. 694–706, 1995.

[230] ISO, IS 8807, *LOTOS a formal description technique based on the temporal ordering of observational behavior*, Feb. 1989.

[231] K. Rompaey, D. Verkest, I. Bolsens, and H. D. Man, "Coware - a design environmenet for heteregeneous hardware/softwa re systems," in *Proceedings of the European Design Automation Conference*, Geneve, September 1996.

[232] R. Klein, "Miami: A hardware software co-simulation environment," in *Proceedings of RSP'96*, pp. 173–177, IEEE CS Press, 1996.

[233] I. Bolsens, B. Lin, K. V. Rompaey, S. Vercauteren, and D. Verkest, "Codesign of dsp systems," in *NATO ASI Hardware/Software Co-Design*, Tremezzo, Italy, June 1995.

[234] D. Ku and G. DeMicheli, "Relative scheduling under timing constraints," *IEEE trans. on CAD*, May 1992.

[235] R. Camposano and R. Tablet, "Design representation for the synthesis of behavioural VHDL models," in *Proceedings of CHDL*, May 1989.

[236] L. Stok, *Architectural Synthesis and Optimization of Digital Systems.* PhD thesis, Eindhoven University of Technology, 1991.

[237] G. Jong, *Generalized Data Flow Graphs, Theory and Applications.* PhD thesis, Eindhoven University of Technology, 1993.

[238] J. Lis and D. Gajski, "Synthesis from VHDL," in *Proceedings of the International Conference on Computer-Aided Design*, pp. 378–381, Oct. 1988.

[239] A. Jerraya, H. Ding, P. Kission, and M. Rahmouni, *Behavioral Synthesis and Component Reuse with VHDL.* Kluwer Academic Publishers, 1997.

[240] R. Gupta, C. C. Jr., and G. DeMicheli, "Synthesis and simulation of digital systems containing interacting hardware and software components," in *Proceedings of the 29th Design Automation Conference (DAC)*, pp. 225–230, IEEE CS Press, 1992.

[241] A. Timmer and J. Jess, "Exact Scheduling Strategies based on Bipartie Graph Matching," in *Proceedings of the European Conference on Design Automation*, (Paris, France), Mars 1995.

[242] A. Kalavade and E. Lee, "A hardware/software codesign methodology for dsp applications," *IEEE Design and Test of Computers*, vol. 10, pp. 16–28, September 1993.

[243] A. Jerraya and K. O'Brien, "Solar: An intermediate format for system-level design and specification," in *IFIP Inter. Workshop on Hardware/software Co-Design*, (Grassau, Allemagne), May 1992.

[244] G. Andrews and F. Schneider, "Concepts and notation for concurrent programming," *Computing Survey*, vol. 15, March 1983.

[245] A. Davis, *Software Requirements: Analysis and Specification.* NY: Elsiever Publisher, 1995.

[246] J. Duley and D. Dietmeyer, "A digital system design language (ddl)," *IEEE trans. on Computers*, vol. C-24, no. 2, 1975.

[247] M. Barbacci, "Instruction set processor specifiocation (isps): The notation and its applications," *IEEE trans. on Computer*, vol. c30, pp. 24–40, january 1981.

[248] P. P. al., "Conlan report," in *Lecture notes in Computer Science 151* (S. Verlag, ed.), (Berlin), 1983.

[249] Institute of Electrical and Electronics Engineers, *IEEE Standard VHDL Language Reference Manual, Std 1076-1993.* IEEE, 1993.

[250] G. Holzmann, *Design and Validation of Computer Protocols.* Englewood Cliffs, N.J: Prentice-Hall,1991.

[251] Computer Networks and ISDN Systems, *CCITT SDL*, 1987.

[252] S. Budowski and P. Dembinski, "An introduction to estelle: A specification language for distributed systems," *Computer Networks and ISDN Systems*, vol. 13, no. 2, pp. 2–23, 1987.

[253] *International Standard, ESTELLE (Formal description technique based on an extended state transition model)*, 1987.

[254] N. Halbwachs, F. Lagnier, and C. Ratel, "Programming and verifying real-time systems by means of the synchronous data-flow lustre," *IEEE Tranactions on Software Enginering*, vol. 18, September 1992.

[255] T. Gautier and P. L. Guernic, "Signal, a declarative language for synchronous programming of real-time systems," *Computer Science, Formal Languages and Computer Architectures*, vol. 274, 1987.

[256] J. L. Peterson, *Petri Net Theory and the modeling of Systems.* Englewood Cliffs, N.J.: Prentice-Hall, 1981.

[257] D. Harel, "Statecharts: A visual formalism for complex systems," in *Science of Computer Programming 8*, North-Holland, 1987.

[258] C. Jones, "Systematic software development using vdm," *C.A.R Hoare Series, Prentice Hall International Series in Computer Science*, 1990.

[259] J. Spivey, "An introduction to Z formal specifications," *Software Engineering Journal*, pp. 40–50, January 1989.

[260] J. R. Abrial, *The B Book. Assigning Programs to Meaning.* Cambridge University Press, 1997.

[261] J. Bruijnin, "Evaluation and integration of specification languages," *Computer Networks and ISDN Systems*, vol. 13, no. 2, pp. 75–89, 1987.

[262] CCITT, *CHILL Language Definition - Recommendation Z.200.* CCITT, May 1984.

[263] M. Romdhani, R. Hautbois, A. Jeffroy, P. de Chazelles, and A. Jerraya, "Evaluation and composition of specification languages, an industrial

point of view," in *Proc. IFIP Conf. Hardware Description Languages (CHDL)*, pp. 519–523, September 1995.

[264] P. Zave and M. Jackson, "Conjunction as composition," *ACM Transactions on Software Engineering and Methodology*, vol. 8, pp. 379–411, October 1993.

[265] D. Wile, "Integrating syntaxes and their associated semantics," Tech. Rep. RR-92-297, Univ. Southern California, 1992.

[266] S. Reis, "Working in the garden environment form conceptual programming," *IEEE Software*, vol. 4, pp. 16–27, November 1987.

[267] C. Valderrama, A. Changuel, P. Raghavan, M. Abid, T. B. Ismail, and A. Jerraya, "A unified model for co-simulation and co-synthesis of mixed hardware/software systems," in *Proc. European Design and Test Conference (EDAC-ETC-EuroASIC)*, IEEE CS Press, March 1995.

[268] W. Loucks, B. Doray, and D. Agnew, "Experiences in real time hardware-software cosimulation," in *Proc. VHDL Int'l Users Forum (VIUF)*, pp. 47–57, April 1993.

[269] C. Valderrama, F. Nacabal, P. Paulin, and A. Jerraya, "Automatic generation of interfaces for distributed C-VHDL,"

[270] D. Thomas, J. Adams, and H. Schmit, "Model and methodology for hardware-software codesign," *IEEE Design and Test of Computers*, vol. 10, pp. 6–15, September 1993.

[271] D. Filo, D. Ku, C. Coelho, and G. DeMicheli, "Interface optimisation for concurrent systems under timing constraints," *IEEE Trans. on VLSI Systems*, vol. 1, pp. 268–281, September 1993.

[272] P. Paulin, C. Liem, T. May, and S. Sutarwala, "Dsp design tool requirements for embedded systems: A telecommunications industrial perspective," *Journal of VLSI Signal Processing (Special issue on synthesis for real-time DSP)*, 1994.

[273] N. Rethman and P. Wilsey, "RAPID: A tool for hardware/software tradeoff analysis," in *Proc. IFIP Conf. Hardware Description Languages (CHDL)*, Elsevier Science, April 1993.

[274] R. Ernst and T. Benner, "Communication, constraints and user directives in COSYMA," Tech. Rep. CY-94-2, Institut für DV-Anlagen, Technische Universität Braunschweig, June 1994.

[275] R. Wilson *et al.*, "The SUIF compiler system." Included in the SUIF compiler system distribution, Stanford University, CA, 1994.

[276] T. Benner, R. Ernst, and A. Österling, "Scalable performance scheduling for hardware-software cosynthesis," in *Proceedings, European Design Automation Conference*, IEEE Computer Society Press, 1995.

[277] T. Benner and R. Ernst, "An approach to mixed systems co-synthesis," in *Proceedings, 5th International Workshop on Hardware/Software Codesign*, pp. 9–14, IEEE Computer Society Press, 1997.

[278] W. Ye, R. Ernst, T. Benner, and J. Henkel, "Fast timing analysis for hardware-software co-synthesis," in *Proceedings, ICCD '93*, pp. 452–457, IEEE Computer Society Press, 1993.

[279] W. Ye and R. Ernst, "Embedded program timing analysis based on program and architecture classification," Tech. Rep. CY-96-3, Institut für DV-Anlagen, Technische Universität Braunschweig, October 1996.

[280] J.Henkel and R. Ernst, "A path-based technique for estimating hardware runtime in HW/SW- cosynthesis," in *8th Intl. Symposium on System Synthesis (ISSS)*, pp. 116–121, Cannes, France, September 13-15 1995.

[281] J. Henkel and R. Ernst, "A hardware/software partitioner using a dynamically determined granularity," in *Proceedings, 1997 Design Automation Conference*, ACM Press, 1997.

[282] D. Herrmann, E. Maas, M. Trawny, R. Ernst, P. Rüffer, M. Seitz, and S. Hasenzahl, "High speed video board as a case study for hardware/software co-design," in *Proceedings, ICCD 96*, IEEE Computer Society Press, 1996.

[283] C. Carreras *et al.*, "A co-design methodology based on formal specification and high-level estimation," in *Proceedings, Fourth International Workshop on Hardware/Software Codesign*, IEEE Computer Society Press, 1996.

[284] Motorola, *TPU: Time Processing Unit Reference Manual.* 1990.

[285] A. Takach and W. Wolf, "Scheduling constraint generation for communicating processes," *IEEE Transactions on VLSI Systems*, vol. 3, pp. 215–230, June 1995.

[286] T. Benner and R. Ernst, "A combined partitioning and scheduling algorithm for heterogeneous multiprocessor systems," Tech. Rep. CY-96-2, Institut für DV-Anlagen, Technische Universität Braunschweig, 1996.

[287] F. Vahid and D. Gajski, "Incremental hardware estimation during hardware/software functional partitioning," *IEEE Transactions on VLSI Systems*, vol. 3, pp. 459–464, September 1995.

[288] W. Ye and R. Ernst, "Worst case timing estimation based on symbolic execution," tech. rep., Institut für DV-Anlagen, Technische Universität Braunschweig, October 1995. COBRA report.

[289] R. Ernst, J. Henkel, and T. Benner, "Hardware/software co-synthesis of microcontrollers," *Design and Test of Computers*, pp. 64–75, Dec. 1992.

[290] G. D. Micheli, "Computer-aided hardware-software codesign," *IEEE Micro*, vol. 14, pp. 10–16, August 1994.

[291] B. G. Hald and J. Madsen, "A flexible architecture representation for High Level Synthesis," in *Second Asian Pacific Conference on Hardware Description Languages*, 1994.

[292] P. V. Knudsen and J. Madsen, "PACE: A dynamic programming algorithm for hardware/software partitioning," in *4th International Workshop on Hardware/Software Codesign, Codes/CASHE'96*, March 1996.

[293] J. Madsen, J. Grode, P. Knudsen, M. Petersen, and A. Haxthausen, "Lycos: the lyngby co-synthesis system," *Design Automation for Embedded Systems*, vol. 2, no. 2, pp. 195–235, 1997.

[294] J. Madsen and B. G. Hald, "A approach to interface synthesis," in *8. International Symposium on System Synthesis*, 1995.

[295] J. Brage, *Foundations of a High-Level Synthesis System.* PhD thesis, Department of Computer Science, the Technical University of Denmark, 1994.

[296] P. G. Paulin and J. P. Knight, "Force-directed scheduling for the behavioral synthesis of ASIC," *IEEE Trans. Computer-Aided Design*, vol. 8, pp. 661–679, 1989.

[297] P. V. Knudsen, "Fine-grain partitioning in codesign," Master's thesis, Technical University of Denmark, 1995.

[298] J. Gong, D. D. Gajski, and S. Narayan, "Software estimation from executable specifications," Tech. Rep. ICS-93-5, Dept. of Information and Computer Science, University of California, Irvine, Irvine, CA 92717-3425, March 8 1993.

[299] J. Grode, "Scheduling of control flow dominated data-flow graphs," Master's thesis, Technical University of Denmark, 1995.

[300] P. V. Knudsen and J. Madsen, "Aspects of System Modelling in Hardware/Software Partitioning," in *Seventh IEEE International Workshop on Rapid Systems Prototyping*, pp. 18–23, 1996.

[301] P. Børn-Jørgensen and J. Madsen, "Critical Path Driven Cosynthesis for Heterogeneous Target Architectures," in *5th International Workshop on Hardware/Software Codesign*, pp. 15–19, 1997.

[302] J. Buck, S. Ha, and E. Lee, "Ptolemy: A framework for simulating and prototyping heterogeneous systems," *International Journal of Computer Simulation*, January 1994.

[303] T. B. Ismail, K. O'Brien, and A. A. Jerraya, "Partif: Interactive system-level partitioning," *VLSI Design*, vol. 3, no. 3-4,, pp. 333–345, 1995.

[304] R. Ernst, J. Henkel, T. Benner, W. Ye, U. Holtmann, D. Herrmann, and M. Trawny, "The COSYMA environment for hardware/software cosynthesis," *Journal of Microprocessors and Microsystems*, 1995.

[305] M. Srivastava and R. Brodersen, "SIERA: A unified framework for rapid-prototyping of system-level hardware and software," *IEEE Transactions on Computer-Aided Design of Integrated Circuits and Systems*, pp. 676–693, June 1995.

[306] W. Wolf, "Object-oriented co-synthesis of distributed embedded systems," *ACM Transactions on Design Automation of Electronic Systems*, vol. 1, July 1996.

[307] M. Theibinger, P. Stravers, and H. Veit, "Castle: an interactive environment for hardware-software codesign," in *Proceedings of International Workshop on Hardware-Software Co-Design*, pp. 203–210, 1994.

[308] E. Barros, W. Rosenstiel, and X. Xiong, "A method for partitioning UNITY language in hardware and software," in *Proceedings, Euro-DAC*, IEEE Computer Society Press, 1994.

[309] A. Kalavade and E. A. Lee, "The extended partitioning problem: Hardware/software mapping, scheduling, and implementation-bin selection," in *Proceedings of Sixth International Workshop on Rapid Systems Prototyping*, pp. 12–18, 1995.

[310] E. D. Lagnese and D. Thomas, "Architectural partitioning of system level synthesis of integrated circuits," *IEEE Trans. CAD/ICAS*, vol. 10, pp. 847–860, July 1991.

[311] F. Vahid and D. D. Gajski, "Specification partitioning for system design," in *Proceedings of the 29$^{th}$ Design Automation Conference*, IEEE Press, 1992.

[312] W. Hardt and R. Camposano, "Specification analysis for HW/SW—partitioning." In 3. GI/ITG Workshop, Passau, Germany, March 1995.

[313] P.Paulin, J. Frehel, E. Berrebi, C. Liem, J.-C. Herluison, and M. Harrand, "High-level synthesis and codesign methods: An appication to a video phone codec," in *Proceredings, EuroDAC/VHDL*, 1995.

[314] P. Marwedel and G. Goessens, *Code Generation for Embedded Processors (DSP)*. Kluwer Academic Publishers, 1995.

[315] A. Alomary, T. Nakata, and Y. Honma, "An asip instruction set optimization algorithm with functional module sharing constraint," in *Proceedings, ICCAD'93*, pp. 526–532, 1993.

[316] J. Madsen and J. Brage, "Codesign analysis of a computer graphics application," *Design Automation for Embedded Systems*, vol. 1, January 1996.

[317] R. K. Gupta and G. de Micheli, "A co-synthesis approach to embedded system design automation," *Design Automation for Embedded Systems*, vol. 1, pp. 69–120, January 1996.

[318] K. Buchenrieder, "A prototyping environment for control-oriented hw/sw systems using statecharts, activity-charts and fpga's," in *Proc. Euro-DAC with Euro-VHDL*, pp. 60–65, IEEE CS Press, 1994.

[319] P. Camurati, F. Corno, P.Prinetto, C. Bayol, and B. Soulas, "System level modeling and verification: A comprehensive design methodology," in *Proc. European Design & Test Conference*, IEEE CS Press, 1994.

[320] M. Chiodo, D.Engels, P. Giusto, H. Hsieh, A. Jurecska, L. Lavagno, K. Suzuki, and A. Sangiovanni-Vincentelli, "A case study in computer aided codesign of embedded controllers," *Design Automation for Embedded Systems*, vol. 1, pp. 51–67, January 1996.

[321] J. Gong, D. Gajski, and S.Narayan, "Software estimation from executable specifications," in *Proc. European Design & Automation Conference*, IEEE CS Press, 1994.

[322] D. Gajski and F. Vahid, "Specification and design of embedded hardware-software systems," *IEEE Design & Test of Computers*, pp. 53–67, Spring 1995.

[323] P.Paulin, C. Liem, T.May, and S. Sutarwala, "Dsp design tool requirements for embedded systems: A telecommunication industrial perspective," *Journal of VLSI Signal Processing*, 1994. Special issue on synthesis for real-time DSP.

[324] T. B. Ismail, G. Marchioro, , and A. Jerraya, "Decoupage de systemes VLSI a partir d'une specification de haut niveau," Technique et Science Informatiques (TSI) Hermes (eds), 1996.

[325] C. Valderrama, A. Changuel, P. V. Raghavan, M. Abid, T. B. Ismail, and A. Jerraya, "A unified model for co-simulation and co-synthesis of mixed hardware/software systems," in *The European Design and Test Confrence ED&TC95*, 1995.

[326] C. W. Krueger, "Software reuse," *ACM Computing Surveys*, vol. 24, pp. 131–183, June 1992.

[327] ITU-T, "Z.100 functional specification and description language," tech. rep., ITU, March 1993. Recommendation Z.100 - Z.104.

[328] R. Saracco and P. A. J. Tilanus, "Ccitt sdl: an overview of the language and its applications," *Computer Networks & ISDN Systems*, vol. 13, no. 2, pp. 65–74, 1987. Special Issue on CCITT SDL.

[329] G. R. Andrews, *Concurrent Programming, Principles and Practice*. Redwood City CA: Benjamin/Cummings, 1991.

[330] Synopsys, Inc., *Synopsys VHDL System Simulator Interfaces Manual: C-language Interface*, June 1993. Version 3.0b.

[331] C. Valderrama, F. Naçabal, P. Paulin, and A. A. Jerraya, "Automatic generation of interfaces for distributed c-vhdl cosimulation of embedded systems: an industrial experience," in *Proc. of Rapid System Prototyping, RSP'96*, p. 72, 1996.

[332] P.Paulin, M. Cornero, C. Liem, F.Naçabal, C. Donawa, S. Sutarwala, T.May, and C. Valderrama, "Trends in embedded system technology: An industrial perspective," in *Hardware/Software Co-design* (G. D. Micheli and M. Sami, eds.), Kluwer Academic Publishers, 1996.

[333] A. Changuel, R. Rolland, and A. A. Jerraya, "Design of an adaptative motors controller based on fuzzy logic using behavioural synthesis," in *Proc. European DAC with EURO-VHDL 96*, pp. 48–52, 1996.

[334] M. Abid, A. Changuel, and A. Jerraya, "Exploration of hardware/software design space through a codesign of robot arm controller," in *Proc. EDA Conference with EURO-VHDL 96*, pp. 42–47, 1996.

# Index